T0367475

GaN TRANSISTORS FOR EFFICIENT POWER CONVERSION

GaN TRANSISTORS FOR EFFICIENT POWER CONVERSION

Second Edition

Alex Lidow

Johan Strydom

Michael de Rooij

David Reusch

Efficient Power Conversion Corporation, El Segundo, California, USA

WILEY

This edition first published 2015

Registered office
John Wiley & Sons Ltd, The Atrium, Southern Gate, Chichester, West Sussex, PO19 8SQ, United Kingdom

For details of our global editorial offices, for customer services and for information about how to apply for permission to reuse the copyright material in this book please see our website at www.wiley.com.

Library of Congress Cataloging-in-Publication Data

Lidow, Alex.
GaN transistors for efficient power conversion / Alex Lidow, Johan Strydom, Michael de Rooij, David Reusch. – Second edition.
1 online resource.
Includes bibliographical references and index.
Description based on print version record and CIP data provided by publisher; resource not viewed.
ISBN 978-1-118-84478-6 (ePub) – ISBN 978-1-118-84479-3 (Adobe PDF) – ISBN 978-1-118-84476-2 (cloth)
1. Field-effect transistors. 2. Gallium nitride. I. Title.
TK7871.95
621.3815′284–dc23

2014023079

A catalogue record for this book is available from the British Library.

ISBN: 978-1-118-84476-2

Set in 10/12 pt TimesLTStd-Roman by Thomson Digital, Noida, India

1 2015

In memory of Eric Lidow,
the original power conversion pioneer.

Contents

Foreword

It is well established that the CMOS inverter and DRAM are the two basic building blocks of digital signal processing. Decades of improving inverter switching speed and memory density under Moore's Law has unearthed numerous applications that were previously unimaginable. Power processing is built upon two similar functional building blocks: power switches and energy storage devices, such as the inductor and capacitor. The push for higher switching frequencies has always been a major catalyst for performance improvement and size reduction.

Since its introduction in the mid-1970s, the power MOSFET, with its greater switching speed, has replaced the bipolar transistor. To date, the power MOSFET has been perfected up to its theoretical limit. Device switching losses can be reduced further with the help of soft-switching techniques. However, its gate drive loss is still excessive, limiting the switching frequency to the low hundreds of kilohertz in most applications.

The recent introduction of GaN, with much improved figures of merit, opens the door for operating frequencies well into the megahertz range. A number of design examples are illustrated in this book and other literatures, citing impressive power density improvements of a factor of 5 or 10. However, I believe the potential contribution of GaN goes beyond the simple measures of efficiency and power density. GaN has the potential to have a profound impact on our design practice, including a possible paradigm shift.

Power electronics is interdisciplinary. The essential constituents of a power electronics system are switches, energy storage devices, circuit topology, system packaging, electromagnetic interactions, thermal management, EMC/EMI, and manufacturing considerations. When the switching frequency is low, these various constituents are loosely coupled. Current design practices address these issues in piecemeal fashion. When a system is designed for a much higher frequency, the components are arranged in close proximity, to minimize undesirable parasitics. This invariably leads to unwanted electromagnetic coupling and thermal interaction.

This increasing intricacy between components and circuits requires a more holistic approach, concurrently taking into account all electrical, mechanical, electromagnetic and thermal considerations. Furthermore, all operations should be executed correctly, both spatially and temporally. These challenges prompt circuit designers to pursue a more integrated approach. For power electronics, integration needs to take place at the functional level or the subsystem level whenever feasible and practical. These integrated modules then serve as the basic building blocks of further system integration. In this manner, customization can be achieved using standardized building blocks, in much the same way as digital electronics systems. With the economy of scale in manufacturing, this will bring significant cost reduction in power electronics equipment and unearth numerous new applications previously precluded due to high cost.

GaN will create fertile ground for research and technological innovations for years to come. Dr. Alex Lidow mentions in this book that it took thirty years for the power MOSFET to reach its current state of maturity. While GaN is still in an early stage of development, a few technical challenges require immediate attention. These issues are recognized by the authors and are addressed in this book.

1. High dv/dt and high di/dt render most of the commercially available gate drive circuits unsuitable for GaN devices, especially for the high-side switch. Chapter 3 offers many important insights in the design of the gate drive circuit.
2. Device packaging and circuit layout are critical. The unwanted effects of parasitics need to be contained. Soft-switching techniques can be very useful for this purpose. A number of important issues related to packaging and layout are addressed in detail in Chapters 4–6.
3. High-frequency magnetic design is also critical. The choice of suitable magnetic materials becomes rather limited when the switching frequency goes beyond 2–3 MHz. Additionally, more creative high-frequency magnetics design practice should be explored. Several recent publications suggest design practices that defy the conventional wisdom and practice, yielding interesting results.
4. The impact of high frequency on EMI/EMC has yet to be explored.

Dr. Alex Lidow is a well-respected leader in the field. Alex has always been in the forefront of technology and a trendsetter. While serving as the CEO of IR, he initiated GaN development in the early 2000s. He also led the team in developing the first integrated DrMOS and DirectFET®, which are now commonly used to power the new generation of microprocessors and many other applications.

This book is a gift to power electronics engineers. It offers a comprehensive view, from device physics, characteristics, and modeling to device and circuit layout considerations and gate drive design, with design considerations for both hard switching and soft switching. Additionally, it further illustrates the utilization of GaN in a wide range of emerging applications.

It is very gratifying to note that three of the four authors of this book are from CPES, joining with Dr. Lidow in an effort to develop this new generation of wide-band-gap power switches – presumably a game-changing device with a scale of impact yet to be defined.

Dr. Fred C. Lee
Director, Center for Power Electronics Systems
University Distinguished Professor, Virginia Tech

Acknowledgments

The authors wish to acknowledge the many exceptional contributions towards the content of this book from our colleagues Jianjun (Joe) Cao, Robert Beach, Alana Nakata, Guang Yuan Zhao, Audrey Downes, Steve Colino, Bhasy Nair, Renee Yawger, Yanping Ma, Robert Strittmatter, Stephen Tsang, Peter Cheng, Larry Chen, F.C. Liu, M.K. Chiang, Winnie Wong, Chunhua Zhou, Seshadri Kolluri, Jiali Cao, Lorenzo Nourafchan, and Andrea Mirenda.

A special thank you is due to Joe Engle who, in addition to reviewing and editing all corners of this work, put all the logistics together to make it happen. Joe also assembled an exceptional group of graphic artists, all of whom worked with endless patience against difficult deadlines.

A note of gratitude to the editors and staff at Wiley who were instrumental in undertaking a diligent review of the text and shepherding the book through the production process.

Finally, we would like to thank Archie Huang and Sue Lin for believing in GaN from the beginning. Their vision and support will change the semiconductor industry forever.

Alex Lidow
Johan Strydom
Michael de Rooij
David Reusch
Efficient Power Conversion Corporation
April 2014

Acknowledgments

1

GaN Technology Overview

1.1 Silicon Power MOSFETs 1976–2010

For over three decades, power management efficiency and cost have improved steadily as innovations in power metal oxide silicon field effect transistor (MOSFET) structures, technology, and circuit topologies have kept pace with the growing need for electrical power in our daily lives. In the new millennium, however, the rate of improvement has slowed as the silicon power MOSFET asymptotically approaches its theoretical bounds.

Power MOSFETs first appeared in 1976 as alternatives to bipolar transistors. These majority-carrier devices were faster, more rugged, and had higher current gain than their minority-carrier counterparts (for a discussion of basic semiconductor physics, a good reference is [1]). As a result, switching power conversion became a commercial reality. Among the earliest high-volume consumers of power MOSFETs were AC-DC switching power supplies for early desktop computers, followed by variable-speed motor drives, fluorescent lights, DC-DC converters, and thousands of other applications that populate our daily lives.

One of the first power MOSFETs was the IRF100 from International Rectifier Corporation, introduced in November 1978. It boasted a 100 V drain-source breakdown voltage and a 0.1 Ω on-resistance ($R_{DS(on)}$), the benchmark of the era. With a die size of over 40 mm^2, and a $34 price tag, this product was not destined to supplant the venerable bipolar transistor immediately. Since then, several manufacturers have developed many generations of power MOSFETs. Benchmarks have been set, and subsequently surpassed, each year for 30-plus years. As of the date of writing, the 100 V benchmark arguably is held by Infineon with the BSC060N10NS3. In comparison with the IRF100 MOSFET's resistivity figure of merit (4 Ωmm^2), the BSC060N10NS3 has a figure of merit of 0.072 Ωmm^2. That is almost at the theoretical limit for a silicon (Si) device [2].

There are still improvements to be made in power MOSFETs. For example, super-junction devices and IGBTs have achieved conductivity improvements beyond the theoretical limits of a simple vertical majority-carrier MOSFET. These innovations may continue for quite some time and certainly will be able to leverage the low cost structure of the power MOSFET and the know-how of a well-educated base of designers who, after many years, have learned to squeeze every ounce of performance out of their power conversion circuits and systems.

GaN Transistors for Efficient Power Conversion, Second Edition.
Alex Lidow, Johan Strydom, Michael de Rooij, and David Reusch.

Companion Website: http://www.wiley.com/go/gan_transistors

1.2 The GaN Journey Begins

Gallium nitride (GaN) high electron mobility transistor (HEMT) devices first appeared in about 2004 with depletion-mode radio frequency (RF) transistors made by Eudyna Corporation in Japan. Using GaN on silicon carbide (SiC) substrates, Eudyna successfully produced transistors designed for the RF market [3]. The HEMT structure was based on the phenomenon first described in 1975 by T. Mimura *et al.* [4], and in 1994 by M. A. Khan *et al.* [5], which demonstrated the unusually high electron mobility described as a two-dimensional electron gas in the region of an aluminum gallium nitride (AlGaN) and GaN heterostructure interface. Adapting this phenomenon to gallium nitride grown on silicon carbide, Eudyna was able to produce benchmark power gain in the multi-gigahertz frequency range. In 2005, Nitronex Corporation introduced the first depletion-mode RF HEMT device made with GaN grown on silicon wafers using their SIGANTIC® technology.

GaN RF transistors have continued to make inroads in RF applications, as several other companies have entered the market. Acceptance outside of this application, however, has been limited by device cost as well as the inconvenience of depletion-mode operation (normally conducting and requires a negative voltage on the gate to turn the device off).

In June 2009, the Efficient Power Conversion Corporation (EPC) introduced the first enhancement-mode GaN on silicon (eGaN®) FETs designed specifically as power MOSFET replacements (since eGaN FETs do not require a negative voltage to be turned off). At the outset, these products were produced in high volume at low cost by using standard silicon manufacturing technology and facilities. Since then, Matsushita, Transphorm, GaN Systems, RFMD, Panasonic, HRL, and International Rectifier, among others, have announced their intention to manufacture GaN transistors for the power conversion market.

The basic requirements for semiconductors used in power conversion are efficiency, reliability, controllability, and cost effectiveness. Without these attributes, a new device structure would not be economically viable. There have been many new structures and materials considered as a successor to silicon; some have been economic successes, others have seen limited or niche acceptance. In the next section, we will look at the comparison between silicon, silicon carbide, and gallium nitride as platform candidates to dominate the next generation of power transistors.

1.3 Why Gallium Nitride?

Silicon has been a dominant material for power management since the late 1950s. The advantages that silicon had over earlier semiconductors, such as germanium or selenium, could be expressed in four key categories:

- silicon enabled new applications not possible with earlier materials
- silicon proved more reliable
- silicon was easier to use in many ways
- silicon devices cost less

All of these advantages stemmed from the basic physical properties of silicon, combined with a huge investment in manufacturing infrastructure and engineering. Let's look at some of those basic properties and compare them with other successor candidates. Table 1.1 identifies five key electrical properties of three semiconductor materials contending for the power management market.

Table 1.1 Material properties of Silicon, GaN, and SiC

Parameter		Silicon	GaN	SiC
Band Gap E_g	eV	1.12	3.39	3.26
Critical Field E_{Crit}	MV/cm	0.23	3.3	2.2
Electron Mobility μ_n	$cm^2/V{\cdot}s$	1400	1500	950
Permittivity ε_r		11.8	9	9.7
Thermal Conductivity λ	W/cm·K	1.5	1.3	3.8

One way of translating these basic crystal parameters into a comparison of device performance is to calculate the best theoretical performance achievable for each of the three candidates. For power devices, there are many characteristics that matter in the variety of power conversion systems available today. Five of the most important are: conduction efficiency (on-resistance), breakdown voltage, size, switching efficiency, and cost.

In the next section, the first four of the material characteristics in Table 1.1 will be reviewed, leading to the conclusion that both SiC and GaN are capable of producing devices with superior on-resistance, breakdown voltage, and a smaller-sized transistor compared to silicon. In Chapter 2, we will look at how these material characteristics translate into superior switching efficiency for a GaN transistor, and in Chapter 11, how a GaN transistor can also be produced at a lower cost than a silicon MOSFET of equivalent performance.

1.3.1 Band Gap (E_g)

The band gap of a semiconductor is related to the strength of the chemical bonds between the atoms in the lattice. These stronger bonds mean that it is harder for an electron to jump from one site to the next. Among the many consequences are lower intrinsic leakage currents and higher operating temperatures for higher band gap semiconductors. Based on the data in Table 1.1, GaN and SiC both have higher band gaps than silicon.

1.3.2 Critical Field (E_{crit})

The stronger chemical bonds that cause the wider band gap also result in a higher critical electric field needed to initiate impact ionization, thus causing avalanche breakdown. The voltage at which a device breaks down can be approximated with the formula:

$$V_{BR} = \tfrac{1}{2}\, w_{drift} \cdot E_{crit} \tag{1.1}$$

The breakdown voltage of a device (V_{BR}), therefore, is proportional to the width of the drift region (w_{drift}). In the case of SiC and GaN, the drift region can be 10 times smaller than in silicon for the same breakdown voltage. In order to support this electric field, there need to be carriers in the drift region that are depleted away at the point where the device reaches the critical field. This is where there is a huge gain in devices with high critical fields. The number of electrons (assuming an N-type semiconductor) between the two terminals can be calculated using Poison's equation:

$$q \cdot N_D = \varepsilon_o \cdot \varepsilon_r \cdot E_{crit} / w_{drift} \tag{1.2}$$

In this equation q is the charge of the electron ($1.6\cdot10^{-19}$ coulombs), N_D is the total number of electrons in the volume, ε_o is the permittivity of a vacuum measured in farads per meter ($8.854\cdot10^{-12}$ F/m), and ε_r is the relative permittivity of the crystal compared to a vacuum. In its simplest form under DC conditions, permittivity is the dielectric constant of the crystal.

Referring to Equation 1.2, it can be seen that if the critical field of the crystal is 10 times higher, and from Equation 1.1, the electrical terminals can be 10 times closer together. Therefore, the number of electrons, N_D, in the drift region can be 100 times greater. This is the basis for the ability of GaN and SiC to outperform silicon in power conversion.

1.3.3 On-Resistance ($R_{DS(on)}$)

The theoretical on-resistance (measured in ohms (Ω)) of this majority-carrier device is therefore

$$R_{DS(on)} = w_{drift}/q\cdot\mu_n\cdot N_D \tag{1.3}$$

Where μ_n is the mobility of electrons. Combining Equations 1.1, 1.2, and 1.3 produces the following relationship between breakdown voltage and on-resistance:

$$R_{DS(on)} = 4\cdot V_{BR}^2/\varepsilon_o\cdot\varepsilon_r\cdot E_{crit}^3 \tag{1.4}$$

This equation can now be plotted as shown in Figure 1.1 for Si, SiC, and GaN. This plot is for an ideal structure. Real semiconductors are not always ideal structures and so it is always a challenge to achieve the theoretical limit. In the case of silicon MOSFETs, it took 30 years.

1.3.4 The Two-Dimensional Electron Gas

The natural structure of crystalline gallium nitride is a hexagonal structure named "wurtzite" (see Figure 1.2). Because this structure is very chemically stable, it is mechanically robust and

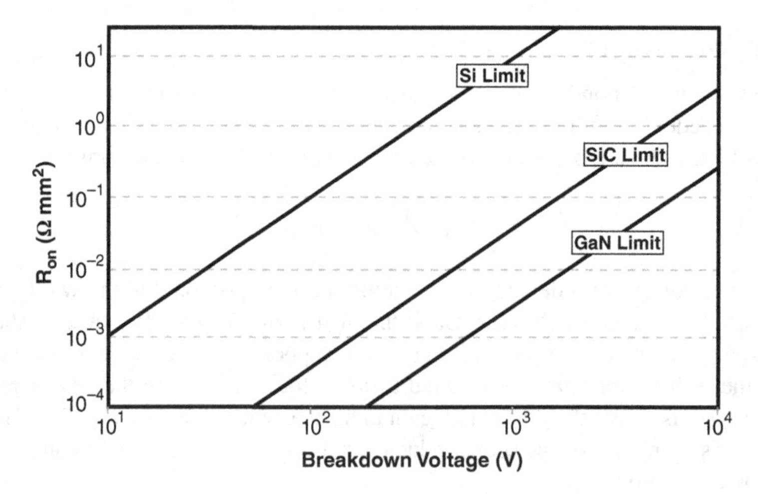

Figure 1.1 Theoretical on-resistance vs. blocking voltage capability for Si, SiC, and GaN based power devices

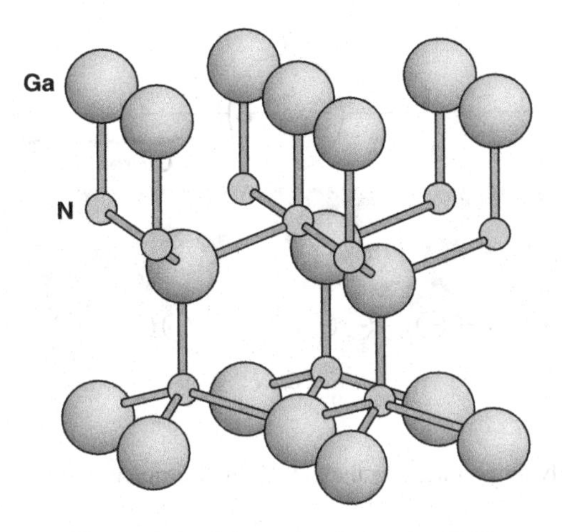

Figure 1.2 Schematic of wurtzite GaN

can withstand high temperatures without decomposition. This crystal structure also gives GaN piezoelectric properties that lead to its ability to achieve very high conductivity compared with other semiconductor materials. Piezoelectricity in GaN is predominantly caused by the displacement of charged elements in the crystal lattice. If the lattice is subjected to strain, the deformation will cause a miniscule shift in the atoms in the lattice that generate an electric field – the higher the strain, the greater the electric field. By growing a thin layer of AlGaN on top of a GaN crystal, a strain is created at the interface that induces a compensating two-dimensional electron gas (2DEG) as shown schematically in Figure 1.3 [6–8]. This 2DEG is used to efficiently conduct electrons when an electric field is applied across it, as in Figure 1.4.

This 2DEG is highly conductive, in part due to the confinement of the electrons to a very small region at the interface. This confinement increases the mobility of electrons from about $1000\,cm^2/V{\cdot}s$ in unstrained GaN to $1500–2000\,cm^2/V{\cdot}s$ in the 2DEG region. The high concentration of electrons with very high mobility is the basis for the high electron mobility transistor (HEMT), the primary subject of this book.

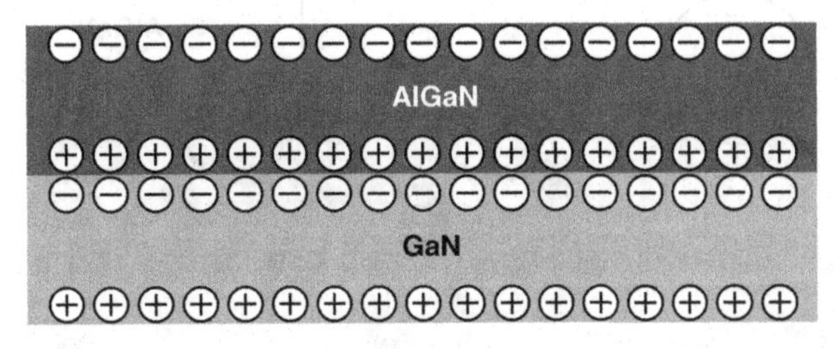

Figure 1.3 Simplified cross section of a GaN/AlGaN heterostructure showing the formation of a 2DEG due to the strain-induced polarization at the interface between the two materials

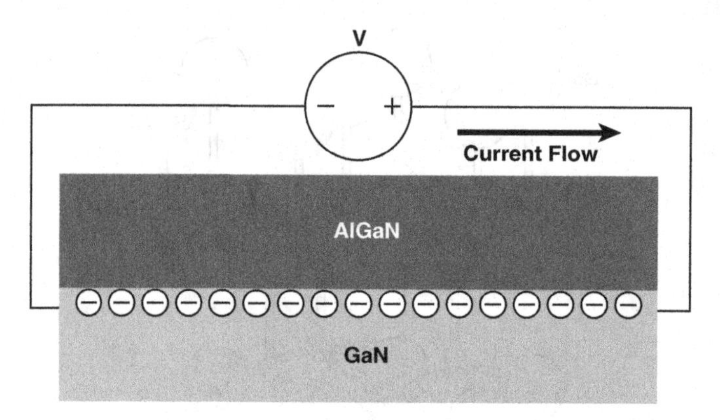

Figure 1.4 By applying a voltage to the 2DEG an electric current is induced in the crystal

1.4 The Basic GaN Transistor Structure

The basic depletion-mode GaN transistor structure is shown in Figure 1.5. As with any power FET, there are gate, source, and drain electrodes. The source and drain electrodes pierce through the top AlGaN layer to form an ohmic contact with the underlying 2DEG. This creates a short circuit between the source and the drain until the 2DEG "pool" of electrons is depleted, and the semi-insulating GaN crystal can block the flow of current. In order to deplete the 2DEG, a gate electrode is placed on top of the AlGaN layer. When a negative voltage relative to both drain and source electrodes is applied to the gate, the electrons in the 2DEG are depleted out of the device. This type of transistor is called a depletion-mode, or d-mode, HEMT.

There are two common ways to produce a d-mode HEMT device. The initial transistors introduced in 2004 had a Schottky gate electrode that was created by depositing a metal layer directly on top of the AlGaN. The Schottky barrier was formed using metals such as Ni-Au or Pt [9–11]. Depletion-mode devices have also been made using an insulating layer and metal gate similar to a MOSFET [12]. Both types are shown in Figure 1.6.

In power conversion applications, d-mode devices are inconvenient because, at the startup of a power converter, a negative bias must first be applied to the power devices. If this negative

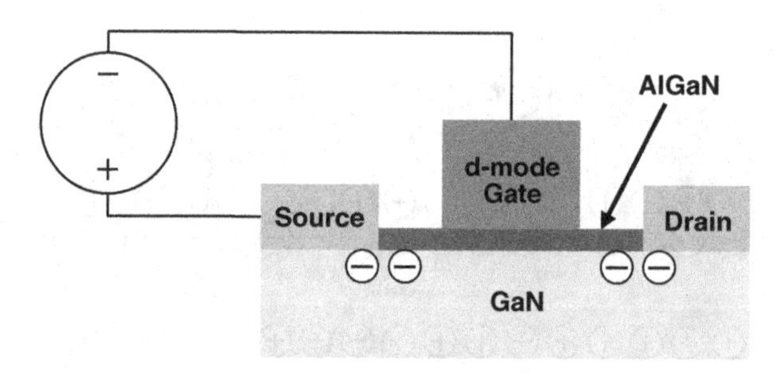

Figure 1.5 By applying a negative voltage to the gate of the device, the electrons in the 2DEG are depleted out of the device. This type of device is called a depletion-mode (d-mode) HEMT

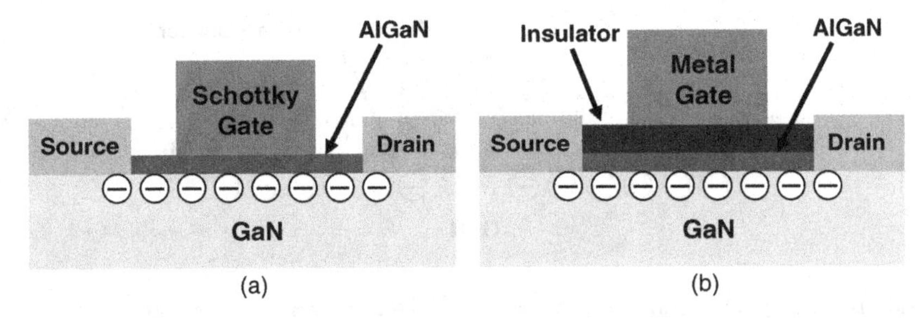

Figure 1.6 Cross section of a basic depletion-mode GaN HEMT with (a) Schottky gate, or (b) insulating gate

bias is not applied first, a short circuit will result. An enhancement-mode (e-mode) device, on the other hand, would not suffer this limitation. With zero bias on the gate, an e-mode device is OFF (Figure 1.7(a)) and will not conduct current until a positive voltage is applied to the gate, as illustrated in Figure 1.7(b).

There are four popular structures that have been used to create enhancement-mode devices: recessed gate, implanted gate, pGaN gate, and cascode hybrid.

1.4.1 Recessed Gate Enhancement-Mode Structure

The recessed gate structure has been discussed extensively in the literature [13] and is created by thinning the AlGaN barrier layer above the 2DEG (see Figure 1.8). By making the AlGaN barrier thinner, the amount of voltage generated by the piezoelectric field is reduced proportionally. When the voltage generated is less than the built-in voltage of the Schottky gate metal, the 2DEG is eliminated with zero bias on the gate. With positive bias, electrons are attracted to the AlGaN interface and complete the circuit between source and drain.

1.4.2 Implanted Gate Enhancement-Mode Structure

Shown in Figure 1.9(a) and (b) is a method for creating an enhancement-mode device by implanting fluorine atoms in the AlGaN barrier layer [14]. These fluorine atoms create a "trapped" negative charge in the AlGaN layer that depletes the 2DEG underneath. By adding a Schottky gate on top, an enhancement-mode HEMT is created.

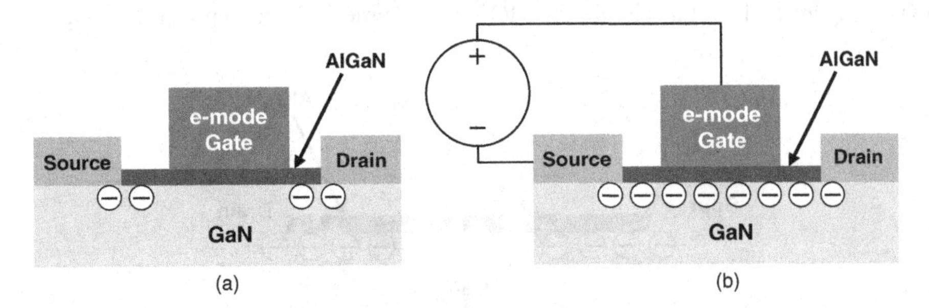

Figure 1.7 (a) An enhancement-mode (e-mode) device depletes the 2DEG with zero volts on the gate. (b) By applying a positive voltage to the gate, the electrons are attracted to the surface, re-establishing the 2DEG

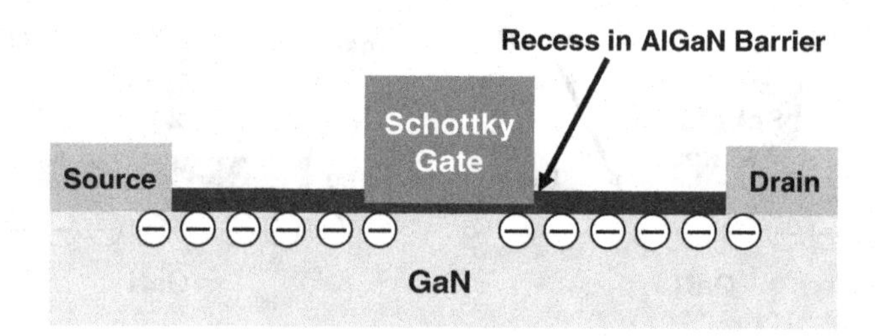

Figure 1.8 By etching away part of the AlGaN barrier layer a recessed gate e-mode transistor can be fabricated

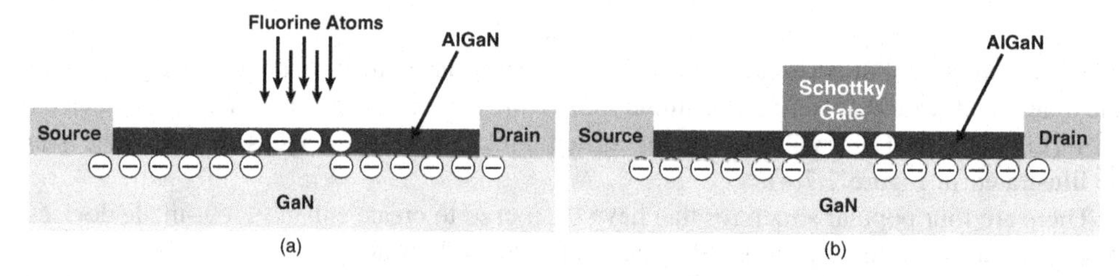

Figure 1.9 (a) By implanting fluorine atoms into the AlGaN barrier layer negative charges are trapped in the barrier. (b) A Schottky gate now can be used to reconstruct the 2DEG when a positive voltage is applied

1.4.3 pGaN Gate Enhancement-Mode Structure

The first enhancement-mode devices sold commercially had a positively charged (p-type) GaN layer grown on top of the AlGaN barrier (see Figure 1.10) [15]. The positive charges in this pGaN layer have a built-in voltage that is larger than the voltage generated by the piezoelectric effect, thus depleting the electrons in the 2DEG and creating an enhancement-mode structure [16].

1.4.4 Cascode Hybrid Enhancement-Mode Structure

An alternative to building a single-chip enhancement-mode GaN transistor is to place an enhancement-mode silicon MOSFET in series with a depletion-mode HEMT device [17,18] as shown in Figure 1.11. In this circuit, the MOSFET is turned on with a positive voltage on the

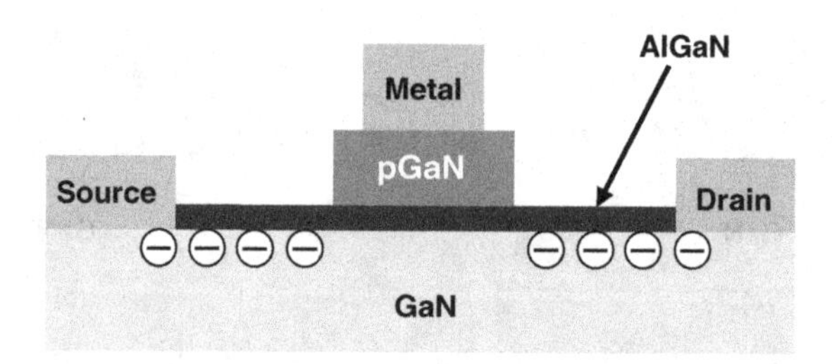

Figure 1.10 By growing a p-type GaN layer on top of the AlGaN the 2DEG is depleted at zero volts on the gate

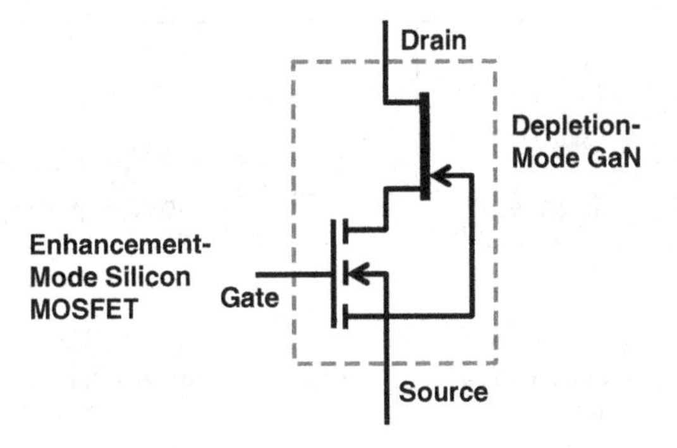

Figure 1.11 Schematic of low-voltage enhancement-mode silicon MOSFET in series with a depletion-mode GaN HEMT

gate when the depletion-mode GaN transistor's gate voltage goes to near-zero volts and turns on. Current can now pass through the depletion-mode GaN HEMT and the MOSFET, which is connected in series with the GaN HEMT. When the voltage on the MOS gate is removed, a negative voltage is created between the depletion-mode GaN transistor gate and its source electrode, turning the GaN device off.

This type of solution for an enhancement-mode GaN "system" works well when the GaN transistor has a relatively high on-resistance compared with the low voltage (usually 30 V rated) silicon MOSFET. Since on-resistance increases with the device breakdown voltage, cascode solutions are most effective when the GaN HEMT is high voltage and the MOSFET is very low voltage. In Figure 1.12 is a chart showing the added on-resistance to the cascode circuit by the enhancement-mode silicon MOSFET. A 600 V cascode device would only have about 3% added on-resistance due to the low-voltage MOSFET. Conversely, as the desired rated voltage goes down, and the on-resistance of the GaN transistor decreases, the MOSFET contribution becomes more significant. For this reason, cascode solutions are only practical at voltages higher than 200 V.

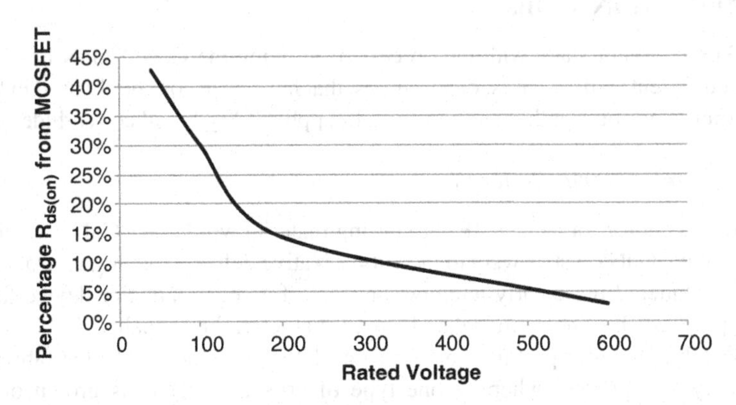

Figure 1.12 At a higher voltage rating the low voltage MOSFET does not add significantly to the on-resistance of the cascode transistor system

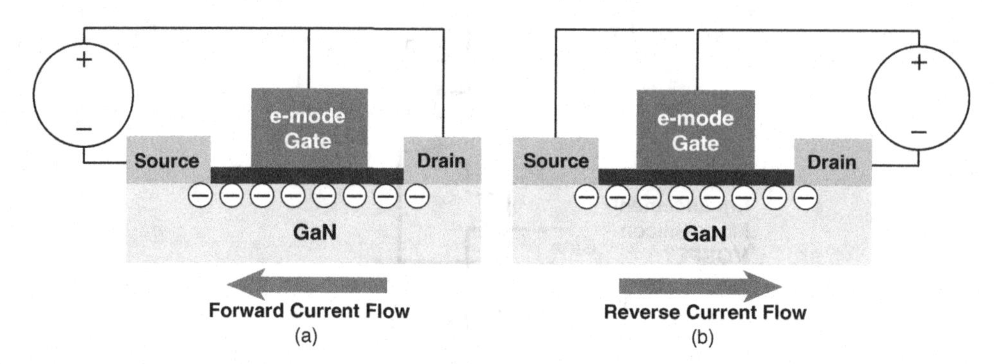

Figure 1.13 Enhancement-mode HEMT transistors can conduct in either the forward or reverse direction

1.4.5 Reverse Conduction in HEMT Transistors

Enhancement-mode GaN transistors can also conduct in the reverse direction when the drain voltage is higher than the gate voltage by at least $V_{GS(th)}$ (see Figure 1.13(b)). In this situation, the 2DEG is again restored under the gate electrode, and current can flow from source to drain. Because the enhancement-mode HEMT has no minority carrier conduction, the device, operating similar to a diode, will turn off instantly when the forward bias is removed between the gate and drain electrodes. This characteristic is quite useful in certain power conversion circuits.

In the reverse direction, the cascode-connected transistor discussed in Section 1.4.4 conducts in the same way as an enhancement-mode GaN HEMT, except that the diode of the MOSFET is conducting the reverse current, which then has to flow through the GaN device. The forward voltage drop of the MOSFET diode creates a slight positive voltage from gate to source in the HEMT which, therefore, is turned on in the forward direction. The HEMT on-resistance is added to the voltage drop of the MOSFET in this configuration. Unlike the enhancement-mode GaN HEMT, the cascode-configured transistor does have a recovery time due to the injection of minority carriers in the silicon-based MOSFET.

1.5 Building a GaN Transistor

Building a GaN transistor starts with the process of growing the GaN/AlGaN heterostructure. There are four different starting bases, or substrates, that have been commonly used in fabricating GaN HEMT transistors: bulk gallium nitride crystal, sapphire (Al_2O_3), silicon carbide, and silicon.

1.5.1 Substrate Material Selection

The most obvious choice for a GaN HEMT starting material would be a GaN crystal. The first attempts at growing GaN crystals were in the 1960s. Native defects from high nitrogen vacancy concentrations rendered these early attempts unusable for semiconductor device fabrication. Since then, progress has been made and small-diameter, high-quality GaN crystals are becoming available, holding promise for use as a platform for active device fabrication.

Heteroepitaxy is a process whereby one type of crystal structure is grown on top of a different crystal. Because GaN crystals have not been readily available, there has been much work focused on growing GaN crystals on top of a more convenient platform such as sapphire,

Table 1.2 Some key properties of Al_2O_3, SiC, and Si [19]

Substrate	Crystal plane	Lattice spacing Å	Lattice mismatch %	Relative thermal expansion 10^{-5}·K^{-1}	Thermal conductivity W/cm·K	Relative Cost
Al_2O_3	(0001)	4.758	16.1	−1.9	0.42	Middle
6H-SiC	(0001)	3.08	3.5	1.4	3.8	Highest
Si	(111)	3.84	−17	3	1.5	Lowest

silicon carbide, or more recently, silicon. The starting point for trying to grow on a dissimilar crystal layer is to find a substrate with the appropriate physical properties.

Referring to Table 1.2, it can be seen that there are tradeoffs between any of the three listed choices for a substrate material. For example, sapphire (Al_2O_3) has a 16.1% mismatch to a GaN crystal lattice and has poor thermal conductivity. Thermal conductivity is especially important in transistors for power conversion because they generate a significant amount of heat flux during operation due to internal power dissipation. Silicon carbide (6H-SiC) substrate, on the other hand, has a reasonably good lattice match and excellent thermal conductivity. The disadvantage comes from the cost of the starting crystal substrate, which can be up to 100 times the cost of a silicon substrate of the same diameter. Silicon is also not an ideal base for a GaN heteroepitaxial structure due to the lattice mismatch and the mismatch of thermal expansion coefficients. Silicon, however, is the least expensive material and there is a large and well-developed infrastructure to process devices on silicon substrates.

For these reasons, silicon carbide is commonly used for devices that require very high power densities, such as linear RF applications; silicon is used for devices in more cost-sensitive commercial applications such as DC-DC conversion, AC-DC conversion, and motion control.

1.5.2 Growing the Heteroepitaxy

There are several types of technologies that have been used to grow GaN on different substrates [20–23]. The two most promising are metal organic chemical vapor deposition (MOCVD) and molecular beam epitaxy (MBE). MOCVD is faster and generally lower cost, whereas MBE is capable of more uniform layers with very abrupt transitions between layers. For GaN HEMT devices in power conversion applications, MOCVD is the dominant technology owing to the cost advantages.

An MOCVD growth occurs in an inductively or radiantly heated reactor. A highly reactive precursor gas is introduced into the chamber where the gas is "cracked" by the hot substrate and reacts to form the desired compound. For GaN growth, the precursors are ammonia (NH_3) and trimethyl-gallium (TMG). For AlGaN growth, the precursors are trimethyl-aluminum (TMA) or triethyl-aluminum (TEA). In addition to the precursors, carrier gases such as H_2 and N_2 are used to enhance mixing and control the flow within the chamber. Temperatures in the range of 900–1100 °C are used for these growths.

A GaN heteroepitaxial structure involves at least four growth stages. Figure 1.14 illustrates this process. The starting material (a), of either SiC or Si, is heated in the reaction chamber. A layer of AlN is then grown (b) to create a seed layer that is hospitable to the AlGaN wurtzite crystal structure. An AlGaN "buffer layer" (c) creates the transition to the GaN crystal (d).

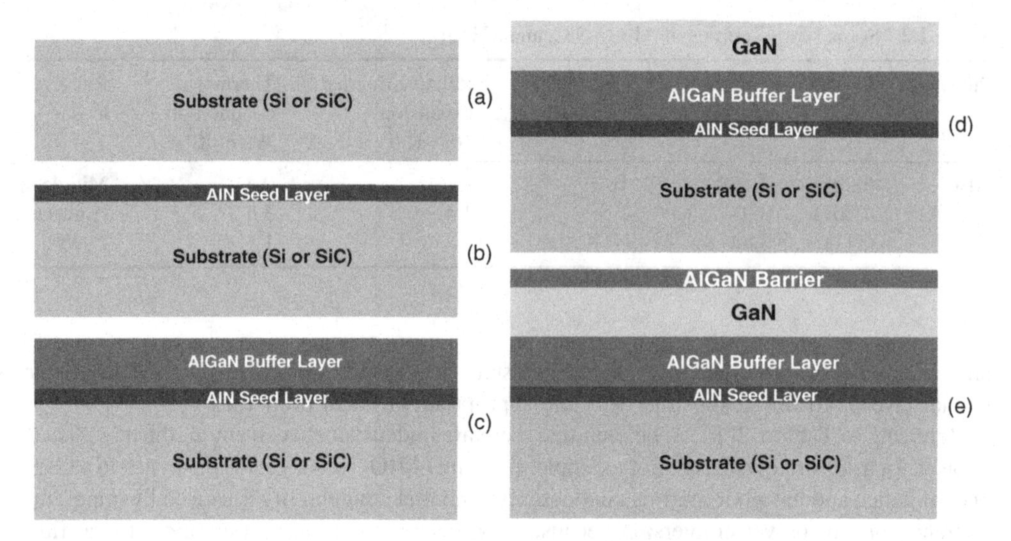

Figure 1.14 An illustration of the basic steps involved in creating a GaN heteroepitaxial structure (not to scale)

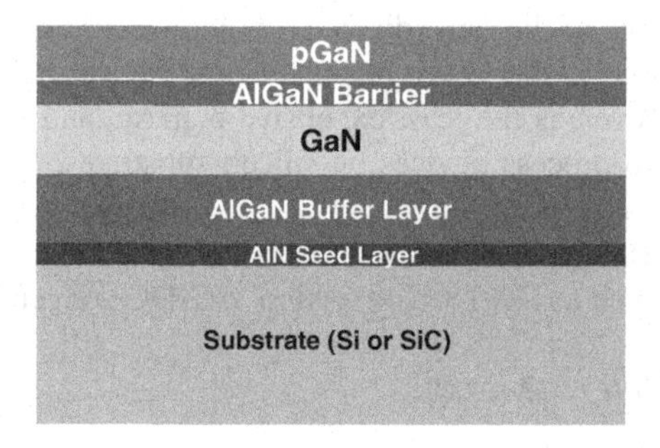

Figure 1.15 An additional GaN layer, doped with p-type impurities can be added to the heteroepitaxy process when producing an enhancement-mode device as illustrated in Figure 1.10

Finally, the thin AlGaN barrier (e) is grown on top of the GaN crystal to create the strain layer that induces the 2DEG formation.

Earlier in this chapter, several methods of making enhancement-mode GaN transistors were described. One method (pGaN enhancement-mode), illustrated in Figure 1.10, includes an additional GaN layer grown on top of the AlGaN barrier. This layer is most commonly doped with p-type impurities such as Mg or Fe. A cross section of this heteroepitaxy structure is shown in Figure 1.15.

1.5.3 Processing the Wafer

Fabricating an HEMT transistor from a heteroepitaxial substrate can be accomplished in variety of sequential steps. One example of a simplified process for making an enhancement-mode HEMT with a pGaN gate is shown in Figure 1.16. A cross section of a completed device using this process is shown in Figure 1.17.

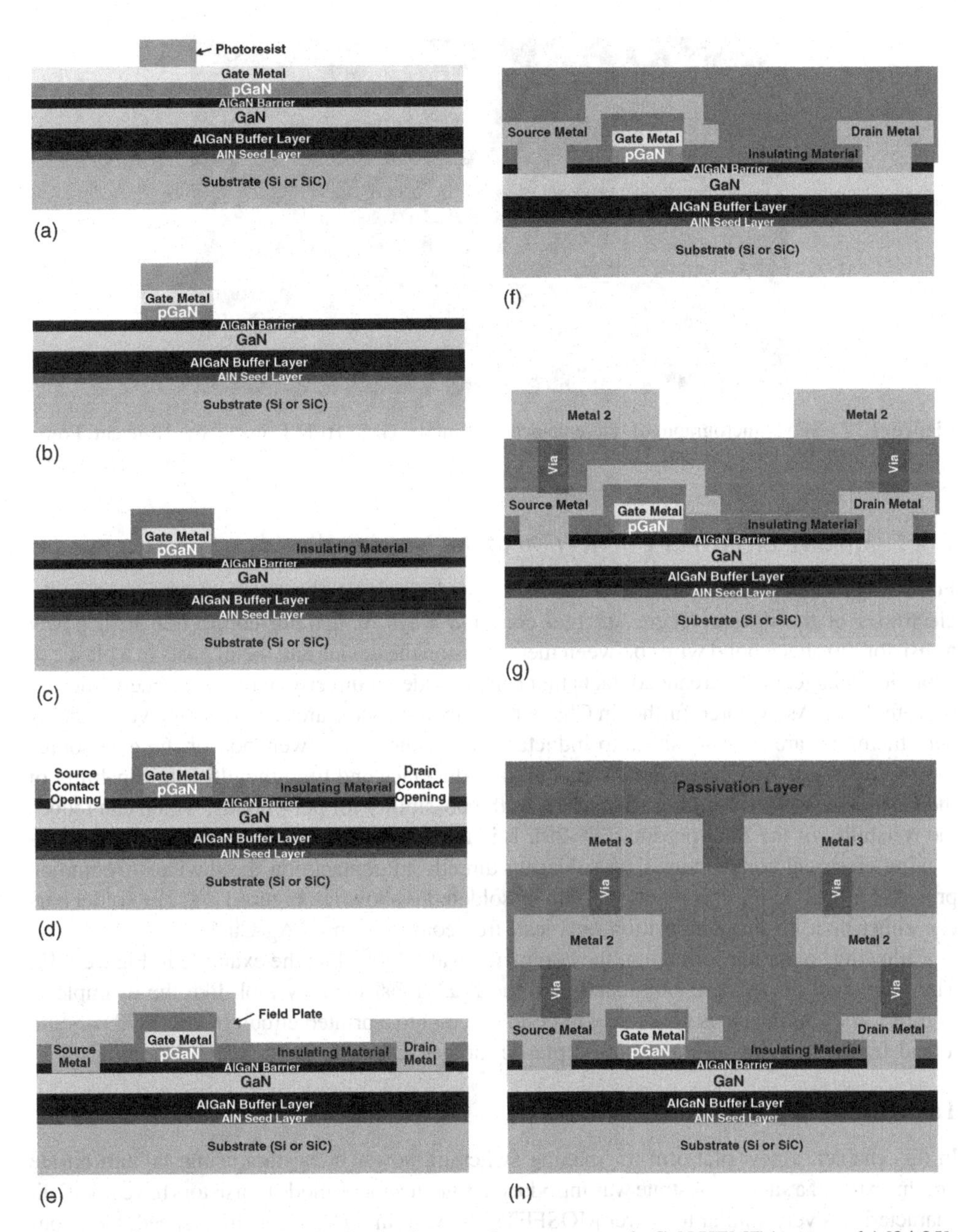

Figure 1.16 A typical process for fabricating an enhancement-mode GaN HEMT (not to scale) [24,25]. The process steps are as follows: (a) Deposit gate metal and define gate pattern using photoresist as a protecting layer. (b) Etch the gate metal and pGaN crystal. (c) Deposit insulating material. (d) Create contact openings to source, drain, and gate (gate contact not pictured). (e) Deposit first aluminum metal layer and define metal pattern. (f) Deposit interlayer dielectric. (g) Cut vias between metal layers, form tungsten via plug, deposit and define second aluminum metal layer. (h) Deposit and define third aluminum layer and deposit final passivation layer

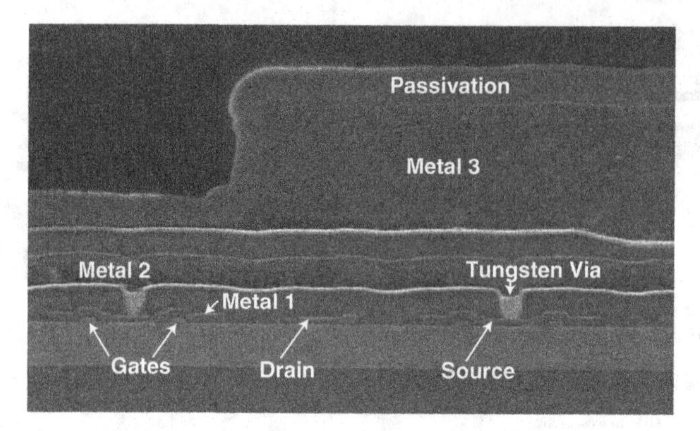

Figure 1.17 SEM micrograph of an enhancement-mode GaN HEMT made by Efficient Power Conversion Corporation [24–26]

1.5.4 Making Electrical Connection to the Outside World

Following the device fabrication, provisions are needed to make the electrical connection to the electrodes of the device. There are two common ways of making connection to a power transistor: (a) attach bond wires between metal pads on the device and metal posts in a plastic or ceramic package, or (b) create contacts that can be soldered directly onto the device while still in wafer form. As explored further in Chapter 3, GaN transistors are able to switch very quickly and, therefore, are very sensitive to inductances in either the power loop or the gate-source loop. Wire bonds have a significant amount of inductance and limit the ultimate capability of the GaN device. Wire bonding also increases the possibility for poor bonds, which can reduce the reliability of the final product [25–29]. It is for this reason that the preferred method for making electrical connection is by soldering directly to contacts on the device. A common process for making these contacts that can be soldered is shown in Figure 1.18. The solder bars can either be a Pb-Sn composition or a lead-free composition of Ag-Cu-Sn.

Following solder bar formation, the completed wafer looks like the example in Figure 1.19. The individual devices are singulated, and the final transistor may look like the example in Figure 1.20. This device is now ready to be soldered onto a printed circuit board (PCB) or onto a lead frame to be incorporated into a plastic molded package.

1.6 Summary

In this chapter, a new platform for making switching power transistors using gallium nitride grown on top of a silicon substrate was introduced. Enhancement-mode transistors have in-circuit characteristics very similar to power MOSFETs, but with improved switching speed, lower on-resistance, and at a smaller size than their silicon predecessors. These new capabilities, married with a step forward in chip-scale, high-density packaging, enable power conversion designers to reduce power losses, reduce system size, improve efficiency, and ultimately reduce system costs.

Chapter 2 will connect these basic physical properties of GaN transistors to the electrical characteristics most important in designing power conversion systems. These electrical characteristics will be compared to state-of-the-art silicon MOSFETs, in order to illustrate both the strong similarities and the subtle differences. In Chapters 3–10, these same electrical

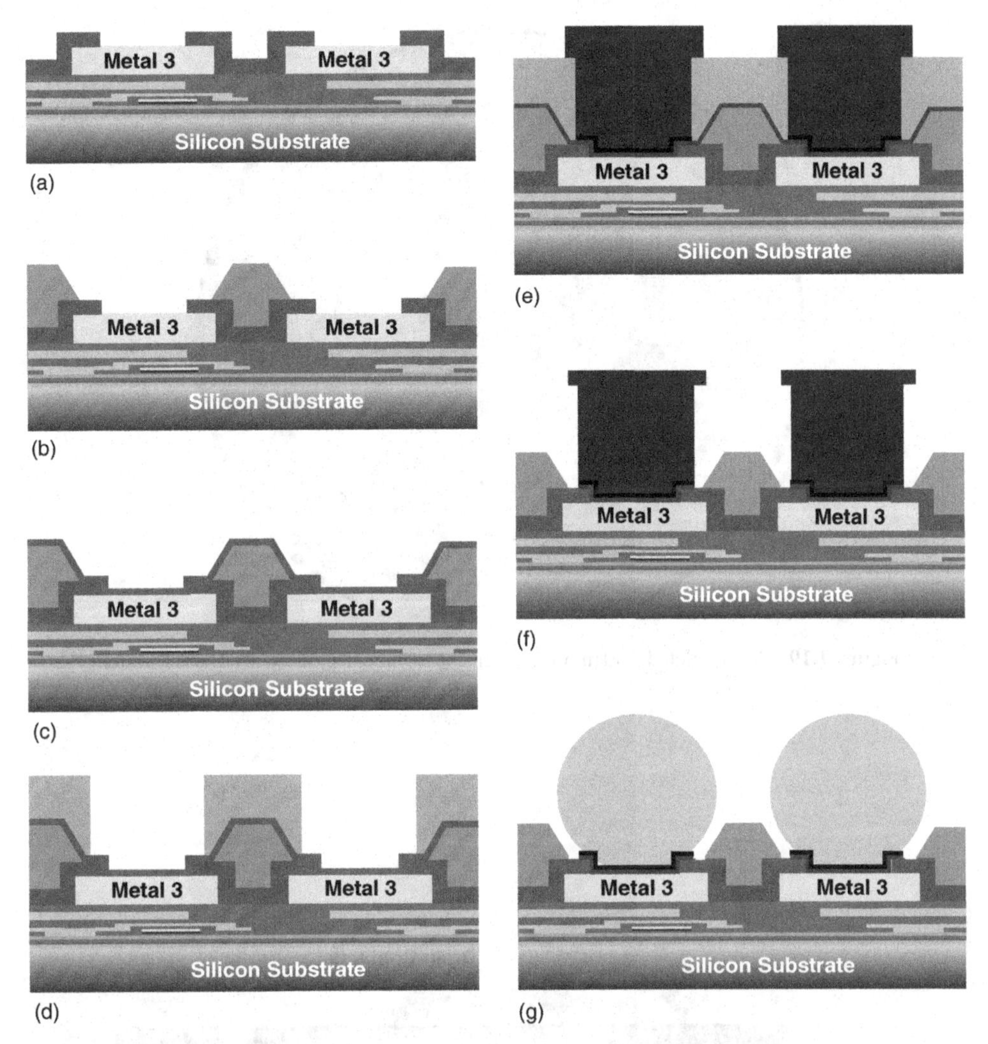

Figure 1.18 A typical process for creating solder bars on an enhancement-mode GaN HEMT (not to scale). The basic process steps are as follows: (a) The finished wafer with openings in the passivation. Metal layer 3 is partially exposed. (b) Photopolyimide is deposited and removed in the area where the solder is desired. (c) An under-bump metal is deposited to create an interface between the aluminum top metal and the solderable material. (d) Photoresist is used to define where the solder will be plated. (e) Copper and solder are plated in the opening. (f) The photoresist is removed and the under-bump metal is etched. (g) The solderable metal is reflowed to form elongated solder bars

characteristics will be related to circuit and system performance in such a way as to give the designer the tools to get the maximum performance from GaN transistors.

Finally, in Chapter 11 we discuss the "why, when, and how" that GaN transistors will displace MOSFETs. Included is a discussion of cost trajectories, reliability, and technology directions for the years 2014 through 2020.

These are the early years of a great new technology.

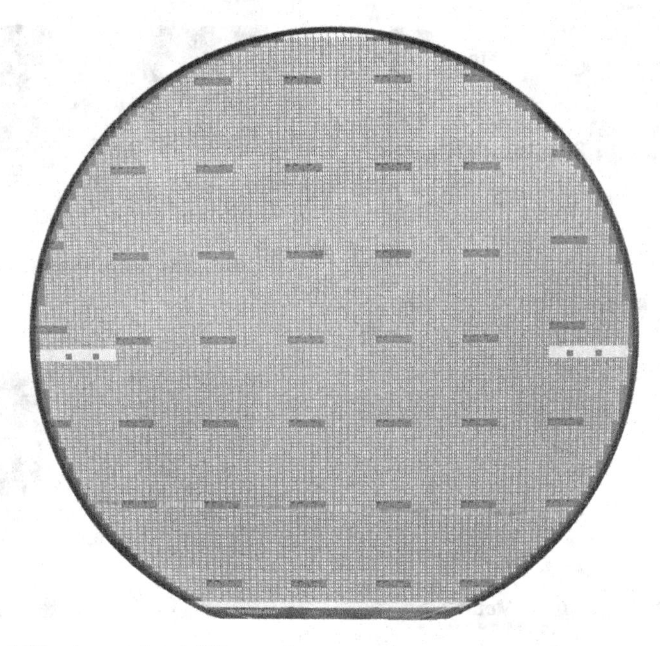

Figure 1.19 A completed 150 mm diameter enhancement-mode GaN HEMT wafer

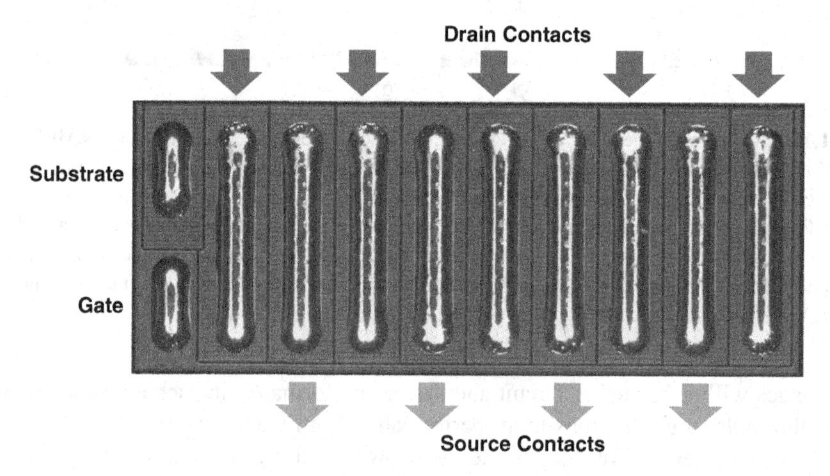

Figure 1.20 A finished device with solder bars in land grid array (LGA). This device measures approximately 4 mm × 1.6 mm

References

1. Sze, S.M. (1981) *Physics of Semiconductor Devices*, 2nd edn, John Wiley and Sons, Hoboken, NJ.
2. Baliga, B.J. (1996) *Power Semiconductor Devices*, PWS Publishing Company, Boston, MA, p. 373.
3. Mitani, E., Haematsu, H., Yokogawa, S. *et al.* "Mass production of high voltage GaAs and GaN devices," *CS Mantech Conference*, Vancouver B.C., Canada, Apr. 24–27, 2006.
4. Mimura, T., Tokoyama, N., Kusakawa, H. *et al.* "GaAs MOSFET for low-power high-speed logic applications," *37th Device Research Conference*, University of Colorado, Boulder, CO, 25–27 June 1979.
5. Khan, M.Asif, Kuznia, J.N., and Olson, D.T. High electron mobility transistor based on a GaN-$Al_xGa_{1-x;}N$ heterojunction. *Applied Physics Letters*, **65** (9), 1121–1123 29.
6. Bykhovski, A., Gelmont, B., and Shur, M. (1993) The influence of the strain induced electric field on the charge distribution in GaNAlNGaN structure. *Journal of Applied Physics*, **74**, 6734.
7. Yu, E., Sullivan, G., Asbeck, P. *et al.* (1997) Measurement of piezoelectrically induced charge in GaN/AlGaN heterostructure field-effect transistors. *Applied Physics Letters*, **71**, 2794.
8. Asbeck, P., Yu, E., Lau, S. *et al.* (1997) Piezoelectric charge densities in AlGaN/GaN HFETs. *Electronics Letters*, **33**, 1230.
9. Liu, Q.Z. and Lau, S.S. (1998) A Review of the Metal-GaN contact technology. *Solid-State Electronics*, **42** (5)
10. Javorka, P., Alam, A., Wolter, M. *et al.* (2002) AlGaN/GaN HEMTs on (111) silicon substrates. *IEEE Electron Device Letters*, **23** (1)
11. Liu, Q.Z., Yu, L.S., Lau, S.S. *et al.* (1997) Thermally stable PtSi Schottky contact on n-GaN. *Applied Physics Letters*, **70** (1)
12. Kordoš, P., Heidelberger, G., Bernát, J. *et al.* (2005) High-power SiO2/AlGaN/GaN metal-oxide-semiconductor heterostructure field-effect transistors. *Applied Physics Letters*, **87**.
13. Lanford, W.B., Tanaka, T., Otoki, Y., and Adesida, I. (2005) Recessed-gate enhancement-mode GaN HEMT with high threshold voltage. *Electronics Letters*, **41** (7)
14. Cai, Y., Zhou, Y., Lau, K.M., and Chen, K.J. (2006) Control of threshold voltage of AlGaN/GaN HEMTs by fluoride-based plasma treatment: from depletion-mode to enhancement-mode. *IEEE Transactions on Electron Devices*, **53** (9)
15. Davis, S., "Enhancement-mode GaN MOSFET Delivers Impressive Performance," *Power Electronics Technology* (March 2010)
16. Hu, X., Simin, G., Yang, J. *et al.* (2000) Enhancement-mode AIGaN/GaN HFET with selectively grown pn junction gate. *Electronics Letters*, **36** (8)
17. Murphy, M., (10 Mar 2009) "Cascode circuit employing a depletion-mode," GaN-based FET, US Patent No. 7,501,670 B2.
18. Huang, X., Liu, Z., Li, Q., and Lee, F.C. (2013) Evaluation and application of 600V GaN HEMT in cascode structure," Proceedings of the 28th Annual IEEE Applied Power Electronics Conference (APEC), Long Beach, CA., March 2013.
19. Strite, S. and Morkoç, H. (1992) GaN, AlN, and InN: a review. *Journal of Vacuum Science and Technology, B*, **10** (4)
20. Nakamura, Shuji. (1991) GaN growth using GaN buffer layer. *Japanese Journal of Applied Physics*, **30** (10A)
21. Nakamura, S., Iwasa, N., Senoh, M., and Mukai, T. (1992) Hole compensation mechanism of p-type GaN films. *Journal of Applied Physics*, **31**, 1258.
22. Amano, H., Sawaki, N., Akasaki, I., and Toyoda, Y. (1986) Metalorganic vapor phase epitaxial growth of a high quality GaN film using an AlN buffer layer. *Applied Physics Letters*, **48**, 353.
23. Hughes, W.C., Rowland, W.H. Jr., Johnson, M.A.L. *et al.* (1995) Molecular beam epitaxy growth and properties of GaN films on GaN/SiC substrates. *Journal of Vacuum Science and Technology, B*, **13** (4)
24. Lidow, A., Beach, R., Nakata, A. *et al.* (26 March 2013) "Enhancement Mode GaN HEMT Device and Method for Fabricating the Same," U.S. Patent 8,404,508.
25. Lidow, A., Beach, R., Nakata, A. *et al.* (8 Jan. 2013) Compensated Gate MOSFET and Method for Fabricating the Same," U.S. Patent No. 8,350,294.
26. Lidow, A., Strydom, J., de Rooij, M., and Ma, Y. (2012) *GaN Transistors for Efficient Power Conversion*, 1st edn, Power Conversion Press, El Segundo, p. 9.

27. Harman, G. (2010) *Wire Bonding in Microelectronics*, 3rd edn, McGraw-Hill Companies Inc., New York.
28. Coucoulas, A. (1970) Compliant bonding," Proceedings 1970 IEEE 20th Electronic Components Conference, Washington, D.C. pp. 380–389.
29. Heleine, T.L., Murcko, R.M., and Wang, S.C. (1991) A wire bond reliability model," Proceedings of the 41st Electronic Components and Technology Conference, Atlanta, GA, 1991.

2

GaN Transistor Electrical Characteristics

2.1 Introduction

In this chapter, the basic physical properties of GaN transistors discussed in Chapter 1 will be connected to electrical characteristics that are important when developing power conversion circuits and systems. These electrical characteristics will be compared to state-of-the-art silicon power MOSFETs in order to explore both their similarities and their differences. Understanding these similarities and differences is fundamental to understanding the extent to which existing power conversion systems can be improved by GaN-based technologies.

2.2 Key Device Parameters

The key operating parameters should give the designer most of the information necessary to design a system with predictable results. For a power-switching transistor, the most basic parameters are: breakdown voltage between source and drain electrodes (BV_{DSS}), on-resistance ($R_{DS(on)}$), and threshold voltage ($V_{GS(th)}$). These three characteristics are enough to get a device to function in a circuit. In order to understand how this device will work when switched on and off, capacitances and reverse-conduction characteristics need to be added. Also, the amount of heat that can be extracted from a device is necessary for developing a full understanding of device and circuit performance. How all of these basic parameters vary over all the operating conditions of a power conversion system is the subject of this chapter.

2.2.1 *Breakdown Voltage (BV_{DSS}) and Leakage Current (I_{DSS})*

The breakdown voltage between the source and drain terminals of a GaN HEMT is determined by several factors [1] including: the fundamental E_{Crit} of GaN discussed in Chapter 1; the specific design of the device; the specifics of the heterostructure; the internal insulating layers in the device structure above the gate, source, and drain electrodes; and the underlying substrate material properties. A semiconductor transistor will break down and conduct current (or potentially destroy itself) when the critical electric field of any of the constituent materials is exceeded.

GaN Transistors for Efficient Power Conversion, Second Edition.
Alex Lidow, Johan Strydom, Michael de Rooij, and David Reusch.

Companion Website: http://www.wiley.com/go/gan_transistors

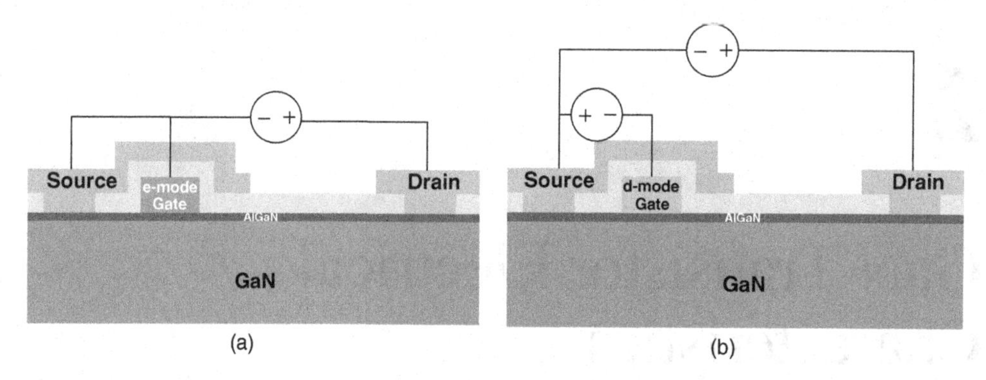

Figure 2.1 An illustration of (a) a basic enhancement-mode GaN transistor with reverse bias applied, and (b) a depletion-mode transistor with the gate turned off and reverse bias applied

To visualize these electric fields in an HEMT transistor, a good starting point is the structure illustrated in Figure 2.1. Applied from drain to source is a voltage such that the transistor is reverse biased. In an enhancement-mode transistor, the transistor is turned off by shorting the gate to the source. In a depletion-mode device, there would need to be a negative bias from the gate to source electrodes to keep the HEMT from conducting current.

A simple two-dimensional analysis of the structure in Figure 2.1(a), shown in Figure 2.2, illustrates the electric fields at any point in the device. Higher electric fields, where the contour lines are closest together, develop near the drain and gate electrodes. When these fields, at any location in the device structure, exceed E_{Crit}, the device will break down and conduct current.

Breakdown can also occur between the metal busing layers in the device. For example, Figure 1.17 showed an enhancement-mode HEMT with three levels of metal used to consolidate drain current and source current and bring them to the outside of the device where they can be connected into a circuit. When the device is reverse biased, sometimes called the "blocking state," one of these layers may be connected to the source potential while an adjacent layer – or perhaps a higher or lower layer in the stack – might be at drain potential. If the E_{Crit} of the dielectric material separating these two layers is exceeded, breakdown will occur. This can be prevented by either increasing the separation of the layers, or by switching to an insulating layer with a higher E_{Crit}.

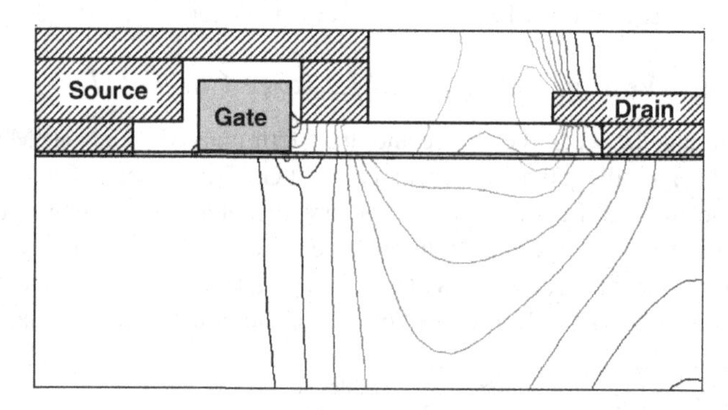

Figure 2.2 The device in Figure 2.1(a) showing the electric fields when voltage is applied from drain to source

If a HEMT device goes into breakdown, either from exceeding the E_{Crit} of the GaN or of an insulating layer, the effect tends to be destructive. In the case where the insulating layers exceed their capacity for blocking voltage, a physical rupture of the dielectric material will develop. The closer the electric field approaches to $E_{Crit\ (insulator)}$, the sooner the rupture will occur. This effect is called “time-dependent dielectric failure” and is discussed extensively in the literature [2]. When the GaN layer exceeds $E_{Crit\ (GaN)}$, a different mechanism causes device failure. When breakdown occurs in the GaN or Al GaN regions, the electrons generated can destroy the 2DEG, causing the device on-resistance to increase greatly [3].

When a transistor of any kind is in the blocking state, there is still a small amount of “leakage current” (I_{DSS}) that will flow between terminals. In the case of an HEMT device, the leakage current can flow from the drain to the source electrodes, from the drain to the gate electrodes, or from the drain to the substrate. The sum of these three leakage currents will be the total I_{DSS} measured between drain and source in a circuit.

These three components of leakage current are shown in Figure 2.3 and are measured on an enhancement-mode HEMT with a breakdown voltage above 700 V. In this example, the substrate material is silicon (Si) and is connected to source potential. The current has been normalized to a 1 mm-wide gate structure.

When designing a power conversion system, I_{DSS} can become a significant source of power loss. For example, if a 100 V device has 100 μA I_{DSS} leakage current, the overall power dissipation due to the leakage current would be 10 mW. In certain applications requiring very low standby power, this amount of loss could become unacceptable.

The drain-to-source leakage current (I_{DSS}) also varies with temperature. Figure 2.4 shows a family of curves from a sample of enhancement-mode transistors with 1 meter gate width, showing leakage current measured at various temperatures. Figure 2.4(a) traces an individual device at various temperatures as a function of V_{DS}. Figure 2.4(b) compares leakage current measurements of six different devices of the same part type plotted against the inverse temperature to illustrate device-to-device variation. Despite the fact that these devices were

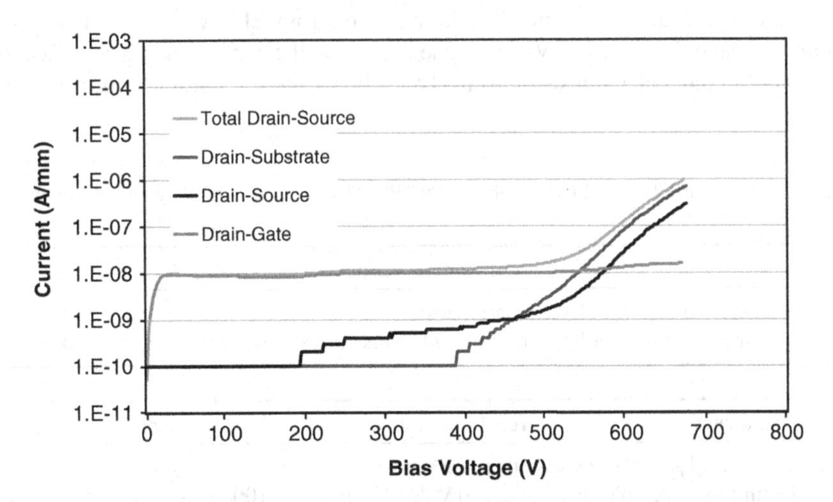

Figure 2.3 There are three main sources of current that add up to the total leakage current between drain and source terminals; drain-gate leakage, drain-source leakage, and drain-substrate leakage

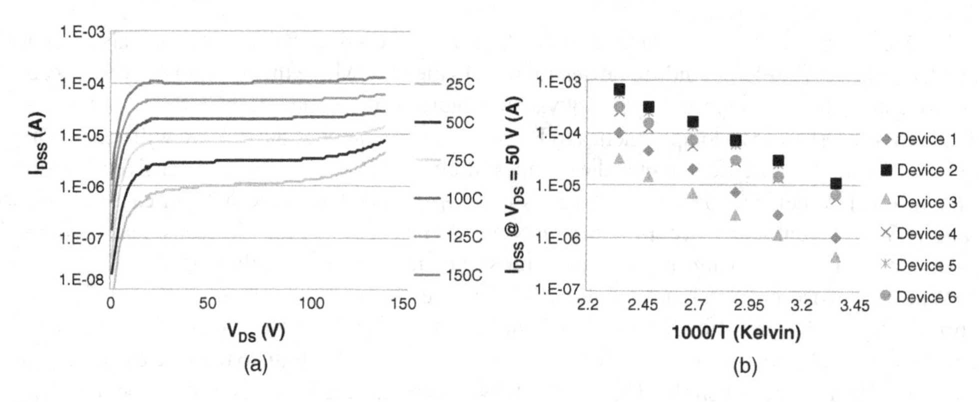

Figure 2.4 I_{DSS} over temperature for an EPC2001 enhancement-mode GaN transistor: (a) I_{DSS} vs. temperature for an individual device, and (b) I_{DSS} (Measured at $V_{DS} = 50$ V) for six different devices showing device-to-device variation

selected to cover a wide range of drain-to-source leakage current, the slope of these measurements for each device gives a consistent activation energy of $E_A = -0.4$ eV. The activation energy for this commercial device lies between values reported in the literature of -0.2 eV due to surface related traps [4,5], and -0.99 eV due to a temperature-assisted tunneling mechanism [4].

Typically, commercial transistors are described using a datasheet. These datasheets vary from supplier to supplier and do not always follow consistent conventions for measurements leading to the numbers listed in the data sheet tables. For BV_{DSS} and I_{DSS}, there are several relevant sections of a datasheet that can be analyzed to deduce device behavior and compare parts from different manufacturers.

In Tables 2.1 and 2.2 are examples of data from two different manufacturers of GaN transistors, Efficient Power Conversion Corporation (EPC) and Transphorm. Table 2.1 gives the specifications for an enhancement-mode transistor from EPC with a nominal maximum drain-source voltage (V_{DS}) of 100 V, and Table 2.2 shows the corresponding specifications for a cascode-configured GaN transistor from Transphorm with a nominal maximum V_{DS} of

Table 2.1 Data from an Efficient Power Conversion EPC2001 datasheet showing sections relating to I_{DSS} and BV_{DSS} [6]

Maximum Ratings			
V_{DS}	Drain-to-source voltage (continuous)	100	V
	Drain-to-source voltage (up to 10,000 5 ms pulses at 125 °C)	120	V

	Parameter	Test Conditions	Min	Typical	Max	Unit
Static Characteristics ($T_j = 25$ °C, unless otherwise stated)						
BV_{DSS}	Drain-to-source voltage	$V_{GS} = 0$ V, $I_D = 300$ µA	100	—	—	V
I_{DSS}	Drain source leakage	$V_{DS} = 80$ V, $V_{GS} = 0$ V	—	100	250	µA

Table 2.2 Data from a Transphorm cascode TPH3006PD datasheet showing sections relating to I_{DSS} and BV_{DSS} [7]

Symbol	Parameter	Limit Value	Unit
Absolute maximum ratings ($T_C = 25\,°C$ unless otherwise stated)			
V_{DSS}	Drain-to-source voltage	600	V
V_{TDS}	Transient drain-to-source voltage[a]	750	V

Symbol	Parameter	Min	Typical	Max	Unit	Test Conditions
Static Characteristics ($T_C = 25\,°C$, unless otherwise stated)						
I_{DSS}	Drain-to-source leakage current, $T_J = 25°C$	—	2.5	90	µA	$V_{DS} = 600\,V$, $V_{GS} = 0\,V$, $T_J = 25\,°C$
I_{DSS}	Drain-to-source leakage current, $T_J = 150\,°C$	—	20	—	µA	$V_{DS} = 600\,V$, $V_{GS} = 0\,V$, $T_J = 150\,°C$

[a] For 1 µsec, duty cycle $D = 0.1$.

600 V. In both cases, there is also a specification for a transient voltage capability above the maximum V_{DS}. This transient capability means that, for short periods of time, the devices can withstand higher voltages than their respective rated maximums. This type of rating is a sign of the relative immaturity of the technology. In the case of a mature 100 V Si MOSFET shown in Table 2.3, there are no transient over-voltage ratings. Instead, an avalanche capability is specified, giving the user license to take the device into full drain-source breakdown with a certain amount of energy (specified in millijoules). Whereas MOSFET users rarely take advantage of this avalanche capability, over time, and with improved device design, GaN transistors will also mature to the point where they can be used in avalanche without catastrophic failure.

Tables 2.1 and 2.2 also list the static (DC) I_{DSS} characteristics for their respective devices. In each case, the test conditions are slightly different. The EPC2001 gives the I_{DSS} at 80 V and the I_D at BV_{DSS}, both at 25 °C, whereas the TPH3006PD lists the I_{DSS} at maximum V_{DS} at both

Table 2.3 Data from an Infineon BSC060N10NS3 G datasheet for an Si MOSFET showing sections relating to I_{DSS} and BV_{DSS} and avalanche energy [8]

Parameter	Symbol	Test Conditons	Value			Unit
Maximum Ratings ($T_J = 25\,°C$, unless otherwise specified)						
Avalanche energy, single pulse	E_{AS}	$I_D = 50\,A$, $R_{GS} = 25\,\Omega$	230			mJ
Static Characteristics						
Drain-source breakdown voltage	$V_{(BR)DSS}$	$V_{GS} = 0\,V$, $I_D = 1\,mA$	100	—	—	V
Zero gate voltage drain current	I_{DSS}	$V_{DS} = 100\,V$, $V_{GS} = 0\,V$, $T_J = 25\,°C$	—	0.01	1	µA
		$V_{DS} = 100\,V$, $V_{GS} = 0\,V$, $T_J = 125\,°C$	—	10	100	µA

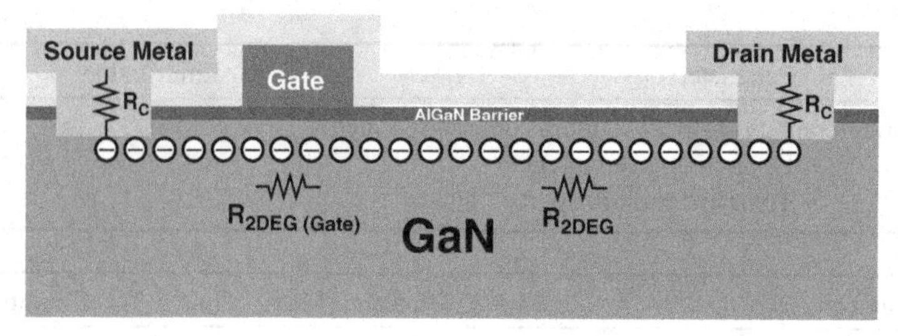

Figure 2.5 Cross section of a GaN HEMT, showing the major components of $R_{DS(on)}$

25 °C and 150 °C. The silicon MOSFET, BSC060N10NS3G from Infineon, has specifications for BV_{DSS} and I_{DSS} under yet a third set of test conditions.

2.2.2 *On-Resistance ($R_{DS(on)}$)*

The on-resistance of a transistor is the sum of all the resistance elements that make up the device. Figure 2.5 shows the major elements that contribute to the $R_{DS(on)}$ of the device. The source and drain metals have to connect to the 2DEG through the AlGaN barrier. This component of resistance is called the contact resistance (R_C). Electrons then flow in the 2DEG with a resistance R_{2DEG}. This resistance is determined by the mobility of the electrons (μ_{2DEG}), the number of electrons created by the 2DEG (N_{2DEG}), the distance the electrons have to travel (L_{2DEG}), the width of the 2DEG (W_{2DEG}), and the universal charge constant, q ($1.6 \cdot 10^{-19}$ coulombs). This resistance can be described by the following formula [9]:

$$R_{2DEG} = L_{2DEG}/(q \cdot \mu_{2DEG} \cdot N_{2DEG} \cdot W_{2DEG}) \tag{2.1}$$

As discussed in Chapter 1, the number of electrons in the 2DEG will depend on the amount of strain induced by the AlGaN barrier. However, under the gate electrode, this 2DEG could have a lower concentration than in the region between the gate and drain electrodes. This electron concentration depends on the type of gate (recessed gate, MOS, Schottky, or pGaN), the particular process used, and the heterostructure deployed. It also depends on the voltage applied to the gate. A fully enhanced gate will have a higher electron concentration than a partially enhanced gate. A good approximation of the resistance of the transistor shown in Figure 2.5 can then be calculated as follows:

$$R_{HEMT} = 2 \cdot R_C + R_{2DEG} + R_{2DEG(Gate)} \tag{2.2}$$

Additional parasitic resistance ($R_{parasitic}$) can come in the form of metal resistance from the multiple metal buses that conduct the current from the individual source and drain electrodes to the terminals of the transistor. In the case of a cascode device, the resistance of the external Si MOSFET would also need to be added to the total resistance of the transistor.

In a power conversion circuit, the conduction losses of the transistor are quite significant and therefore the device is typically used either fully turned on, or fully turned off. For this reason, a key parameter for specifying any power transistor is the on-resistance, $R_{DS(on)}$, and can be defined as:

$$R_{DS(on)} = R_{HEMT(Fully\ Enhanced)} + R_{parasitic} \tag{2.3}$$

Each of these components of $R_{DS(on)}$ will vary with temperature. The metal layers typically are made of combinations of copper and aluminum and have resistivity temperature coefficients in the range of $3.8 \cdot 10^{-3}$/°C [10] for copper, to $3.9 \cdot 10^{-3}$/°C [11] for aluminum. In contrast, a 2DEG's resistivity temperature coefficient is significantly higher in the range of $1.3 \cdot 10^{-2}$/°C [12,13], and the contact resistance, R_C, has a temperature coefficient in the range of $4.7 \cdot 10^{-3}$/°C [12]. The transistor's $R_{DS(on)}$, as a function of temperature, would be expected to be somewhere in between these numbers and can be approximated as:

$$R_{DS(on)}(t) = R_{parasitic}(t) + (2 \cdot R_C(t)) + (R_{2DEG} + R_{2DEG(Gate)})(t) \tag{2.4}$$

The final temperature variation of $R_{DS(on)}$ will depend on the design of the device: how much of the $R_{DS(on)}$ comes from 2DEG, contact resistance, or parasitic metal resistance. Devices in commercial use, however, demonstrate a variation of $R_{DS(on)}$ with temperature that is somewhat less than silicon MOSFETs as shown in Figure 2.9. The 100 V-rated enhancement-mode GaN HEMT in this example has a temperature coefficient of approximately $6.5 \cdot 10^{-3}$/°C compared with the Si MOSFET's temperature coefficient of $20 \cdot 10^{-3}$/°C. For devices designed for higher BV_{DSS}, the 2DEG will be a greater fraction of the total $R_{DS(on)}$ and, since the temperature coefficient of the 2DEG is higher than that of the parasitic elements and the contact resistance, the temperature coefficient will be higher (see Figure 2.6).

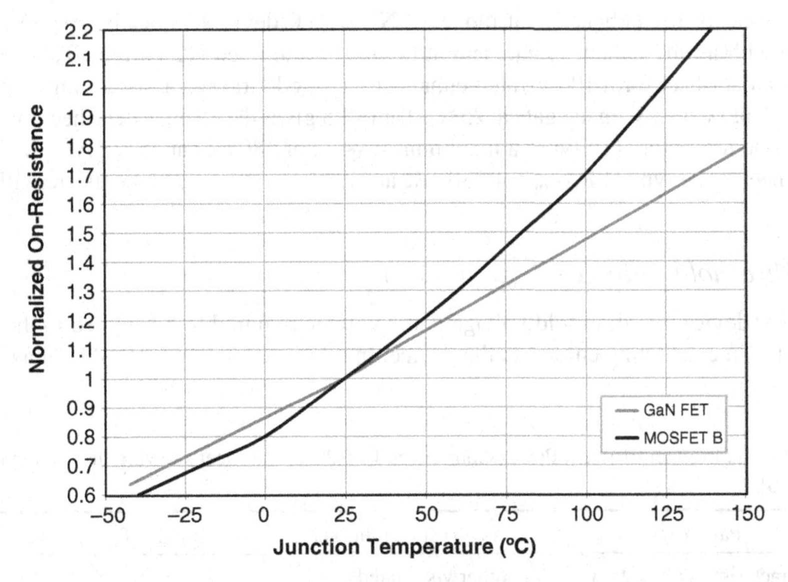

Figure 2.6 Normalized $R_{DS(on)}$ vs. temperature for a 100 V enhancement-mode GaN transistor (EPC2001) compared with an Si power MOSFET with similar ratings [6,8]

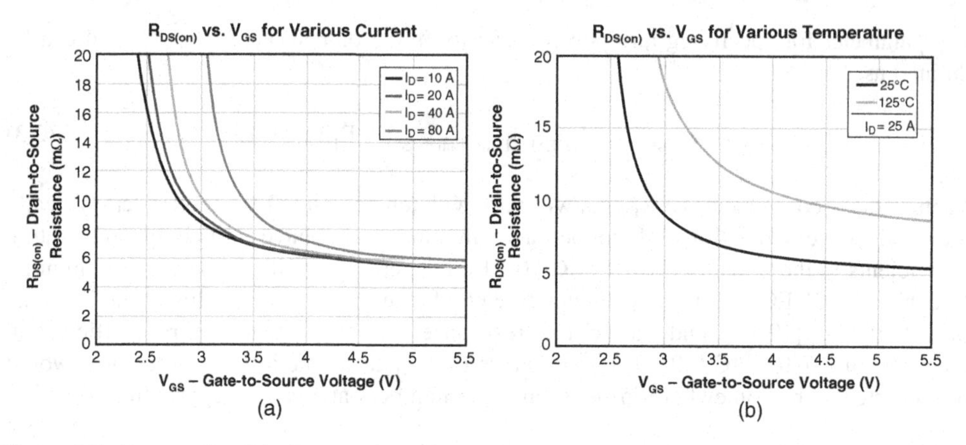

Figure 2.7 Two graphs of the $R_{DS(on)}$ of an enhancement-mode GaN HEMT (EPC2001) as a function of gate voltage for various drain currents (a) and at various temperatures (b) [6]

Equation 2.2 breaks down the resistance of the 2DEG into two components, the 2DEG between the drain and gate and the 2DEG under the gate. The resistance of the 2DEG under the gate changes from a very high value when the device is in the off state, to a very low value when the device is in the on-state. This transition depends on the specific design of the gate and has a significant impact on device performance in a switching circuit.

In Figure 2.7 is a graph of the $R_{DS(on)}$ of an enhancement-mode GaN HEMT as a function of gate-to-source voltage for various drain currents (I_D) and at various temperatures. In this example, the resistance of the gate 2DEG rapidly decreases until it is fully enhanced at about $V_{GS} = 4$ V. Beyond that voltage there is only a nominal decrease in overall device $R_{DS(on)}$.

The datasheets for enhancement-mode GaN HEMT devices typically specify this on-resistance with a fully enhanced gate and at the maximum rated DC current. Table 2.4 gives data from a datasheet for a 100 V-rated enhancement-mode transistor specifying a maximum $R_{DS(on)}$ of 7 mΩ with 5 V on the gate at 25 °C. Table 2.5 gives data from a datasheet for a 600 V-rated cascode transistor specifying a maximum $R_{DS(on)}$ of 180 mΩ at 25 °C. Also indicated in this datasheet is the typical $R_{DS(on)}$ of 330 mΩ at 175 °C. Both measurements are with 8 V on the gate.

2.2.3 *Threshold Voltage ($V_{GS(th)}$ or V_{th})*

For a power device, the threshold voltage is the voltage required to be applied to the gate-to-source to begin conducting current in the device. In other words, the threshold voltage defines

Table 2.4 Data from an Efficient Power Conversion EPC2001 datasheet showing the section relating to $R_{DS(on)}$ [6]

Parameter		Test Conditions	Min	Typ	Max	Unit
Static Characteristics ($T_J = 25$ °C, unless otherwise stated)						
$R_{DS(on)}$	Drain-source On-Resistance	$V_{GS} = 5$ V, $I_D = 25$ A	—	5.6	7	mΩ

Table 2.5 Data from a Transphorm cascode TPH3006PD datasheet showing the section relating to $R_{DS(on)}$ [7]

Symbol	Parameter	Min	Typical	Max	Unit	Test Conditions
Electrical Characteristics (T_C = 25 °C, unless otherwise stated)						
Static						
$R_{DS(on)}$	Drain-source On-Resistance T_J = 25 °C	—	0.15	0.18	Ω	V_{GS} = 8 V, I_D = 11 A, T_J = 25 °C
$R_{DS(on)}$	Drain-source On-Resistance T_J = 175 °C	—	0.33	—	Ω	V_{GS} = 8 V, I_D = 11 A, T_J = 175 °C

the voltage below which the device is off. An enhancement-mode or cascode device has a positive threshold voltage, and a depletion-mode device has a negative threshold voltage.

For a GaN power device, the threshold voltage is when the 2DEG underneath the gate is fully depleted by the voltage generated by the gate electrode [14]. This occurs when the voltage of the gate balances the voltage generated by the piezoelectric strain in the AlGaN/GaN barrier. This voltage has two components, the voltage applied externally to the gate (defined as V_{th}), plus the built-in voltage due to the specifics of the gate metallurgy. In the case of a Schottky gate device, this built-in voltage is the Schottky barrier height [15] of the gate metal on top of the AlGaN barrier. In the case of a MOS HEMT device, it is the voltage generated by the gate metal across the insulator as well as the AlGaN barrier. In the case of a pGaN gate, it is the voltage generated by the built-in field caused by a p-type semiconductor material next to an n-type material.

Because the strain in the AlGaN barrier is relatively constant with temperature, as are the voltages generated by the internal metallurgy, the threshold voltage in a GaN HEMT is relatively constant with temperature as shown in Figure 2.8. The threshold voltage of a cascode device, however, would track the change in threshold of the Si MOSFET in series with the GaN d-mode HEMT, and will decline with increasing temperature.

Table 2.6 shows data from the datasheet of a 100 V-rated enhancement-mode GaN HEMT. The typical value of 1.4 V is measured at 5 mA, a small amount of current compared with the 25 A DC rating for this same transistor.

Table 2.7 presents data from a 600 V-rated cascode transistor. In this case the V_{th} is specified at 1 mA, which is a small value compared with the 17 A DC rating. No information is given on the change of this device's threshold voltage as a function of temperature. Figure 2.9, however, shows a graph of the threshold voltage versus temperature for a 100 V Infineon BSC060N10NS3 [8] power MOSFET. The V_{th} for this silicon device drops about 38% from about 2.75 V to about 2 V, when temperature changes from 25 °C to 125 °C compared with a nominal 3% increase in the enhancement-mode GaN transistor.

2.3 Capacitance and Charge

A transistor's capacitance is a significant factor in determining the energy that will be lost in the device during a transition from the on-state to the off-state, or vice versa. The capacitance (C) determines the amount of charge (Q) that needs to be supplied to various terminals of the device to change the voltage across those terminals ($Q = C \cdot V$). The faster this charge is supplied, the faster the device will change voltage.

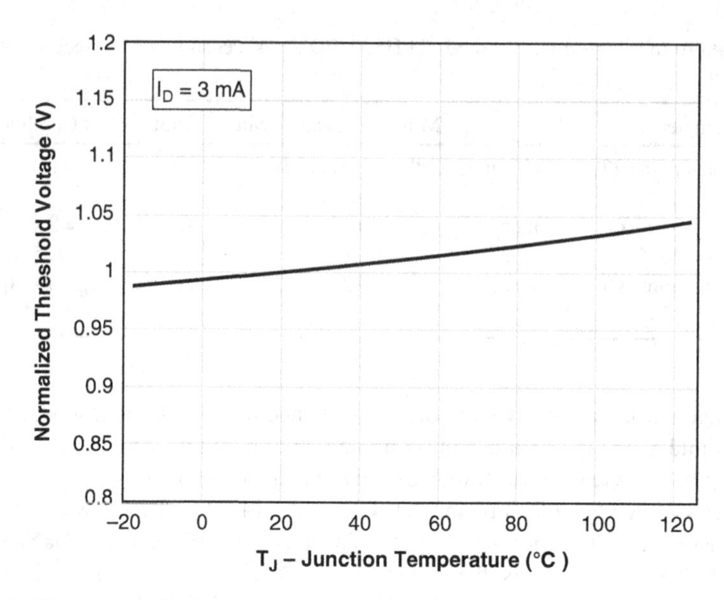

Figure 2.8 EPC2010 normalized threshold voltage vs. temperature showing only a 3% change over the normal operating range of this device

Table 2.6 Data from an Efficient Power Conversion EPC2001 datasheet showing the section relating to V_{th} [6]

	Parameter	Test Conditions	Min	Typ	Max	Unit
Static Characteristics ($T_J = 25\,°C$, unless otherwise stated)						
$V_{GS(th)}$	Gate threshold voltage	$V_{DS} = V_{GS}$, $I_D = 5\,mA$	0.7	1.4	2.5	V

Table 2.7 Data from a Transphorm cascode TPH3006PD datasheet showing the section relating to V_{th} [7]

Symbol	Parameter	Min	Typical	Max	Unit	Test Conditions
Electrical Characteristics ($T_C = 25\,°C$, unless otherwise stated)						
$V_{GS(th)}$	Gate threshold voltage	1.35	1.8	2.35	V	$V_{DS} = V_{GS}$, $I_D = 1\,mA$

There are three main elements of capacitance related to a FET: (1) gate-to-source capacitance (C_{GS}), (2) gate-to-drain capacitance (C_{GD}), and (3) drain-to-source capacitance (C_{DS}). Figure 2.10 illustrates the physical origin of each of these capacitances. Occasionally designers need to look just at the total capacitance seen at either the input terminals ($C_{ISS} = C_{GD} + C_{GS}$), or output terminals ($C_{OSS} = C_{GD} + C_{DS}$).

These capacitances are a function of the voltage applied to various terminals. Figure 2.11 shows how the values change for an enhancement-mode HEMT, as the voltage from drain-to-source increases. The reason for the drop in capacitance as V_{DS} increases is that the free electrons in the GaN are depleted. For example, the initial step down in C_{OSS} is caused by the depletion of the 2DEG near the surface. Higher V_{DS} values extend the depletion region laterally

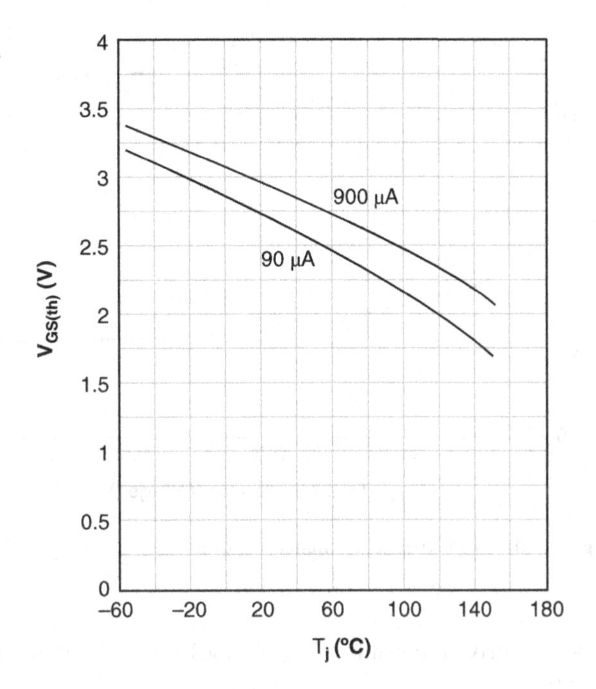

Figure 2.9 Data from an Infineon BSC060N10NS3 G datasheet showing the change in V_{th} vs. temperature [8]

from the field plate edge to the drain, further depleting the 2DEG and eliminating its capacitive component.

The result of integrating the capacitance between two terminals across the range of voltage applied to the same terminals is the amount of charge (Q) that is stored in the capacitor. Since current-integrated-over-time equals charge, it is often very convenient to look at the amount of charge necessary to change the voltage across various terminals in the GaN HEMT. Figure 2.12 shows the amount of gate charge, Q_G, that must be supplied to increase the voltage from gate-to-source to the desired voltage. Q_G is the integrated value of C_{ISS} from the starting voltage on the gate to the ending voltage. Referring to Figure 2.12, it can be seen that about 5 nC is needed to achieve 5 V on the gate, a value that will ensure that the device is

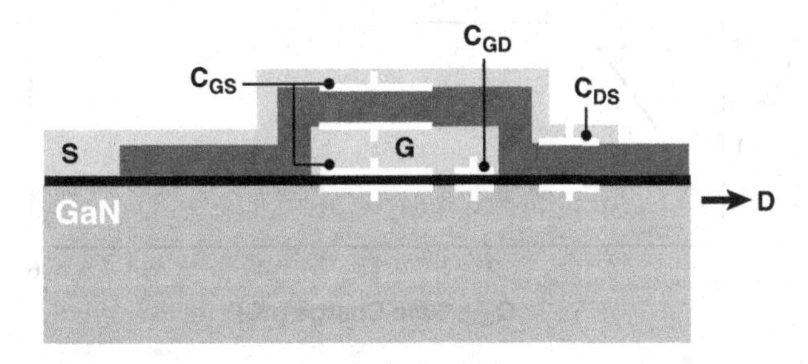

Figure 2.10 Schematic of GaN transistor capacitive sources

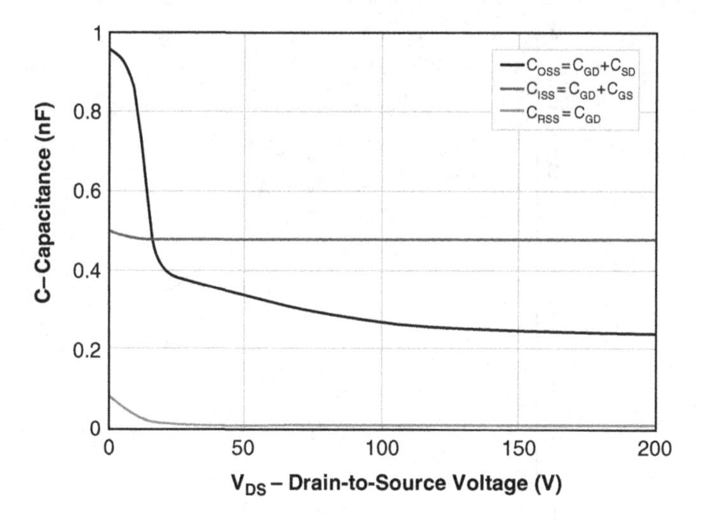

Figure 2.11 EPC2010 capacitance vs. drain-source voltage [16]

fully turned on. If the gate drive is capable of supplying 1 A of current, it will take about 5 ns to achieve this voltage.

The Q_{GD} and Q_{GS} are also specified separately because they impact the voltage- and current-switching transition speeds, respectively. Also, the ratio of these two values, Q_{GD}/Q_{GS}, called the Miller ratio, is often an important metric to determine the point at which a device might turn on due to a voltage transient applied across the drain and source. The Miller ratio will be discussed in greater detail in Chapter 3.

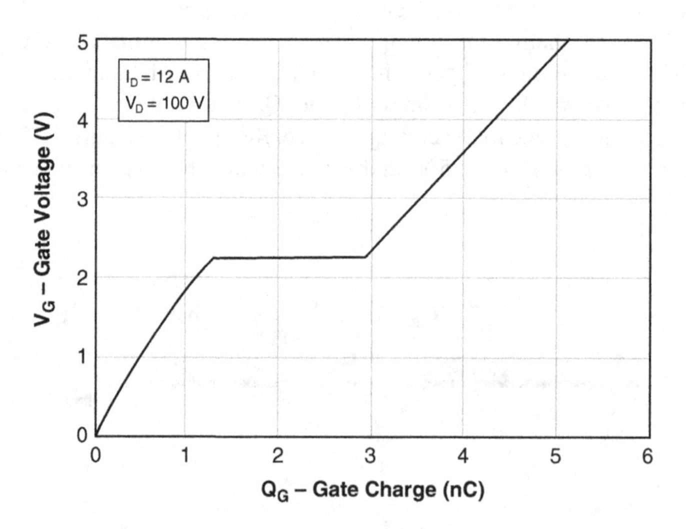

Figure 2.12 EPC2010 gate charge vs. gate voltage [16]

Table 2.8 Data from an Efficient Power Conversion EPC2001 datasheet showing the section relating to capacitance and charge [6]

	Parameter	Test Conditions	Min	Typ	Max	Unit
Dynamic Characteristics ($T_J = 25\,°C$, unless otherwise stated)						
C_{ISS}	Input capacitance	$V_{DS} = 50\,V$, $V_{GS} = 0\,V$		850	950	pF
C_{OSS}	Output capacitance			450	525	
C_{RSS}	Reverse transfer capacitance			20	30	
Q_G	Total gate charge ($V_{GS} = 5\,V$)	$V_{DS} = 50\,V$, $I_D = 25\,A$		8	10	nC
Q_{GD}	Gate-to-drain charge			2.2	2.7	
Q_{GS}	Gate-to-source charge			2.3	2.8	
Q_{OSS}	Output charge	$V_{DS} = 50\,V$, $V_{GS} = 0\,V$		35	40	

The capacitances and charges for an enhancement-mode GaN HEMT are shown in Table 2.8. The gate-to-drain charge, and its corresponding capacitance C_{RSS} (or C_{GD}), will change with drain-to-source voltage. In this example, the values are given at 50 V, which is half the rated BV_{DSS}. This convention is used because, historically, the operating point for transistors in power conversion designs was about half the maximum rated voltage to provide safety margins for overshoot caused during switching transients.

2.4 Reverse Conduction

V_{SD} is the voltage drop across the device when voltage is applied from source-to-drain. This is the reverse direction from the normal forward FET conduction. In an Si MOSFET, there is a p-n junction that forms a diode from the body of the channel to the drain of the transistor. It is, therefore, called a body-drain diode. As discussed in Chapter 1, enhancement-mode GaN HEMT transistors do not have a p-n diode, but they do conduct in a way similar to a diode in the reverse direction. Figure 2.13 shows how this "body diode" forward voltage drop varies with source-drain current. It should be noted that, because this body diode is formed by turning on the 2DEG in the reverse direction using the drain-gate voltage to enhance the channel, if the gate voltage is lowered below 0 V, the forward drop will increase proportionately. For example, if the gate drive of a circuit turns off the GaN HEMT by applying a *negative* 1 V to the gate, the V_{SD} at 0.5 A will be 2.8 V instead of 1.8 V with 0 V_{GS}.

Because the reverse conduction in a GaN transistor is due to the turning on of the 2DEG, the forward voltage drop will change with temperature in much the same way as the $R_{DS(on)}$ changes with temperature in forward conduction. In an Si MOSFET, the reverse conduction is due to a p-n junction diode. Contrary to a GaN HEMT, the diode forward drop goes down with temperature in the Si MOSFET.

The voltage drop across an enhancement-mode GaN HEMT rated at 100 V when conducting in the reverse direction is shown in Table 2.9 [6].

For a cascode device, the reverse-conduction mechanism is somewhat different. The body diode of the Si MOSFET is conducting in series with the depletion-mode HEMT. The voltage across the MOSFET body diode helps to further enhance the depletion-mode HEMT, but the channel of the HEMT is now in series with the Si diode, and the voltage drop is the sum of the two. The reverse conduction is therefore a combination of the MOSFET diode and a depletion-

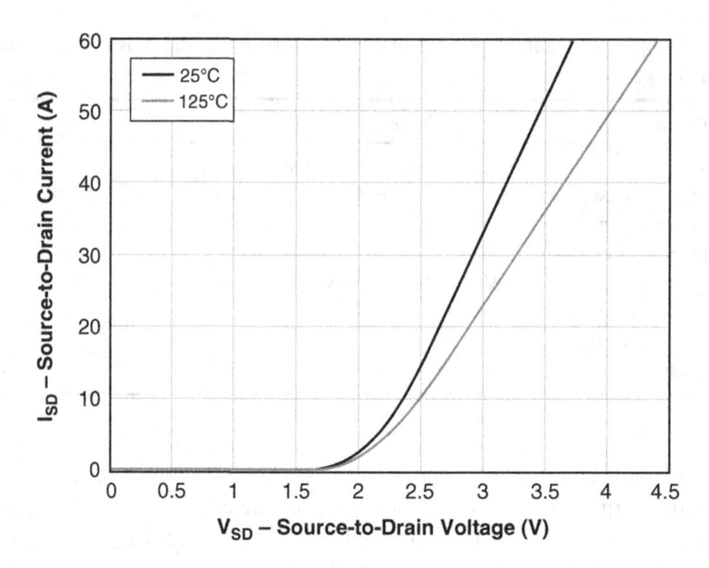

Figure 2.13 EPC2010 body-diode forward drop vs. source-drain current and temperature [16]

Table 2.9 Excerpt from an Efficient Power Conversion EPC2001 datasheet showing the section relating to the voltage drop when the transistor is conducting in the reverse direction [6]

Parameter		Min	Typical	Max	Unit	Test Conditions
Electrical Characteristics ($T_C = 25\,°C$, unless otherwise stated)						
V_{SD}	Source-drain forward voltage	—	1.75	—	V	$I_S = 0.5\,A$, $V_{GS} = 0\,V$, $T = 25\,°C$
		—	1.8	—	V	$I_S = 0.5\,A$, $V_{GS} = 0\,V$, $T = 125\,°C$

mode GaN HEMT, and could have either a positive or negative temperature coefficient to the V_{SD} depending on the specifics of the cascode design.

There is one other charge element, Q_{RR}, which does not directly relate to a device capacitance in an enhancement-mode GaN HEMT. This is the amount of charge dissipated when a body diode is turned off. This charge comes from the minority carriers left over during diode conduction in a MOSFET. Because there are no minority carriers involved in conduction in an enhancement-mode GaN HEMT, there are no reverse recovery losses. Therefore, Q_{RR} is zero, which is a significant advantage compared with power MOSFETs, and will be discussed in greater detail in Chapters 3 and 6.

A cascode transistor, however, does have the stored charge that has to be swept out of the MOSFET before the diode from the series-connected Si MOSFET turns off, as shown in Table 2.10. Because the cascode MOSFET is a low-voltage device, the amount of stored charge is small compared to a comparable 600 V MOSFET. To illustrate this difference, Table 2.11 gives data from an Infineon 650 V CoolMOS™ datasheet [17]. This CoolMOS™ MOSFET has a maximum $R_{DS(on)}$ of 130 mΩ. In comparison, the Transphorm cascode device has a maximum $R_{DS(on)}$ of 180 mΩ – about 50% higher. Nevertheless, the MOSFET has a stored charge, Q_{rr}, that is more than 100 times greater than the cascode GaN transistor.

Table 2.10 Data from a Transphorm cascode TPH3006PD datasheet showing the section relating to conduction in the reverse direction [7]

Symbol	Parameter	Min	Typical	Max	Unit	Test Conditions
Electrical Characteristics ($T_C = 25$ °C, unless otherwise stated)						
Reverse Operation						
I_S	Reverse current	—	—	11	A	$V_{GS} = 0$ V, $T_J = 100$ °C
V_{SD}	Reverse voltage	—	3.8	4.7	V	$V_{GS} = 0$ V, $I_S = 11$ A, $T_J = 25$ °C
V_{SD}	Reverse voltage	—	2.3	2.6	V	$V_{GS} = 0$ V, $I_S = 5.5$ A, $T_J = 25$ °C
t_{rr}	Reverse recovery time	—	30	—	ns	$I_S = 11$ A, $V_{DD} = 480$ V, di/dt = 450 A/μs, $T_J = 25$ °C
Q_{rr}	Reverse recovery charge	—	54	—	nC	

Table 2.11 Data from an Infineon 650 V CoolMOS™ datasheet showing the section relating to conduction in the reverse direction [17]

Parameter	Symbol	Values			Unit	Note/Test Conditions
		Min	Typical	Max		
Diode forward voltage	V_{SD}	—	0.8	—	V	$V_{GS} = 0$ V, $I_F = 44$ A, $T_J = 25$ °C
Reverse recovery time	t_{rr}	—	630	—	ns	$V_R = 400$ V, $I_F = 15$ A, $di_F/dt = 55$ A/μs
Reverse recovery charge	Q_{rr}	—	6.4	—	μC	$V_R = 400$ V, $I_F = 15$ A, $di_F/dt = 55$ A/μs
Peak reverse recovery current	I_{rrm}	—	20	—	A	$V_R = 400$ V, $I_F = 15$ A, $di_F/dt = 55$ A/μs

2.5 Thermal Resistance

The power consumed in a device during operation is dissipated in the form of heat. It is therefore important to understand the ability of a device to transfer the heat to the surrounding environment. Devices interface to their environment through a variety of packaged and "package-less" formats, but the specifications for the transfer of heat have two of the same component elements: junction-to-case thermal resistance ($R_{\theta JC}$) and junction-to-ambient thermal resistance ($R_{\theta JA}$). In the case of a double-sided device that can be cooled from the top and the bottom surfaces, a third parameter, junction-to-board ($R_{\theta JB}$), needs to be added to the list.

Figure 2.14 illustrates the first of these two thermal resistance parameters for a conventional single-sided device. In a single-sided device, the transistor is mounted onto a copper leadframe. This leadframe can be connected to a larger heatsink using thermal interface material, such as thermal grease, to improve the interface contact efficiency if the surfaces are not perfectly flat. The $R_{\theta JC}$ for this configuration would be the thermal resistance from the top die surface in contact with the leadframe through the copper leadframe or "case" – the larger the area of the

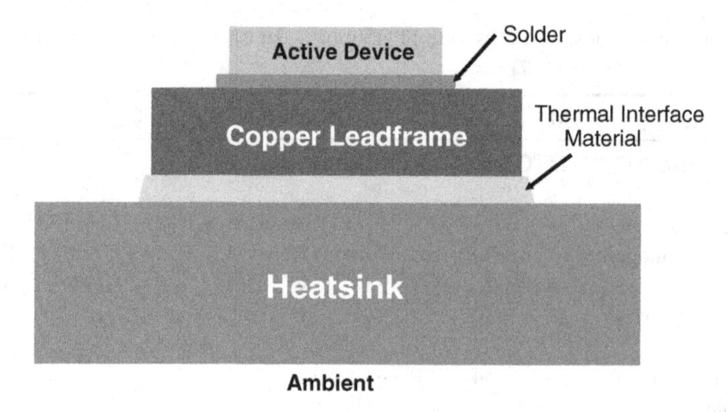

Figure 2.14 Cross section of a transistor mounted in a typical package including a copper leadframe or "case"

transistor, the lower this resistance. The thicker the substrate or the case, the longer the thermal path to the heatsink, and the higher $R_{\theta JC}$ will become. The heat flux then has to be transferred into the environment. The parameter that describes this is $R_{\theta JA}$, which includes the junction-to-case resistance and makes an assumption about the area of the back surface that is exposed to the ambient environment.

Figure 2.15 gives a simple schematic model describing the thermal resistance in steady state for the structure shown in Figure 2.14. The figure shows the thermal resistance broken down by location in the structure starting from the junction to the bottom of the leadframe ($R_{\theta JC}$), the thermal resistance through the thermal interface material ($R_{\theta TIM}$), and the thermal resistance from the interface with the heatsink to the ambient ($R_{\theta SA}$). The sum of these three is the thermal resistance from the junction to the ambient ($R_{\theta JA}$).

Table 2.10 shows data from a previously cited MOSFET datasheet (BSC060N10NS3G). In this example, there are two values given for $R_{\theta JA}$: the first value is for the resistance of the back surface of the package radiating into the environment without an added heatsink, and the second value makes the assumption there is a 6 cm^2 cooling layer of 70 μm thick copper embedded in an FR4 printed circuit board (PCB) acting as the heatsink.

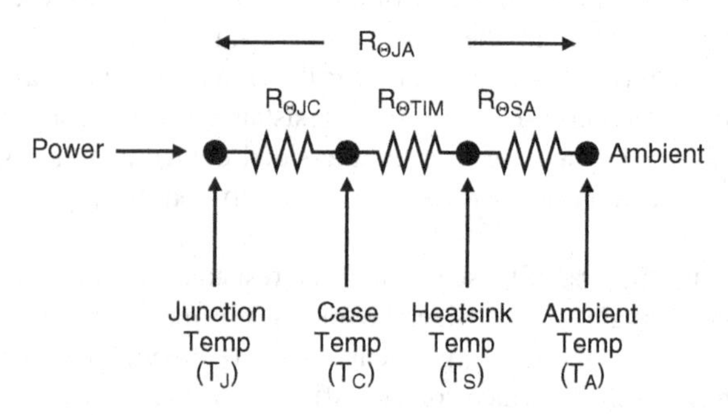

Figure 2.15 Steady-state thermal resistance schematic model for the physical device structure given in Figure 2.14

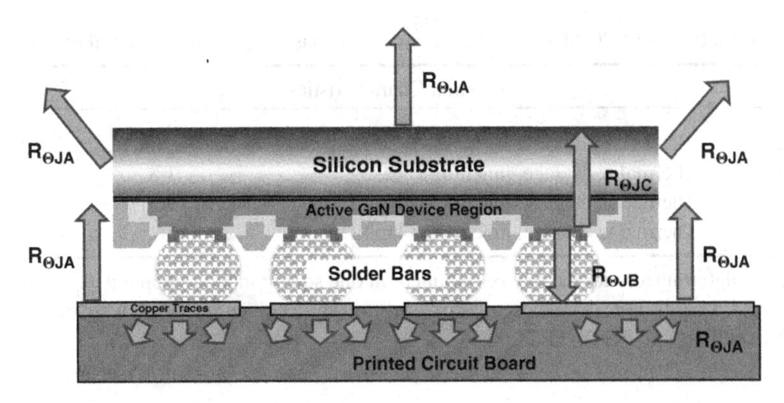

Figure 2.16 Side view of a LGA device mounted to a printed circuit board illustrating the three basic components of thermal resistance

Figure 2.16 illustrates all three thermal parameters in the case of the land grid array (LGA) format device first shown in Figure 1.20 of Chapter 1. In this example, the HEMT is mounted with the active transistor facing the PCB and separated by the solder bars that form the device terminals. $R_{\theta JC}$ is still defined as the resistance from the active surface of the die to the case. For a GaN HEMT in an LGA format, the case is the surface of the bare Si substrate (facing upwards in this illustration). The new thermal parameter, $R_{\theta JB}$, is the thermal resistance from the active surface of the transistor, through the solder bars to the copper traces on the PCB. The third thermal parameter, $R_{\theta JA}$ is the thermal resistance standardized to a specific PCB area (typically one square inch) with no heatsink, and gives an estimate of the overall thermal resistance between junction and ambient for this case. This differs from $R_{\theta JB}$, as it also includes the specific PCB-to-ambient thermal resistance, $R_{\theta BA}$, in series, as well as some direct case-to-ambient thermal resistance, $R_{\theta CA}$, due to radiation and convection.

The data in Tables 2.12 and 2.13 show that the resistance between the junction and the ambient, $R_{\theta JA}$, is the largest resistor and could therefore limit the amount of heat the device can dissipate. For this reason, it is common to add a heatsink to the device to facilitate heat removal. Figure 2.17 shows an assembly of two GaN HEMT devices with the addition of a heatsink and thermal interface material. The overall system thermal resistance is now the

Table 2.12 Data from an Infineon BSC060N10NS3 G datasheet for an Si MOSFET showing sections relating to steady state thermal resistance [8]

Parameter	Symbol	Test Conditions	Values			Unit
			Min	Typical	Max	
Thermal Characteristics						
Thermal resistance, junction-case	$R_{\theta JC}$	—	—	—	1	K/W
Thermal resistance, junction-ambient	$R_{\theta JA}$	Minimal footprint	—	—	62	
		6 ms cooling area	—	—	50	

Table 2.13 Data from an EPC2010 datasheet showing the basic steady state thermal parameters [16]

Thermal Characteristics			
		Typical	
$R_{\theta JC}$	Thermal resistance, junction-to-case	1.8	°C/W
$R_{\theta JB}$	Thermal resistance, junction-to-board	16	°C/W
$R_{\theta JA}$	Thermal resistance, junction-to-ambient[1]	56	°C/W

Note 1: $R_{\theta JA}$ is determined with the device mounted on one square inch of copper pad, single layer 2 oz. copper on FR4 board. For details, see http://epc-co.com/epc/documents/product-training/Appnote_Thermal_Performance_of_eGaN_FETs.pdf.

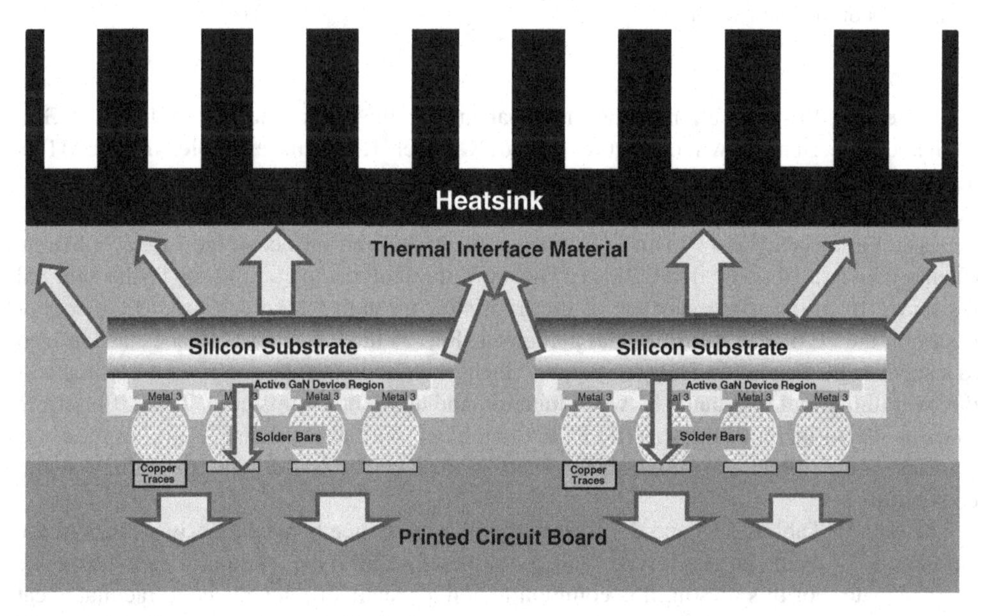

Figure 2.17 Side view of two LGA devices mounted to a printed circuit board and connected to a heatsink via thermal interface material

sum of the parallel resistances of all the paths illustrated including the thermal resistance of the thermal interface material and the heatsink. This system can be modeled using a simple resistor network [18].

2.6 Transient Thermal Impedance

Rarely is a transistor used in steady state with a continuous DC flow of current through the device. In order to give designers a measure of the thermal impact of a shorter pulse, or a repetitive pulse of various duty cycles, datasheets also have a transient thermal impedance graph, such as the one shown in Figure 2.18 [16].

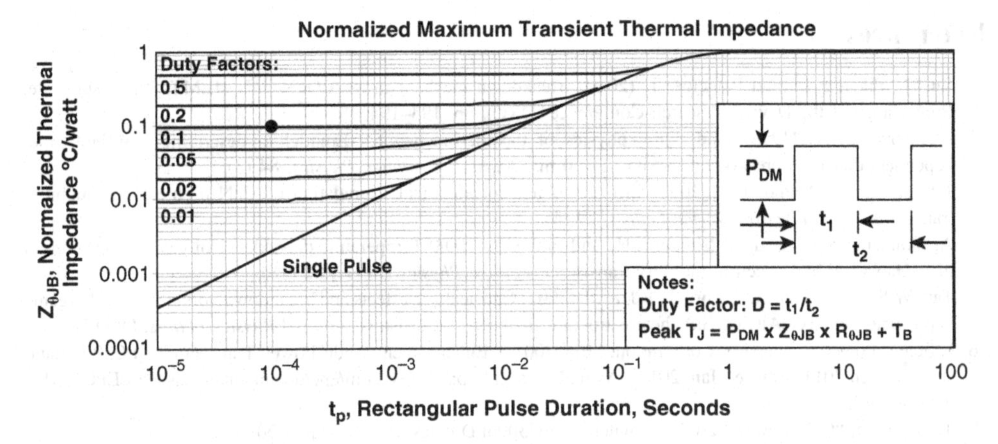

Figure 2.18 Transient thermal response curve for the EPC2010 eGaN FET [16]

To illustrate how these data in the graph can be used, take the example of the EPC2010 [16] mounted on a PCB without a heatsink on the back. If the circuit pulses the part with a 10 W instantaneous pulse (PDM) at a 10% duty cycle, and the pulses are 100 μs long, and the board is sufficient to absorb the heat with little temperature impact, the effective thermal resistance and junction temperature rise would be as shown below. The effective junction-to-board temperature can be given by Equation 2.5, with the normalized thermal impedance $Z_{\theta JB}$ determined from Figure 2.18 for a duty cycle of 10%:

$$R_{\theta JB(\text{effective})} = R_{\theta JB} \times Z_{\theta JB} = 16\,°\text{C/W} \times 0.1 = 1.6\,°\text{C/W} \tag{2.5}$$

$$\Delta T_J = \text{PDM} \times R_{\theta JB(\text{effective})} = 10\text{W} \times 1.6\,°\text{C/W} = 16\,°\text{C} \tag{2.6}$$

Note the need to add to the ΔT_J in Equation 2.6 any rise in the temperature of the surface of the PCB to get a more accurate gauge of the actual rise in device temperature.

2.7 Summary

In this chapter, the basic electrical and thermal characteristics of GaN transistors were discussed and related to the physical and design characteristics of the devices.

The three characteristics that make a device function in a circuit are: breakdown voltage between source and drain electrodes (BV_{DSS}), on-resistance ($R_{DS(on)}$), and threshold voltage ($V_{GS(th)}$). Capacitance and reverse-conduction characteristics were discussed to show how a power device will work when switched on and off. However, to develop a more complete understanding of device and circuit performance, the characteristics and specifications associated with the amount of heat that can be extracted from a device were also discussed.

The next two chapters are about circuit and layout techniques for GaN transistors. With a step-function improvement in switching speed and power density, designers need to take extra care to properly drive the gate of the device and reduce parasitic elements in the surrounding circuits.

References

1. Lu, B., Piner, E.L., and Palacios, T. (2010) Breakdown mechanism in AlGaN/GaN HEMTs on Si substrate, Proceedings of the Device Research Conference (DRC), pp. 193–194.
2. McPherson, J.W. (1998) Underlying physics of the thermochemical Emodel in describing low-field time-dependent dielectric breakdown in SiO_2 thin films. *Journal of Applied Physics*, **84** (3).
3. Joh, J. and del Alamo, J.A. (2008) Critical voltage for electrical degradation of GaN high-electron mobility transistors. *IEEE Electron Device Letters*, **29** (4).
4. Arulkumaran, S., Egawa, T., Ishikawa, H., and Jimbo, T. (2003) Temperature dependence of gate-leakage current in AlGaN/GaN high-electron-mobility transistors. *Applied Physics Letters*, **82** (18).
5. Tan, W.S., Houston, P.A., Parbrook, P.J. *et al.* (2002) Gate leakage effects and breakdown voltage in metal organic vapor phase epitaxy AlGaN/GaN heterostructure field-effect transistors. *Applied Physics Letters*, **80** (17).
6. Efficient Power Conversion Corporation, "EPC2001 – Enhancement-mode Power Transistor," EPC2001 datasheet, March 2011 [Revised Jan. 2013]. Available from http://epc-co.com/epc/documents/datasheets/EPC2001_-datasheet.pdf.
7. Transphorm, "GaN Power Low-loss Switch," TPH3006PD datasheet, 27 March 2013.
8. Infineon, "OptiMOS^{TM} Power-Transistor," BSC060N10NS3 G datasheet, 21 Oct. 2009.
9. Sze, S.M. (1981) *Physics of Semiconductor Devices*, 2nd edn, John Wiley and Sons, Hoboken, NJ, pp. 31.
10. Giancoli, Douglas. (2009) [1984]. 25. Electric Currents and Resistance, in: *Physics for Scientists and Engineers with Modern Physics*, 4th edn (ed. Phillips Jocelyn), Prentice Hall, Upper Saddle River, New Jersey, p. 658.
11. Serway, R.A. (1998) Principles of Physics, in *Fort Worth*, 2nd edn, Saunders College Pub, Texas, London, p. 602.
12. Cuerdo, R., Pedros, J., Navarro, A. *et al.* (2008) High temperature assessment of nitride-based devices. *Journal of Materials Science-Materials in Electronics*, **19** (2), pp. 189–193.
13. Vitanov, S., Palankovski, V., Maroldt, S., and Quay, R. (2010) High-temperature modeling of AlGaN/GaN HEMTs. *Solid-State Electronics*, **54**, 1105–1112.
14. Rashmi, Abhinav Kranti, Haldar, S., and Gupta, R.S. (2002) An accurate charge control model for spontaneous and piezoelectric polarization dependent two-dimensional electron gas sheet charge density of lattice-mismatched. *Solid-State Electronics*, **46**, 621–630.
15. Liu, Q.Z. and Lau, S.S. (1998) A review of the metal-GaN contact technology. *Solid-State Electronics*, **42** (5).
16. Efficient Power Conversion Corporation, "EPC2010 – Enhancement-mode Power Transistor," EPC2010 datasheet, July 2011 [Revised Feb. 2013]. Available from http://epc-co.com/epc/documents/datasheets/EPC2010_-datasheet.pdf.
17. Infineon, "CoolMOS^{TM} Power-Transistor," IPL65R130C7 datasheet, April 2013.
18. Strydom, J. (May 2012) "The eGaN® FET–Silicon Power Shoot-Out Volume 8: Envelope Tracking," Power Electronics Technology.

3

Driving GaN Transistors

3.1 Introduction

This chapter discusses the basic techniques for using GaN transistors in high performance power conversion circuits. GaN transistors generally behave like power MOSFETs, but at much higher switching speeds and power densities. A good understanding of these similarities and differences is fundamental to understanding by how much existing power conversion systems can be improved by using GaN-based device technologies. The next three chapters highlight the benefits of GaN technology, design techniques for maximum performance, and ways to avoid common pitfalls that can result from the new GaN performance capabilities. Techniques to be addressed include:

- how to drive a GaN transistor
- how to layout a high-efficiency GaN transistor circuit
- how to model and measure, both thermally and electrically, a high power-density GaN transistor-based circuit.

To understand the differences in, and opportunities offered by these faster switching devices, the alternative GaN transistor structures need to be considered individually. Two of the structures shown in Chapter 1 will be examined: enhancement-mode transistors using a p-GaN type gate, and the two-transistor cascode configuration. The gate electrodes of both types of structures have very high input impedance, and control of the device is therefore accomplished by supplying or removing a certain amount of charge from the gate electrode.

Table 3.1 gives definitions of the various charge components that will be used to examine the switching properties of GaN and MOSFET devices. In Figure 3.1(a), these definitions are used to segment transistor switching into four regions: (1) the charge required to bring the gate electrode up to device threshold, (2) the charge required to complete the current rise transition time (t_{CR}) and reach the plateau voltage (V_{pl}), (3) the charge required to complete the voltage fall transition time (t_{VF}), and (4) the charge supplied to drive the gate to the steady-state gate voltage. In Figure 3.1(b) a gate charge curve for a GaN transistor is shown with the various gate charges.

GaN Transistors for Efficient Power Conversion, Second Edition.
Alex Lidow, Johan Strydom, Michael de Rooij, and David Reusch.

Companion Website: http://www.wiley.com/go/gan_transistors

Table 3.1 Gate charge components and their definitions

Gate Charge Components	Definitions
Q_{GS}	Charge required to increase gate voltage from zero to the plateau voltage.
Q_{GS1}	Charge required to increase gate voltage from zero to the threshold voltage of the device.
Q_{GS2}	Charge required to commutate the device current.
$Q_{GS} = Q_{GS1} + Q_{GS2}$	
Q_{GD}	Charge required to commutate the device voltage, at which point the device enters the linear region.
Q_G	Total gate charge required to drive a device from zero to rated gate voltage, including Q_{GD}

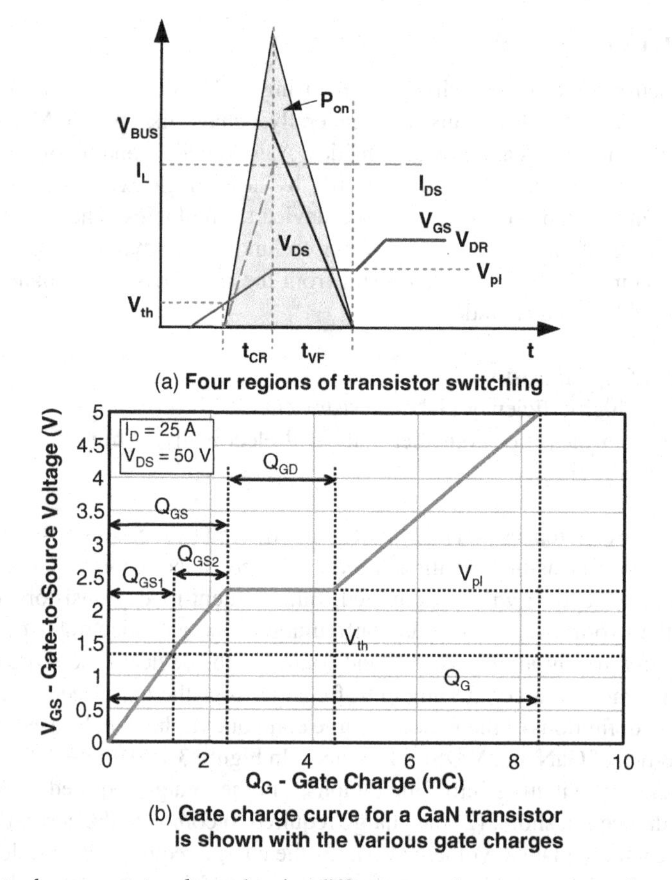

Figure 3.1 Gate charge vs. gate voltage showing different gate charge components for an EPC2010 GaN transistor [1]

To better understand why GaN transistors switch so much faster than MOSFETs, these two transistor technologies can be compared quantitatively using figures of merit (FOM). As mentioned in Chapter 1, the theoretical on-resistance versus blocking voltage of a GaN transistor is at least three orders of magnitude lower than that of silicon [2], and the

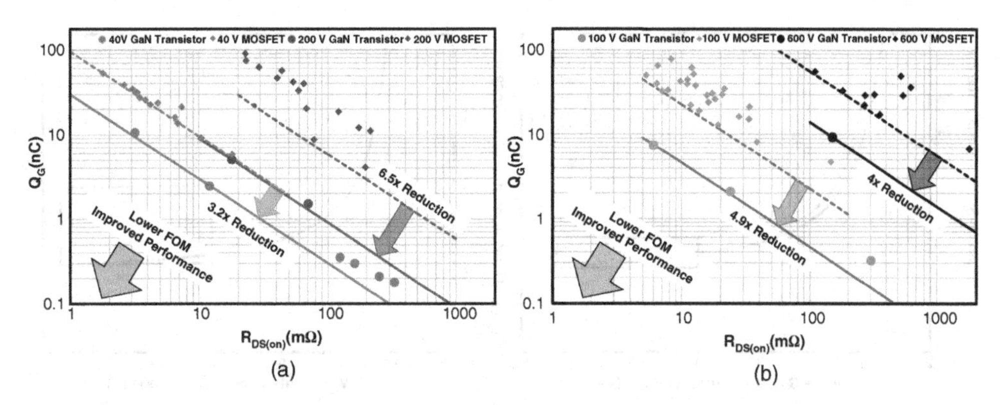

Figure 3.2 On-resistance vs. total gate charge comparison for Si- and GaN-based power devices showing (a) 40 V and 200 V, and (b) 100 V and 600 V devices

first-generation production devices were already beyond the silicon limit [3]. In general, these smaller devices have less capacitance when compared to silicon MOSFETs. Although more specialized FOMs will be discussed in the Chapters 6 and 7, the $R_{DS(on)} \times Q_G$ product is commonly used for comparing different MOSFET technologies. Using this FOM to compare GaN to silicon (Figure 3.2) shows an improvement of three to seven times.

3.2 Gate Drive Voltage

Both enhancement-mode and cascode GaN transistors have maximum voltage limits that can be applied to the gate electrode with respect to the source. For a cascode device, such as the TPH3006PD [4], this maximum voltage is ±18 V. For an enhancement-mode device, such as the EPC2010, this maximum voltage is +6 V/−5 V. (Note that different technologies and different manufacturers will rate their devices differently.) Exceeding these limits may damage devices permanently, and therefore must be avoided. Fortunately, the amount of gate voltage required to drive the device to the on-state is significantly lower than the maximum voltage allowed. With very fast switching speed, however, care must be taken to avoid overshoot that might inadvertently take the gates above the maximum voltage limit. For the enhancement-mode transistor, the device $R_{DS(on)}$ is specified in the datasheet at a recommended 5 V_{GS}, which is 1 V below the absolute maximum rating.

Because of this relatively tight requirement, we will discuss the gate drive requirements for an enhancement-mode GaN device first, with the understanding that the same basic principles apply to a transistor in a cascode configuration.

It is possible to drive these enhancement-mode devices with gate voltages as low as 4 V without a significant increase in $R_{DS(on)}$, as shown by the rectangular box area in Figure 3.3. Furthermore, it is recommended to keep the gate driver voltage below 5.25 V to allow enough margin between the gate voltage and the absolute maximum gate voltage. These recommended gate voltage limits can be readily achieved by near-critical damping of the gate drive turn-on power loop.

The gate driver, the transistor, the gate drive bypass capacitor (C_{VDD}), and the inductance of the interconnections between them (L_G), form an LCR-series resonant tank as shown in Figure 3.4(a). This equivalent resistance value includes the transistor gate resistance (R_G), gate drive pull-up resistance (R_{Source}), the high-frequency interconnect resistance between the

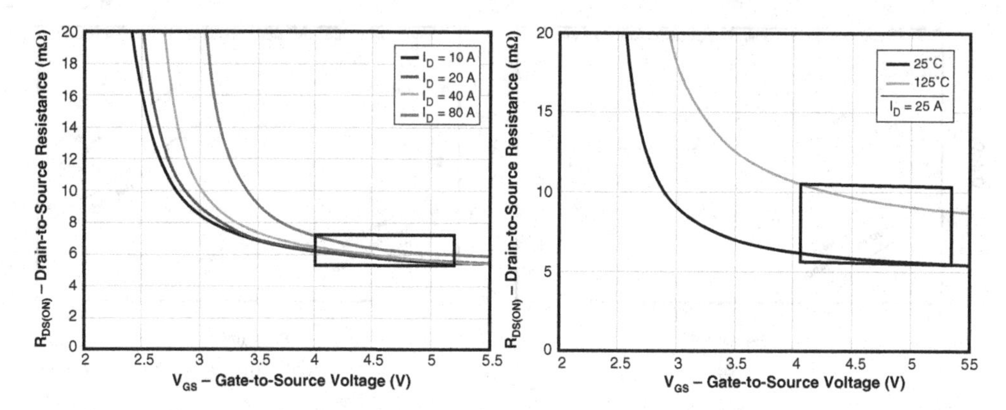

Figure 3.3 Two graphs of the $R_{DS(on)}$ of an enhancement-mode GaN HEMT (EPC2001) as a function of gate voltage for various drain currents (a) and at various temperatures (b) showing the recommended gate drive voltage range (boxed area) [6]

components, as well as the equivalent series resistance (ESR) of the gate driver supply capacitor. To critically damp this loop, the overall gate loop resistance $R_{G(eq)}$ ($R_{G(eq)} = R_G + R_{Source}$) must be larger than the value given in Equation 3.1. This is achieved by minimizing the gate loop inductance (L_G) (layout techniques for minimal gate inductance will be discussed in Chapter 4) and adjusting the series gate resistance to limit overshoot. The resultant damped gate voltage turn-on [5] is shown in Figure 3.5.

$$R_{G(eq)} \geq \sqrt{\frac{4L_G}{C_{GS}}} \tag{3.1}$$

For the gate drive voltage falling edge, the minus 5 V minimum does not present any practical limitations. It is, therefore, possible to drive the GaN transistor faster at turn-off and allow some negative ringing. Furthermore, the gate drive turn-off power loop, shown in Figure 3.4(b), will have smaller inductance, as this loop does not include the gate drive bypass capacitor. However, care should still be taken to avoid subsequent positive gate voltage ringing beyond the gate threshold of the device, as this will cause the device to turn on again. Figure 3.5 shows an under-damped turn-off with the subsequent positive ringing of less than half a volt.

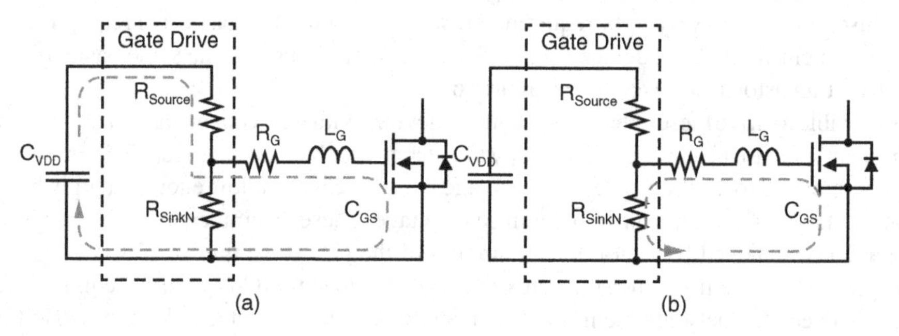

Figure 3.4 Resonant loop formed between the gate driver and eGaN FET during (a) turn-on and (b) turn-off

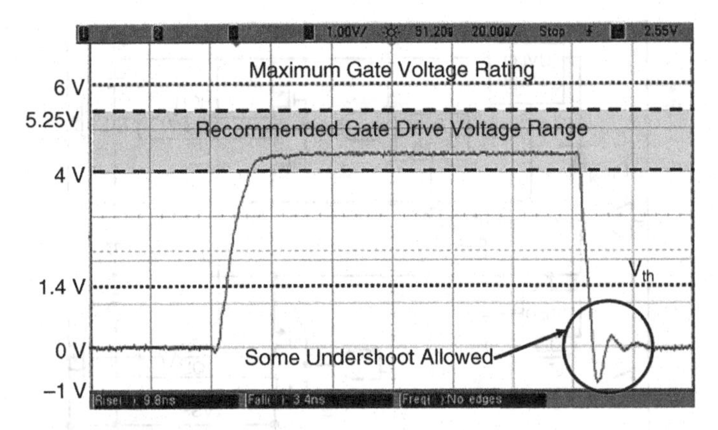

Figure 3.5 Example of an enhancement-mode transistor gate drive voltage showing a critically damped voltage rise and a slightly under-damped voltage fall

Since the turn-on and turn-off damping requirements are different, the minimum gate loop resistance values will also differ. These differences are best addressed by separating the pull-up and pull-down gate driver resistances at the driver output (creating two separate driver outputs), thus allowing the use of two separate gate resistors to independently adjust the turn-on and turn-off gate loop damping.

3.3 Bootstrapping and Floating Supplies

The limited gate drive supply voltage range for the enhancement-mode GaN transistor also has an impact on the generation of a floating high-side supply for half-bridge applications. An almost ubiquitous solution for generating this supply is the use of a bootstrap circuit as shown in Figure 3.6. This circuit operates by charging the floating high-side supply capacitor (C_{Boot})

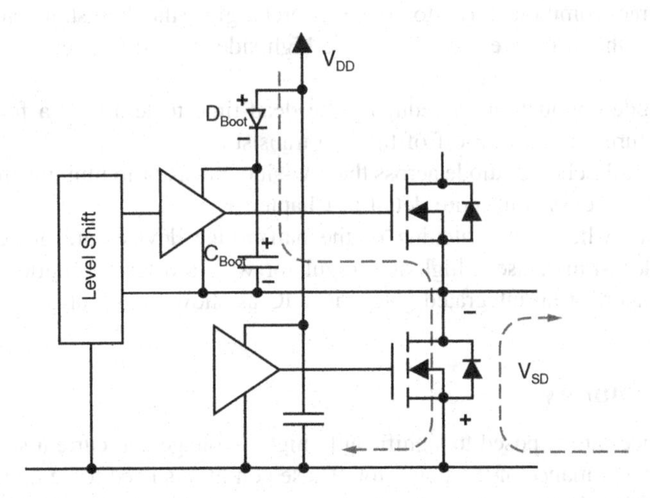

Figure 3.6 Half-bridge circuit with bootstrap supply showing impact of low-side diode conduction

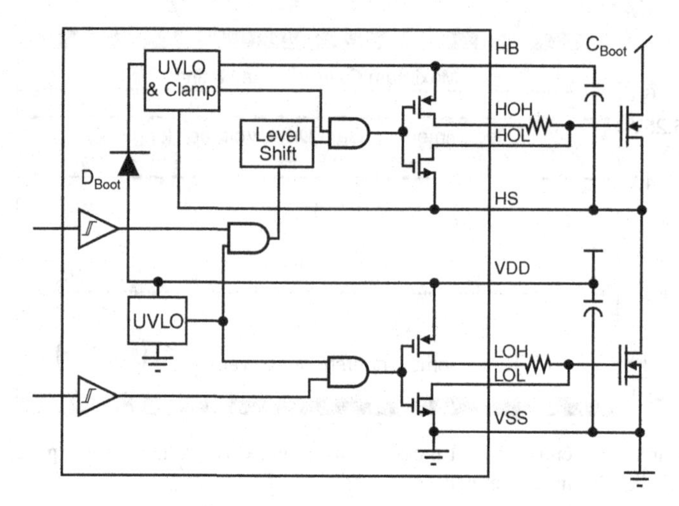

Figure 3.7 Block diagram of a half-bridge GaN transistor driver with integrated high-side supply regulation [7]

during the on-state of the low-side device from the fixed low-side supply through a high-speed bootstrap diode (D_{Boot}). During the high side on state, the bootstrap diode blocks the full bus voltage, and the floating capacitor supplies the required gate drive energy at the voltage of the bootstrap capacitor.

In practice, the bootstrap capacitor is charged to the low-side supply voltage minus the drop across the bootstrap diode, typically around a half a volt, resulting in a lower high-side supply voltage. When the bootstrap capacitor is fully charged, the diode will block current and end the charging period. However, due to the body diode drop of the GaN transistor, prolonged diode conduction of the low-side device will cause the bootstrap supply to charge up to the low-side bus voltage (V_{DD}) plus the reverse conduction voltage drop (V_{SD}) – minus the bootstrap diode, which altogether could be higher that the maximum allowable gate voltage.

There are three common ways to avoid overcharging the bootstrap capacitor, which increases the likelihood of over-voltage on the high-side transistor gate:

1. Minimize diode conduction by reducing the dead-time to less than a few nanoseconds between the turn-on and turn-off of the two transistors.
2. Place an external Schottky diode across the low-side transistor to limit the diode drop. This option will be discussed in more detail in Chapter 6.
3. In applications where the switching of the half-bridge devices are not complementary (prolonged dead-time), use a high-side regulator with a discrete solution using a series voltage regulator, or an integrated gate driver IC as shown in Figures 3.7 and 3.8 [7,8].

3.4 dv/dt Immunity

GaN power devices are exposed to significantly higher voltage and current slew rates, which can impact the performance of the transistor. These conditions need to be understood well in order to fully utilize the technology.

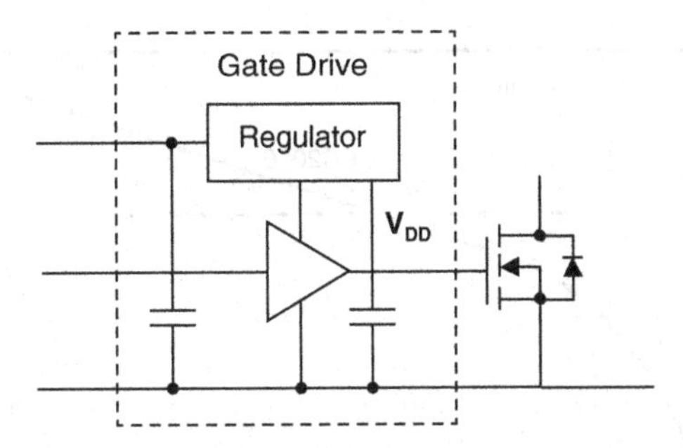

Figure 3.8 Block diagram of a single GaN transistor driver with integrated supply regulation [8]

A high, positive-voltage slew rate (dv/dt) on the drain of an off-state device can occur in both hard- and soft-switching applications, and is characterized by a quick charging of the device's capacitances as depicted in Figure 3.9. During this dv/dt event, the drain-source capacitance (C_{DS}) is charged. Concurrently, the gate-drain (C_{GD}) and gate-source (C_{GS}) capacitors in series also are charged. The concern is that, unless addressed, the charging current through the C_{GD} capacitor will flow through and charge C_{GS} beyond V_{th} and turn the device on. This event, sometimes called Miller turn-on, can be very dissipative.

Such unintended turn-on can be avoided by supplying an alternative parallel path across C_{GS} through which the C_{GD} charging current can then flow. With the addition of a gate driver pull-down keeping the device off, some of the current flowing through C_{GD} can be diverted from C_{GS} through the series gate resistor (R_G) to the gate driver pull-down resistor (R_{Sink}). This

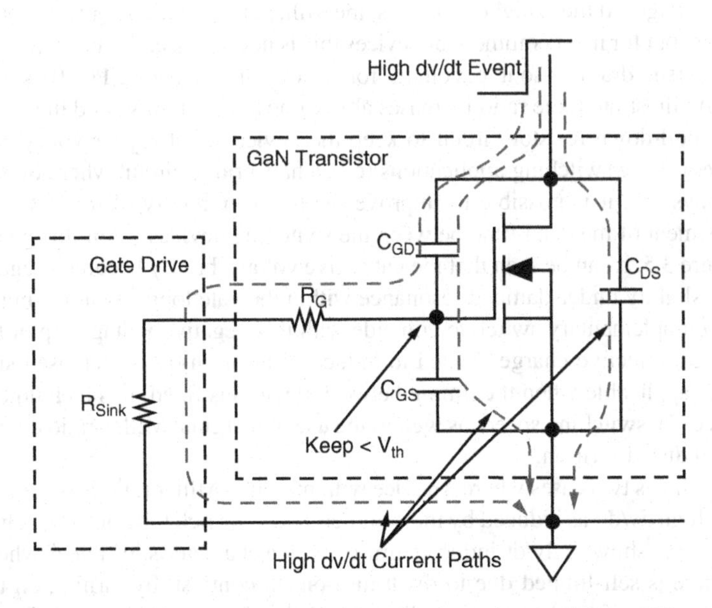

Figure 3.9 Effect of dv/dt on a device in the off state and requirements for avoiding Miller-induced shoot-through

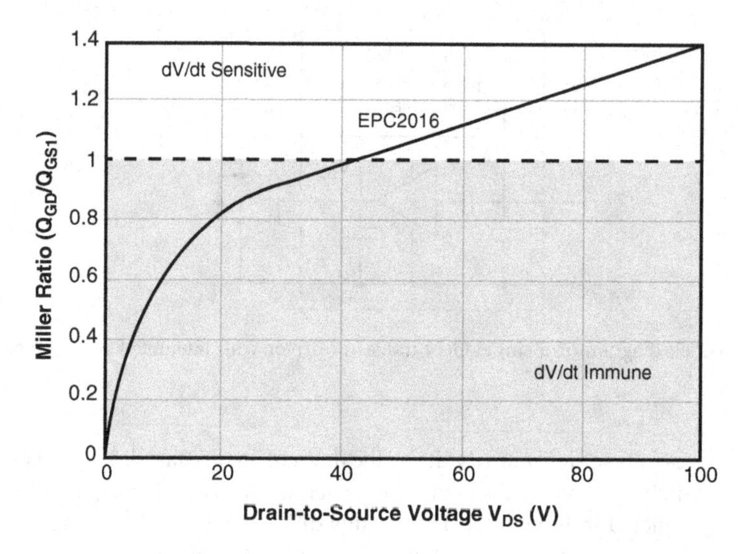

Figure 3.10 Example of Miller charge ratio vs. drain-to-source voltage using an EPC2016 [10]

additional path allows the efficient operation of devices that would otherwise be sensitive to dv/dt turn-on.

To determine the dv/dt susceptibility of a power device, a Miller charge ratio (Q_{GD}/Q_{GS1}), as function of drain-to-source voltage, needs to be evaluated. A Miller ratio of less than one will guarantee dv/dt immunity [9]. GaN transistors, like MOSFETs, are typically operated up to 80% of rated voltage. At these higher voltages, the Miller charge ratio should preferably remain at less than one, but for most commercial devices this is not the case. As an example, the Miller charge ratio versus drain-to-source voltage for a 100 V-rated part, EPC2016, is plotted in Figure 3.10. As illustrated, the ratio increases above one at about 40 V, and therefore, requires at least some pull-down resistor circuit to keep the device off at higher voltages.

For complementary switching applications (e.g. a half-bridge circuit where one or the other switch is always on), it is possible to improve the dv/dt immunity of the device artificially through adjustment of the dead-time between the switching devices. From the gate waveform shown in Figure 3.5, it can be seen that the gate drive voltage briefly becomes negative at turn-off due to the slightly under-damped resonance within the gate loop. By adjusting the turn-on timing of the complementary switch to coincide with this negative voltage dip of the device's gate voltage, the effective charge needed to induce Miller turn-on is increased significantly. Although only applicable when the timing between devices is fixed, this technique allows for an increase in dv/dt switching speed, as well as the use of marginal Miller-ratio devices without fear of dv/dt induced turn-on.

Figure 3.11 shows two cases where a device with marginal Miller ratio is turned off and then subjected to a high dv/dt, as induced by the complementary switch turn-on. The solid line in the drain voltage curve shows a dv/dt-induced turn-on with a characteristic "knee" where the drain voltage rise time is self-limited due to dv/dt turn-on. In contrast, by turning on the complementary device during the gate voltage dip (dotted line V_{DS}), dv/dt turn-on is avoided and higher dv/dt edge rates are achieved.

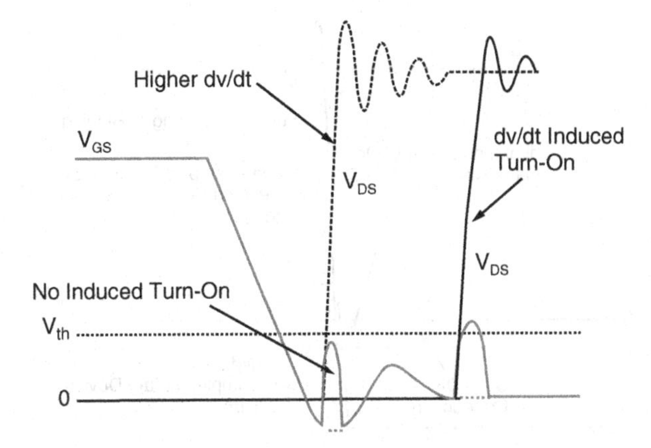

Figure 3.11 Improvement of dv/dt turn-on immunity through controlled gate timing using under-damped gate turn-off

3.5 di/dt Immunity

A rising current through an off-state device, as shown in Figure 3.12, will induce a step voltage across the common-source inductance (CSI). The CSI is the inductance on the source side of a device that is common to both the power loop (drain-to-source current) and the gate drive loop (gate-to-source current). This positive voltage step will induce an opposing voltage across C_{GS}. For a rising current, this causes the gate voltage to be driven to a negative value and with insufficient damping of the off-state gate loop LCR resonant tank, this initial negative voltage

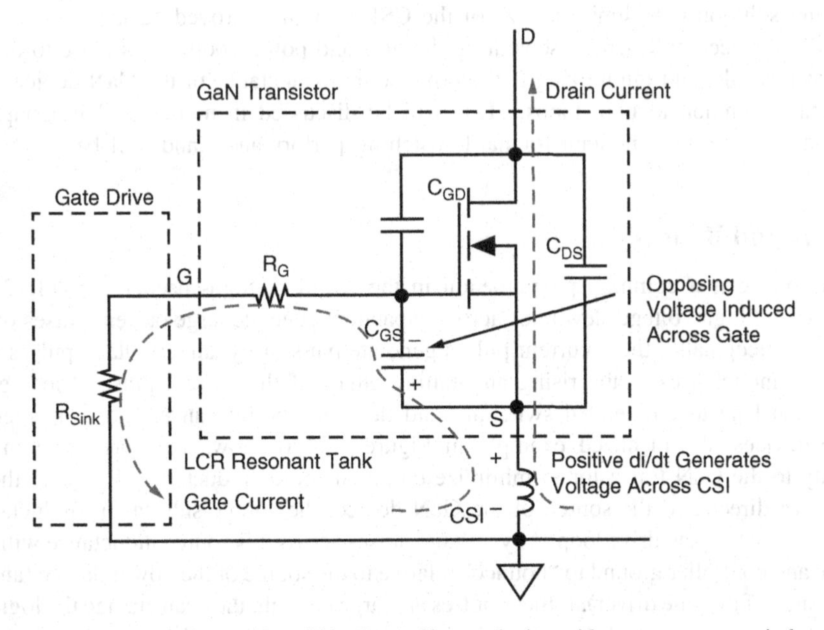

Figure 3.12 Impact of a positive di/dt of an off-state device with common-source inductance

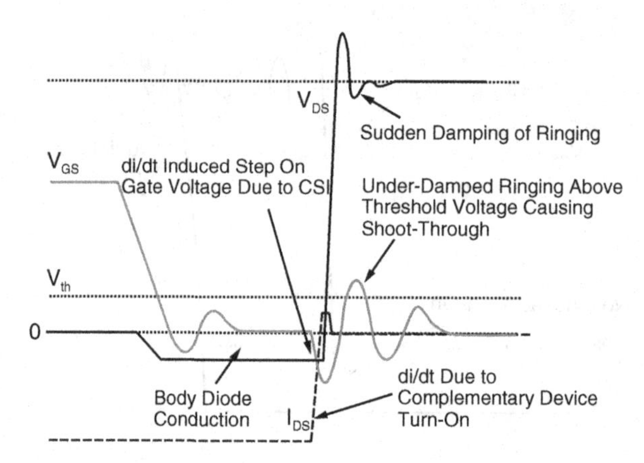

Figure 3.13 di/dt-induced turn-on (shoot-through) of an off-state device with under-damped gate turn-off power loop

step across the gate could induce positive ringing and cause an unintended turn-on and shoot-through as shown in Figure 3.13.

It is possible to avoid this type of di/dt turn-on by sufficiently damping the gate turn-off loop, although some level of undershoot may be preferred, as described in the dv/dt immunity case above. However, increasing the gate turn-off power loop damping through an increase in gate pull-down resistance would negatively impact dv/dt immunity. Thus, adjusting gate resistance alone for devices with marginal Miller charge ratios may not be enough to avoid di/dt and/or dv/dt turn-on.

A better solution is to limit the size of the CSI through improved packaging and device layout. This is accomplished by separating the gate and power loops to as close to the GaN device as possible, and minimizing the internal source inductance of the GaN device, which will remain common to both loops. (This will be discussed in more detail in Chapter 4.) Reducing CSI is also beneficial for hard-switching performance, and will be discussed in Chapter 6.

3.6 Ground Bounce

Ground bounce is a common phenomenon in the world of high-speed logic [11,12]. The concept is that high voltage slew rates across capacitors generate large current pulses of short duration. Conceptually, these current pulses generate pairs of dynamic voltage pulses across any layout inductances at the rising and falling edges of the current pulse. These ground bounces can lead to unintended switching and degraded performance, and can potentially damage devices. An idealized example in Figure 3.14(a) shows the gate drive in close proximity to the GaN transistor to minimize common-source inductance. By tying the gate drive return directly to the source of the GaN device, the source-side layout inductance is pushed outside the gate drive loop. Any voltage pulses across this source inductance will cause the logic and controller ground to "bounce" relative to the source of the power device (and thus the "ground" of the gate driver). If these pulses are large enough, they can change the logic state of the gate driver input, and thus, negatively impact a GaN power device.

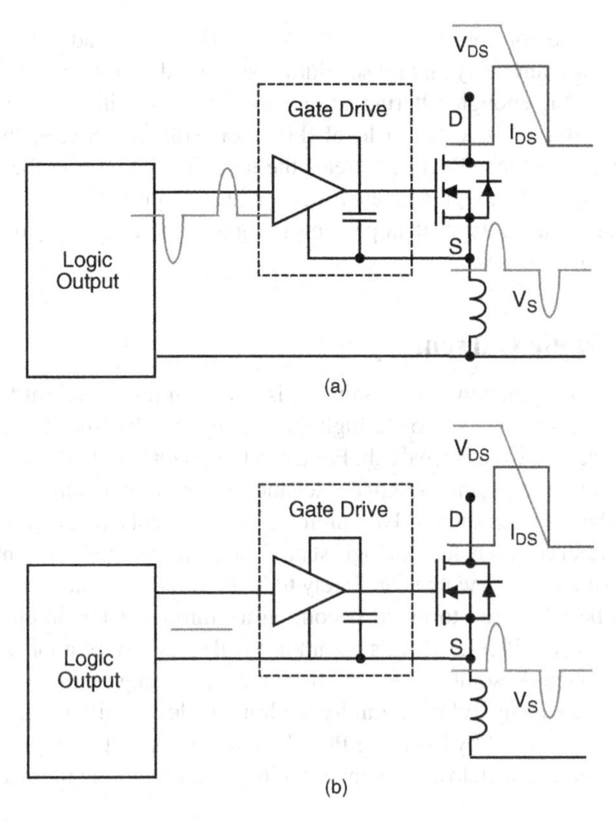

Figure 3.14 Inductance between the gate drive and the power ground causes a "bounce" of the gate driver ground. (a) Tying controller to power ground (b) Tying controller to gate driver ground

The best way to avoid ground bounce is to place the controller on the same ground as the gate driver, as shown in Figure 3.14(b), something that may not be practical with multiple low-side switching devices. In those cases, there are two ways to address ground bounce as shown in Figure 3.15. First, the ground bounce noise can be filtered out by placing a small RC low-pass

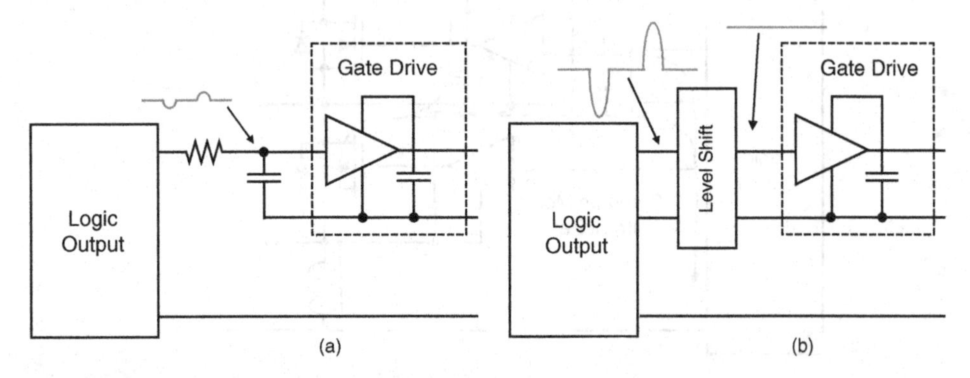

Figure 3.15 Solutions for filtering out ground "bounce" noise from gate drive input. (a) RC filter and (b) level-shifter or isolator

filter (LPF) between the controller and the gate driver. There is a tradeoff between too much filtering causing significant delay and pulse width distortion due to variation in the gate driver input thresholds, or not enough filtering maintaining susceptibility to logic glitches. The second alternative solution is to use a level-shifter or isolator between the controller and the gate driver. This approach effectively treats the low-side gate driver the same way as the floating high-side driver. Although a level-shifter increases complexity and component count, it does have the added advantage of improving the gate driver propagation delay matching between the high side and low side.

3.7 Common Mode Current

Another mechanism for generating logic glitches is common mode current through the level-shifter or isolator. This can happen to the high-side device in a half-bridge application during high positive or negative switch-node dv/dt. For a positive dv/dt event, as shown in Figure 3.16, the high voltage slew rate across the isolator capacitance causes the generation of a common mode current flowing in the loops, as shown. The common mode current causes ground bounce within the level-shifter and can cause changes in logic state if the common mode current is large enough.

With GaN transistors, the slew rates are likely to be hundreds of volts per nanosecond. This issue will need to be addressed to avoid becoming a limiting factor on circuit performance and, since this is a level-shifter issue, it is common to all GaN device applications that have a high-side floating device. Resolution requires minimizing floating high-side-to-ground capacitance as well as increasing dv/dt immunity within the level-shifter. High-side-to-ground capacitance can be minimized by avoiding PCB layout overlap between ground and the high side, selecting components with low inherent capacitance, and limiting the size of the high-side copper area.

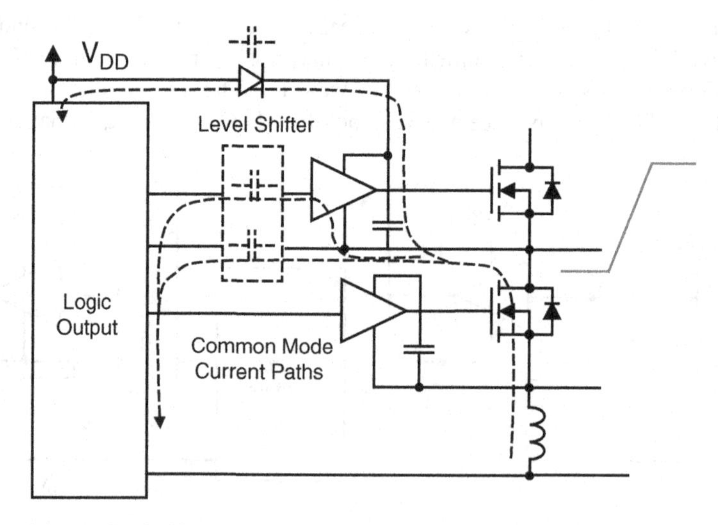

Figure 3.16 High speed switching causes large common mode current across level-shifter and bootstrap diode capacitance

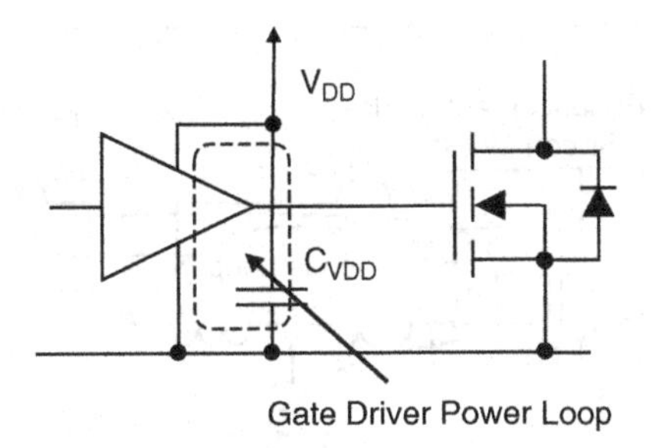

Figure 3.17 Schematic of GaN transistor and driver showing gate drive power loop

3.8 Gate Driver Edge Rate

As GaN devices continue to improve, their relative capacitances and figures of merit will continue to decrease. This means that the required resistance for damping in the equation described earlier (Equation 3.1) will increase with decreasing die size (higher $R_{DS(on)}$) and also with improvements in GaN technology. It also means that the gate charge time will decrease with decreasing gate capacitance, and the resulting decrease in theoretical switching time. To achieve this, however, the actual gate driver rise time must become even faster. This requires the minimization of the gate drive loop inductance as well as the gate driver power loop, as shown in Figure 3.17. Implementation means that the gate driver packaging inductance must be minimized through the use of chip-scale packaging, or similar low inductance packaging, and the integration of some of the V_{DD} bus capacitance, C_{VDD}, within the driver. Furthermore, the gate drive and the high-speed GaN device need to be closely connected, while the interconnection impedance is minimized. This interconnection would require complementing pin-outs and packaging options for both the gate driver and GaN transistor.

3.9 Driving Cascode GaN Devices

Cascode devices have a number of unique driving requirements. One of the main characteristics of the cascode device is the fact that it is a hybrid design with two discrete devices, a depletion-mode GaN transistor in-series with a MOSFET, each made on dissimilar processes that require external connections.

Driving the cascode device through the MOSFET gate has a number of advantages and disadvantages compared to an enhancement-mode device.

Some advantages are as follows:

- The cascode device gate terminal is that of a MOSFET. It has the same MOSFET gate voltage ratings and can be driven, in concept, with traditional MOSFET drivers. It doesn't necessarily need as much attention to avoid gate overshoot.
- Turning off the GaN transistor with a positive current is self-commutating, once the MOSFET turns off. In other words, the load current itself is responsible for generating the necessary depletion-mode gate voltage to turn the device off through charging up the MOSFET output capacitance. The higher the load current, the faster this switching occurs, thus resulting in a turn-off energy that is largely load current independent.

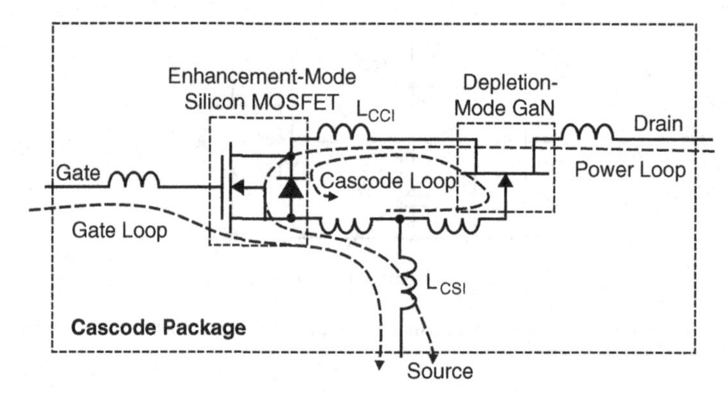

Figure 3.18 Schematic of cascode GaN device showing parasitic inductances and the different high-frequency loops between transistors

Some disadvantages are as follows:

- There are two loops where common inductance is important. Referring to Figure 3.18, there is the common-source inductance (L_{CSI}), similar to a single enhancement-mode device, and the common cascode inductance (L_{CCI}) in the loop between the MOSFET and the depletion-mode GaN transistor. Both of these loops will negatively impact switching speed.
- The use of two discrete devices means the interconnect parasitics are larger than for a single enhancement-mode device. As with most parasitics, this can be addressed through higher levels of package integration and complexity [13].
- The turn-on speed of the device is limited by the speed of the low-voltage silicon MOSFET and the CSI between the driver and MOSFET. One method to address both of these issues is to integrate the MOSFET and driver into a single module using a low-voltage LDMOS process [14].
- The size of the entire cascode device is at least twice that of the enhancement-mode GaN transistor.
- The addition of a lower-voltage series MOSFET device negatively impacts the resultant $R_{DS(on)}$ of the cascode device. This impact decreases with increasing device voltage rating, as was shown in Figure 1.12. This makes the cascode structure less suitable for lower-voltage applications.

Another of the concerns with using a cascode structure is the static voltage sharing during the off-state, and the dynamic voltage sharing at turn-off and turn-on. For static sharing, the depletion-mode GaN transistor and MOSFET devices must have similar I_{DS} leakage currents. If they are not well matched, the drain-to-source voltage drop across the MOSFET will either keep increasing or decreasing. If it keeps increasing, the low-voltage MOSFET's maximum voltage will be reached, at which point its I_{DS} will start increasing due to avalanche breakdown, until equilibrium is reached. The voltage across the depletion-mode GaN FET gate-to-source will then be equal to the MOSFET's rated breakdown voltage. On the other hand, if the MOSFET leakage current is higher than the depletion-mode transistor, the MOSFET drain-to-source voltage will collapse to near zero, at which point the GaN device starts to turn on, increasing its leakage current, and restoring drain leakage equilibrium.

Dynamically, the total output capacitance charge ratio between the MOSFET and the depletion-mode GaN transistor should be similar to the ratio of their rated drain voltages. What complicates matters is the non-linearity of these capacitances, coupled with the additional parasitic inductances between the devices. These factors can generate significant voltages during current rise and fall intervals. It may even be possible under certain circumstances to dynamically over-voltage the source-to-gate of the depletion-mode GaN transistor.

3.10 Summary

In this chapter the driver considerations for high-speed GaN transistors were addressed, including the following:

- Gate power loop inductance minimization: The gate driver should be designed to minimize the inductance between the V_{DD} supply capacitor and the actual gate driver power devices (sink and source devices). This will minimize gate driver rise time and maximize driver di/dt.
- Ground bounce immunity: The gate driver design should be made with the assumption that the driver ground and the controller ground can differ significantly, and the input logic pin must be immune to noise-induced changes in logic state.
- High dv/dt immunity for high-side drivers: Logic-isolators or level-shifters used to transfer the control logic signal to the floating high-side device need to be immune to high dv/dt rise and fall times without changing the logic state.
- Optimized driver packaging and pin-out: The gate drive and high-speed GaN device need to be closely connected with the interconnection impedance minimized. This requires pin-outs and packaging options that complement the GaN transistor.
- Separate control of the turn-on and turn-off: For a general purpose GaN driver, the speed of the driver needs to be matched to the size and speed of the device being driven. This flexibility requires a low-resistance gate driver with the option of additional external resistors. Furthermore, to adjust both the turn-on and turn-off separately, it is preferred to have separate pins for turn-off and turn-on.
- Regulation of gate drive supply voltage: For enhancement-mode transistors in particular, both low-side drivers, and especially high-side drivers, need to regulate the gate drive supply voltage to avoid an over-voltage condition on the transistor gate.

The next chapter will focus on layout techniques and ways to minimize the parasitic inductances that have increased importance due to the higher switching speed of GaN transistors.

References

1. Efficient Power Conversion Corporation, "EPC2010 – Enhancement-mode Power Transistor," EPC2010 datasheet, March 2011 [Revised Feb. 2013]. Available from http://epc-co.com/epc/documents/datasheets/EPC2010_datasheet.pdf.
2. Baliga, B.J. (1989) Power semiconductor device figure-of-merit for high frequency applications. *IEEE Electron Device Letters*, **10**, 455–457.
3. Beach, R. "Master the fundamentals of your gallium-nitride power transistors," *Electronic Design Europe*, 29 April 29, 2010.

4. Transphorm, "TPH3006PD – GaN Power Low-loss Switch," Transform datasheet, 27 March 2013.
5. Reusch, D., Gilham, D., Su, Y., and Lee, F. (2012) "Gallium nitride based 3D integrated non-isolated point of load module," in *Applied Power Electronics Conference and Exposition (APEC)*, 2012, Twenty-Seventh Annual IEEE, Orlando, FL, Feb. 2012, pp. 38–45.
6. Efficient Power Conversion Corporation, "EPC2001 – Enhancement-mode Power Transistor," EPC2001 datasheet, March 2011 [Revised Jan. 2013]. Available from http://epc-co.com/epc/documents/datasheets/EPC2001_datasheet.pdf.
7. Texas Instruments, "LM5113 5 A, 100 V Half-Bridge Gate Driver for Enhancement Mode GaN GETs," LM5113 datasheet, June 2011 [Revised April 2013].
8. Texas Instruments, "4-A and 6-A High-Speed 5-V Drive, Optimized Single-Gate Driver," UCC27611 datasheet, Dec. 2012.
9. Wu, T. *Cdv/dt Induced Turn-On In Synchronous Buck Regulators*, white paper, International Rectifier Corporation.
10. EPC Corporation, "EPC2016 – Enhancement-mode Power Transistor," EPC2016 datasheet, Sep. 2013 [Revised Sept. 2013].
11. King, P. "Ground Bounce Basics and Best Practices," Agilent Technologies, Available from http://www.home.agilent.com/upload/cmc_upload/All/Ground_Bounce.pdf.
12. Fairchild Semiconductor, "Understanding and Minimizing Ground Bounce," Appl. Note AN-640, Available from http://www.fairchildsemi.com/an/AN/AN-640.pdf.
13. Patterson, G. "GaN Switching for Efficient Converters," *Power Electronics Europe*, Issue 5, 2013, pp. 18–21, Available from http://www.power-mag.com/pdf/issuearchive/63.pdf.
14. Roberts, J. and Klowak, G. "GaN Transistors – Drive Control, Thermal Management, and Isolation," Power Electronics Magazine, Feb. 2013, Available from http://powerelectronics.com/gan-transistors/gan-transistors-drive-control-thermal-management-and-isolation. pp. 24–28.

4

Layout Considerations for GaN Transistor Circuits

4.1 Introduction

The previous chapter discussed the driver requirements for GaN transistors, which are capable of much higher switching speeds than Si MOSFETs. The faster switching speeds also magnify the impact of parasitic inductances on performance. As GaN matures and becomes capable of even higher switching speeds, the minimization of parasitics will be even more critical to fully utilizing GaN transistors. In this chapter, the focus will be on layout techniques and ways to minimize these parasitics. In subsequent chapters, the impact of these parasitics will be quantified along with their relative importance for performance in various applications.

For a half-bridge configuration, which is used in about 90% of power converters, there are two main power loops to consider: (1) the high-frequency power loop formed by the two power switching devices along with the high-frequency bus capacitor, and (2) the gate drive loop formed by the gate driver, power device, and high-frequency gate drive capacitor. The common source inductance (CSI) is defined by the part of the loop inductance that is common to both gate loop and power loop. It is represented by the inductance shown in Figure 4.1 where the power and gate loops coincide.

4.2 Minimizing Parasitic Inductance

The minimization of all parasitic inductances is vital when considering the layout of high-speed power devices. It is not possible to reduce all components of inductance equally, and therefore they must be addressed in order of importance, starting with common source inductance, then power loop inductance, and, lastly, gate loop inductance. The importance of CSI was discussed in Chapter 3, but the actual layout implementation varies with the packaging of GaN transistors. Chapter 6 will further quantify the impact of CSI on circuit performance.

For high-voltage PQFN (Power Quad Flat No lead) MOSFET packages, the need for a separate gate-return source pin is well known [1], and is implemented in high-voltage GaN PQFN structures [2,3]. When these separate pins are available, the gate drive loop and the power loop are

GaN Transistors for Efficient Power Conversion, Second Edition.
Alex Lidow, Johan Strydom, Michael de Rooij, and David Reusch.

Companion Website: http://www.wiley.com/go/gan_transistors

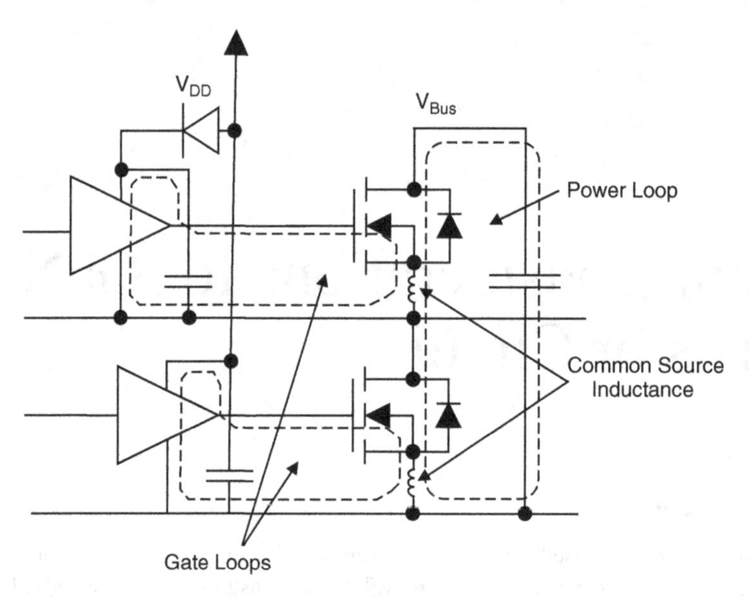

Figure 4.1 Schematic of a half-bridge power stage showing power and gate drive loops with common source inductance defined by where the loops coincide

separated within the package and must not be connected externally. As stated in the previous gate driver chapter, the reduction in common source inductance comes at the expense of external source inductance, pushed outside the gate loop. This external inductance can lead to increased ground bounce due to the improved speed of the device once CSI is removed [4].

Enhancement-mode transistors are available in a wafer level chip-scale package (WLCSP) with terminals in a land grid array (LGA). Some of these devices do not offer a separate gate-return source pin, but rather a number of very low inductance LGA solder bars, as shown in Figure 4.2. These parts can be treated in the same way as one provided with a dedicated gate-return pin or bar, by allocating the source pads closest to the gate to act as the "star" connection point for both gate loop and power loop. The layout of the gate and power loops are then separated by having the currents flow in opposite or orthogonal directions, as shown in Figure 4.2.

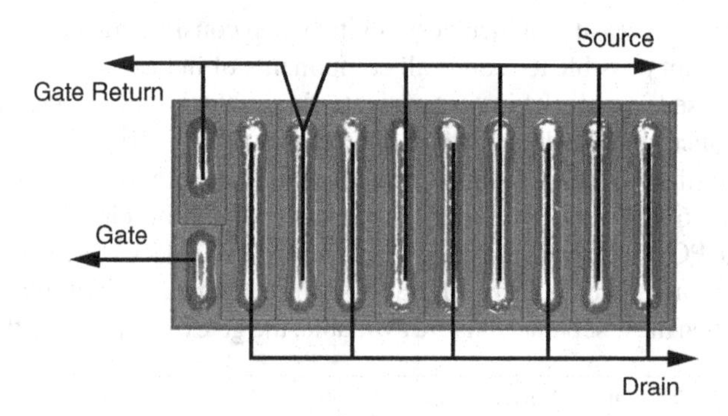

Figure 4.2 GaN transistor in an LGA format showing the direction of device current flow that minimizes common-source inductance

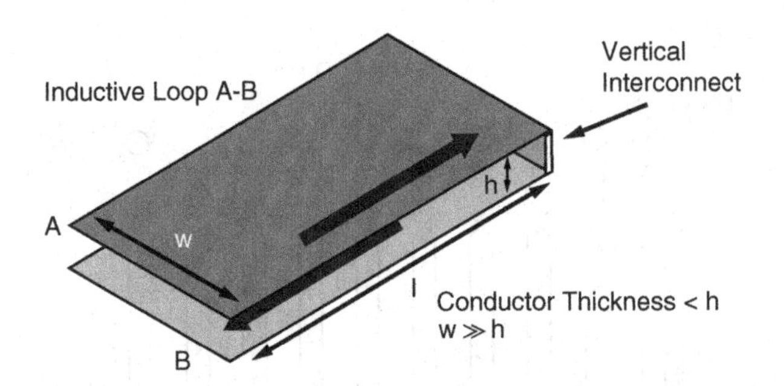

Figure 4.3 Theoretical parallel plate transmission line with end termination forming an inductive loop

Both of the remaining inductance loops can be minimized in a similar manner. While minimizing the inductance of the individual elements that make up the loop (i.e. capacitor ESL, device lead inductance, PCB interconnect inductance) is important, the designer must also focus on minimizing the total loop inductance. Because the inductance of the loop is determined by the magnetic energy that is stored within, it is possible to further minimize the overall loop inductance by using the coupling between adjacent conductors to induce magnetic field self-cancellation. For such a coupled structure, consider a theoretical parallel plate transmission line with end terminations as shown in Figure 4.3. The inductance from A to B in this loop is given by equation 4.1.

$$L_{A\text{-}B} = \mu_R \cdot \mu_0 \cdot (h \cdot l)/w \qquad (4.1)$$

In this equation μ_0 is the permeability of free air, and μ_R is the relative permeability of the PCB material. It shows that the inductance is proportional to the cross-sectional area of the loop ($h \times l$) and is inversely proportional to the width of the conductors, w. To form this loop, the current has to flow in opposite directions in the two adjacent layers, which causes magnetic field self-cancellation and reduces the inductance. The loop inductance will increase linearly with an increase in conductor spacing (h). It is therefore recommended to make all high-frequency loops as short and wide as possible, with the return path directly adjacent and as close as possible.

It should be noted that Equation 4.1 does not include the vertical interconnection between the layers, or even require that there be one. This approach, therefore, can be implemented on any incremental loop inductance, provided that there are two equal and opposing currents in adjacent layers that form the area of the loop. It is especially useful for LGA devices, as shown in Figure 4.4.

By interleaving the drain and source terminals on one side of the device, a number of small loops with opposing currents are generated that will decrease the overall inductance through magnetic field self-cancellation. This is not only true for the PCB traces shown in Figure 4.4(a), but also for the vertical LGA solder bars and the interlayer connection vias shown in Figure 4.4(b). With multiple small magnetic field canceling loops formed, the total magnetic energy, and therefore inductance, is significantly reduced [5]. A further reduction in partial loop inductance is possible by bringing both drain and source currents out on both sides of the device from the centerline and duplicating the magnetic field cancellation effect. This works by reducing the current in each conductor, thus further reducing the energy stored, and the shorter current path yields a lower inductance as previously discussed.

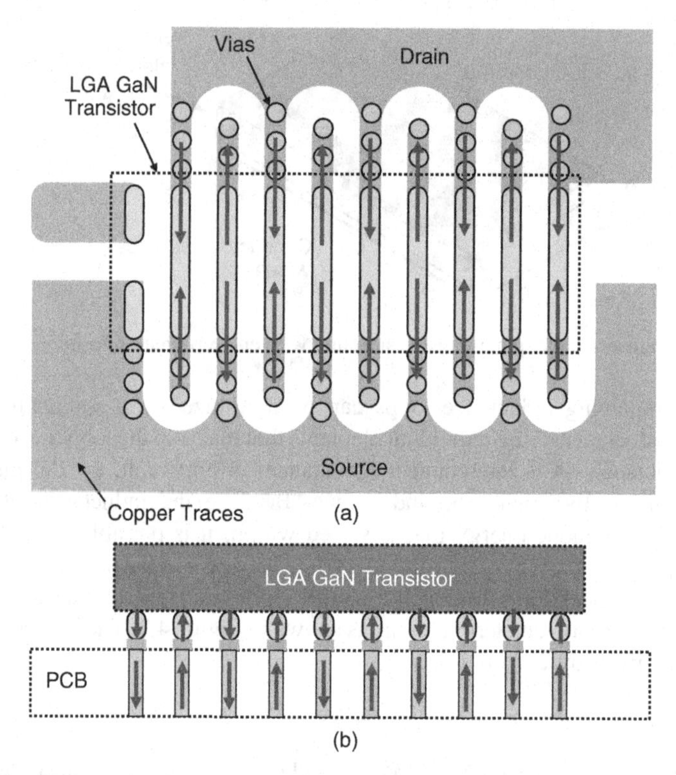

Figure 4.4 LGA GaN transistor mounted on a PCB showing alternating current flow (a) top view (b) side view

4.3 Conventional Power Loop Designs

To see how this inductance minimization is realized in an actual layout, two conventional approaches to power loops are presented for comparison. These two approaches will be called "lateral" and "vertical" respectively. The lateral layout places the input capacitors and devices on the same side of the PCB, in close proximity to minimize the size of the high-frequency power loop. The high-frequency loop for this design is contained on the same side of the PCB and is considered a lateral power loop as a result of the power loop flowing laterally on a single PCB layer. An example of the lateral layout using an LGA transistor design is shown in Figure 4.5 with the high-frequency loop highlighted.

While minimizing the physical size of the loop is important for reducing parasitic inductance, the design of the inner layers is also critical. For the lateral power loop design, the first inner layer serves as a "shield layer." This layer has a critical role in shielding the circuit from the fields generated by the high-frequency power loop. The power loop generates a magnetic field that induces a current in the shield layer that flows in the opposite direction from the power loop. The current in the shield layer generates a magnetic field to counteract the original power loop's magnetic field. The end result is a cancellation of magnetic fields that translates into a reduction in parasitic power loop inductance.

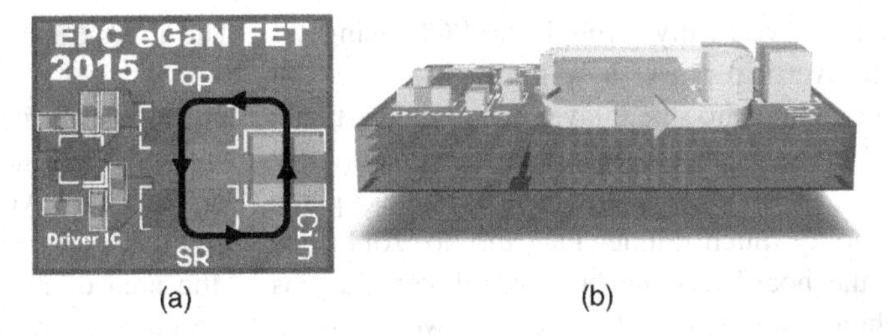

Figure 4.5 Conventional lateral power loop for LGA GaN transistor-based converter: (a) top view (b) side view

Having a complete shield plane in close proximity to the power loop yields the lowest power loop inductance. For the lateral power loop design, the high-frequency loop inductance shows little dependence on board thickness because the power loop is completely contained on the top layer. The lateral design, however, is very dependent on the distance from the power loop to the shield layer contained on the first inner layer [6].

The second conventional layout, shown in Figure 4.6, places the input capacitors and transistors on opposite sides of the PCB, with the capacitors located directly underneath the devices to minimize the physical loop size. This layout is called a vertical power loop because

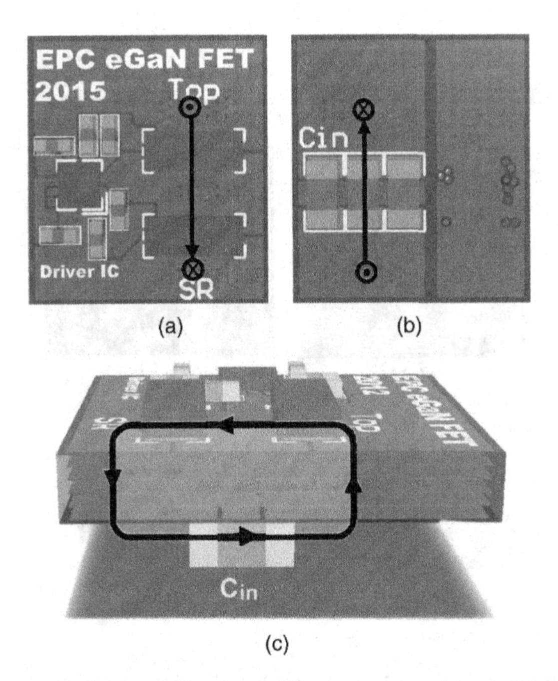

Figure 4.6 Conventional vertical power loop for LGA transistor-based converter: (a) top view (b) bottom view (c) side view

the loop is connected vertically through the PCB using vias. The LGA transistor design of Figure 4.6 has the vertical power loop highlighted.

For this design, there is no shield layer due to the vertical structure. The vertical power loop uses a magnetic field self-cancellation method (with currents flowing in opposite directions) to reduce inductance, as opposed to the use of a shield plane. For the PCB layout, the board thickness is generally much thinner than the horizontal length of the traces on the top and bottom side of the board. As the board thickness decreases, the area of the loop shrinks significantly when compared to the lateral power loop, and the current that is flowing in opposing directions on the top and bottom layers begins to provide magnetic field self-cancellation. For a vertical power loop to be most effective, the board thickness must be minimized.

4.4 Optimizing the Power Loop

An improved layout technique that provides the benefits of reduced loop size, magnetic field self-cancellation, consistent inductance independent of board thickness, a single-sided component PCB design, yielding high efficiency for a multilayer structure, is shown in Figure 4.7. The design utilizes the first inner layer, shown in Figure 4.7(b), as the power loop return path. This return path is located directly underneath the top layer's power loop, shown in Figure 4.7(a), allowing for the smallest physical loop size combined with magnetic field self-cancellation. The side view (Figure 4.7(c)) illustrates the concept of creating a low profile magnetic field self-canceling loop in a multilayer PCB structure.

The characteristics of the conventional and optimal designs are compared in Table 4.1.

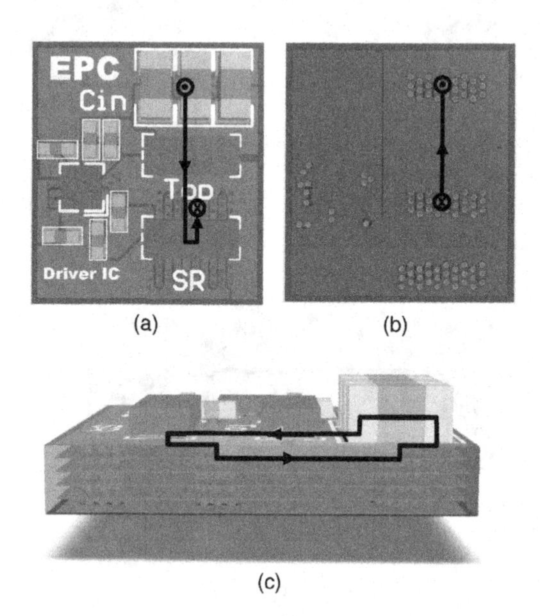

Figure 4.7 Optimum power loop for LGA transistor-based converter: (a) top view (b) top view of inner layer 1 (c) side view

Table 4.1 Characteristics of conventional and optimal power loop designs

	Lateral loop	Vertical loop	Optimal loop
Single-sided PCB capability	Yes	No	Yes
Magnetic field self-cancellation	No	Yes	Yes
Inductance independent of board thickness	Yes	No	Yes
Shield layer required	Yes	No	No

This improved layout places the input capacitors in close proximity to the top device, with the positive input voltage terminals located next to the drain connections of the top transistor. The GaN devices are located in the same arrangement as the lateral and vertical power loop cases. Located between the two transistors is a series of interleaved inductor node and ground vias arranged to match the LGA fingers. The interleaved inductor node and ground vias are duplicated on the bottom side of the synchronous rectifier.

These interleaved vias provide three advantages:

- The interleaving of the vias with current flowing in opposing direction reduces magnetic energy storage and helps generate magnetic field cancellation. This results in reduced eddy and proximity effects, reducing AC conduction losses.
- The vias located in between the two transistors provide a shorter high-frequency loop inductance path, leading to lower parasitic inductance.
- The vias located beneath the lower transistor reduces resistance and accompanying conduction losses during the transistor freewheeling period.

In Chapter 10, some comparative results for these different power loops in a hard-switching buck converter application will demonstrate the significant efficiency improvement that can be achieved through proper layout.

4.5 Paralleling GaN Transistors

The layout considerations presented above assume that a single GaN device will be used per switching element. For higher power applications, it may be necessary to place multiple transistors in parallel, and have them behave like a single device. Alternatively, multiple devices for a switching element may be required in more complex structures such as a half-bridge, where additional current paths need to be considered.

4.5.1 Paralleling GaN Transistors for a Single Switch

Figure 4.8 shows an example with three parallel devices with their interconnections between them, indicated as series resistor and inductor elements. To achieve the best possible overall resistance matching, the drains and sources of each of the devices are connected in a diagonal symmetry such that any additional resistance mismatch in the drain paths are mirrored and compensated for in the source paths. This configuration is typically used in hot-swap and similar slow switching applications where DC and low frequency current sharing is critical. The reason that this implementation cannot be used for fast-switching converters becomes clear when considering where to place the gate driver return connection. Choosing a geometrically symmetrical point – for example "E" in Figure 4.8 – results in a significant

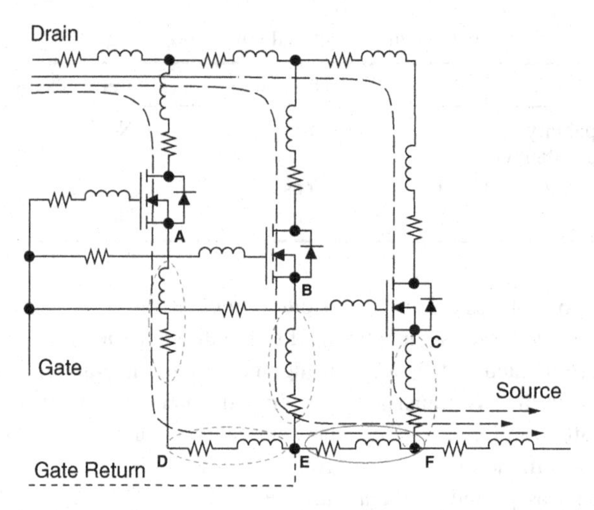

Figure 4.8 Schematic diagram showing three parallel devices with interconnect parasitics. The three drain current paths (dashed lines) are designed for matched drain-to-source impedance. Positive common-source inductance circled in dotted oval and negative CSI circled in solid oval

mismatch in CSI between the loops (dotted circles and solid circle respectively). At high speeds, CSI is the most important parasitic element, so any high-speed layout must solve the CSI symmetry issue.

The requirement for symmetry, in order to place GaN devices in parallel efficiently, is illustrated conceptually in Figure 4.9. Symmetry of the power loops, the CSI components, as well as the gate loop inductances is key to effectively paralleling GaN transistors. Even with improved symmetry, the CSI when using multiple devices will be higher than for a single device as the gate return connection point will be pushed further away.

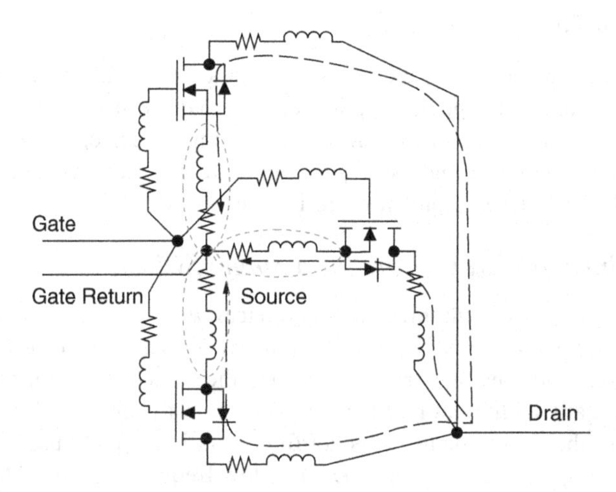

Figure 4.9 Conceptual schematic diagram showing three parallel devices with interconnect parasitics designed for complete symmetry. Common-source inductance circled in dotted oval and drain current paths shown in dashed lines

A layout with complete symmetry can be challenging. When considering the priority of these different parasitic components, the need for symmetry in CSI is foremost. Second would be the power loop (or, for a single switching element, the common drain-to-source inductance outside the transformer). Last is the gate drive loop inductance, as the gate drive speed and power always will be lower than that of the switching device itself. Following the above requirements for complete symmetry and applying the magnetic field self-canceling approach through interleaved vias and opposing currents in neighboring PCB layers, it is possible to generate an adequate solution using LGA GaN transistors, as shown in Figure 4.10.

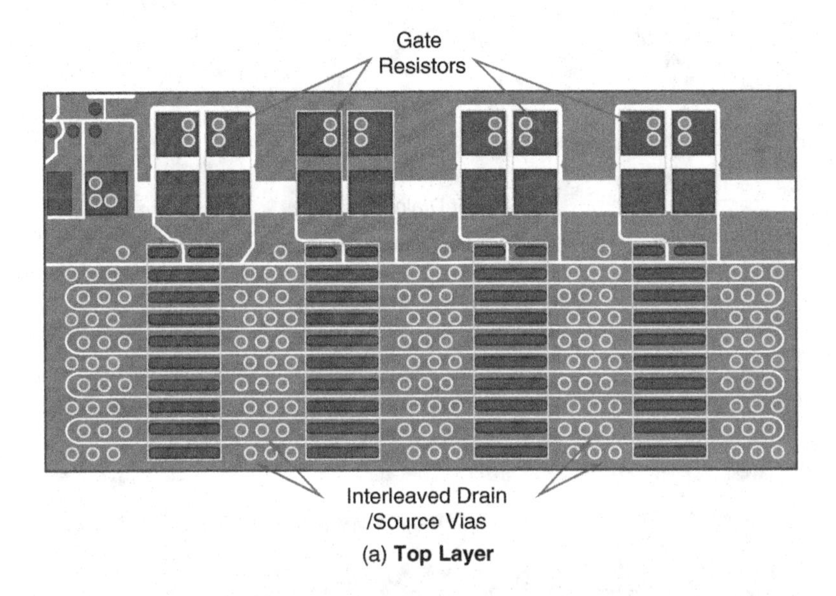

(a) **Top Layer**

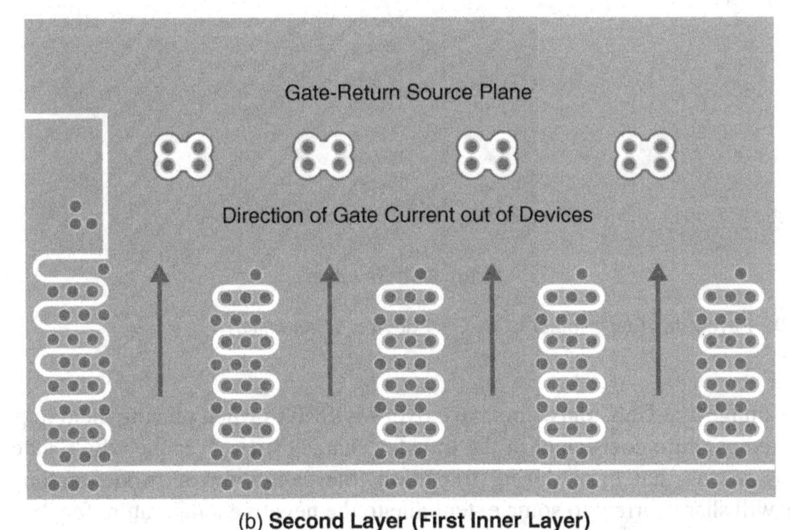

(b) **Second Layer (First Inner Layer)**

Figure 4.10 Layout for four parallel LGA GaN devices

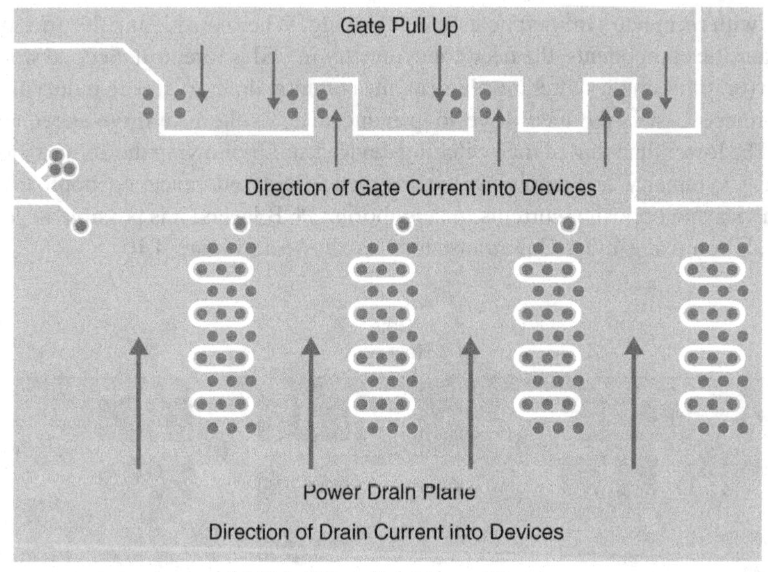

(c) **Third layer (Last Inner Layer)**

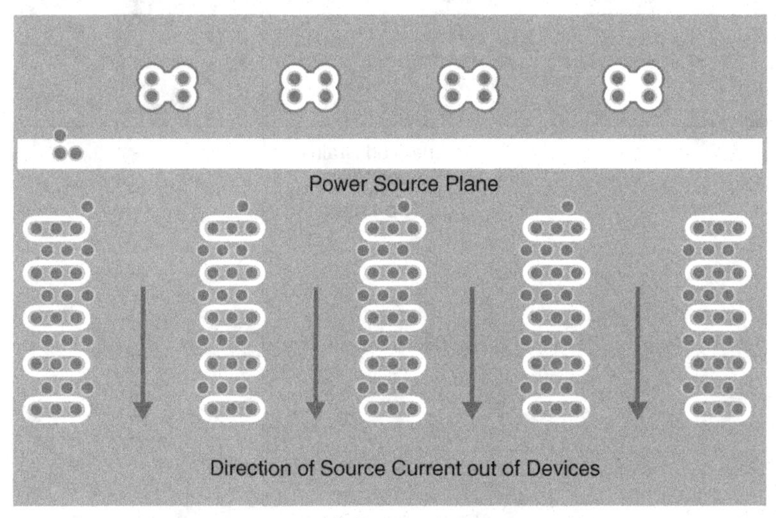

(d) **Bottom Layer**

Figure 4.10 Layout for four parallel LGA GaN devices (*Continued*)

One advantage that GaN transistors have over MOSFETs when placing them in parallel is a positive temperature coefficient in the transfer characteristic over the whole gate voltage range. This means that even during transition, the saturation or triode region of GaN transistors will share current to some extent due to the negative temperature feedback effect. Thus, if one device carries too much current it will heat up relative to the other devices, its on-resistance and plateau voltage will go up, and thus the current in the device is reduced automatically. In a MOSFET during parallel switching, the reduction of threshold voltage with increasing temperature can cause the current to increase, and thus an imbalance between devices can occur.

The key elements that have been integrated in a design with four GaN transistors connected in parallel to generate a single switching element can be summarized as follows:

- On the top layer, the devices are placed in a row, with four being the maximum number one would place in parallel before mirroring this design around another axis.
- All the devices have their drain and source terminals extending on both sides of the device to minimize drain and source inductance. These traces are then connected immediately down to all the subsequent PCB layers through a number of parallel and interleaved vias to further reduce inductance as much as possible.
- The gate of each of the devices is connected through separate pull-up and pull-down resistors, one per device. This allows for independent adjustment of the device switching speed as needed.
- On the second layer, the gate-return source connection is made directly to all the source vias and is not externally connected to any other part of the power circuit. This minimizes the CSI and isolates the gate drive from the power loop.
- On the third layer, both pull-up and pull down gate driver outputs are bused across to each of the devices, and then through the vias up to the gate resistors on the top layer. Thus, the gate-return path on layer two is sandwiched between the gate drive conductors on both adjacent layers.
- The power loop drain connection is also made on the third layer. The drain current flows up towards the devices before distributing laterally in both directions from the drain vias to the top layer.
- On the bottom layer, the power loop source connection is made. The source current flows down, away from the devices, after distributing laterally in both directions once through the source vias from the top layer. The third and bottom layers are adjacent, and the drain and source currents in these layers are of equal and opposite magnitude, allowing for magnetic field self-cancellation and minimizing loop inductance.

4.5.2 Paralleling GaN Transistors for Half-Bridge Applications

For paralleling devices in half-bridge applications, the above layout approach may work, but it does not provide an optimal solution due to some practical limitations. To form a half-bridge, consider another switching element with multiple parallel devices placed in mirror image below that of Figure 4.10. This will result in a layout with gate drivers on opposite edges of the power device layout. This configuration is not suitable for a single half-bridge gate driver, but could be implemented with a separate floating gate driver for each group of parallel devices, as long as a suitable symmetrical layout for bringing out the switch node can be achieved. Alternatively, the other switching element should be mirrored along the left or right side of the layout as shown in Figure 4.10. This would result in a single gate driver along the top edge of the layout, but the high-frequency power loop will run left to right (or right to left), resulting in a mismatch in common-source inductance similar to the equivalent circuit shown in Figure 4.8.

Although both above approaches are possible, a better alternative would be to place in parallel complete half-bridge power loops, rather than individual devices. A conceptual

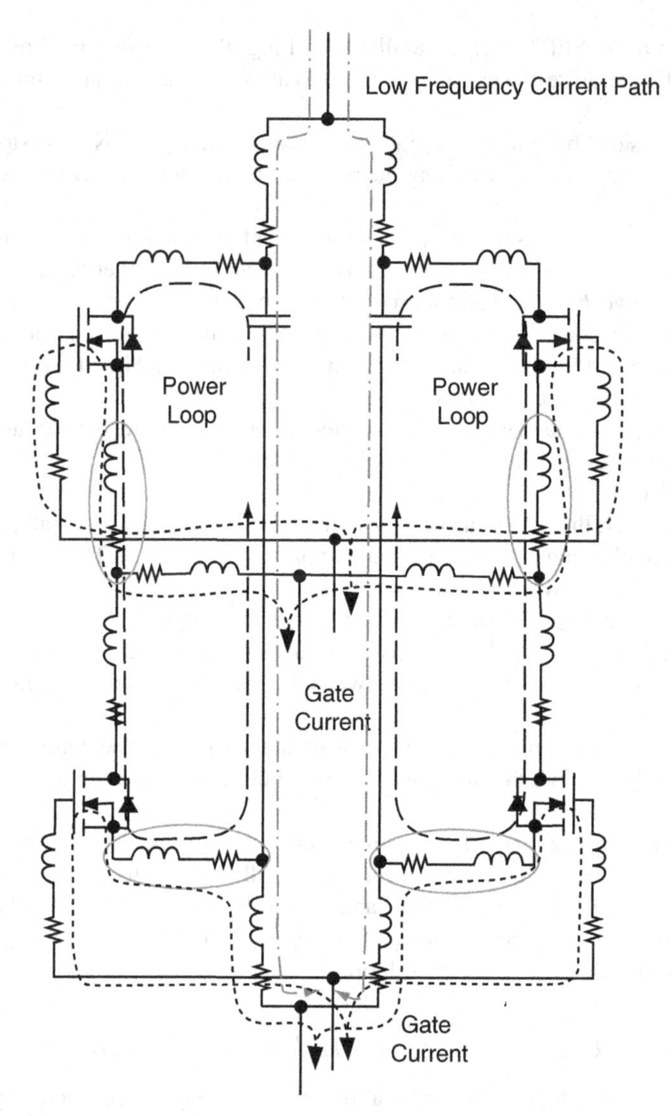

Figure 4.11 Schematic diagram showing two parallel half-bridge power loops with common gate driver. High-frequency current paths are shown in dashed lines for power loops and dotted lines for gate drive loops. Low-frequency current path are shown in dash-dot lines. Common-source inductance indicated in ovals

schematic for two such loops is shown in Figure 4.11. At first glance this may seem counterintuitive, as this solution will result in larger inductances between the DC supply terminals, but these additional inductances do not carry any high-frequency currents. Following the same order of layout requirements as before, a complete half-bridge layout with four parallel power loops is shown in Figure 4.12. An example comparing these paralleling approaches and verifying the performance advantage of the parallel loop approach is shown in Chapter 10.

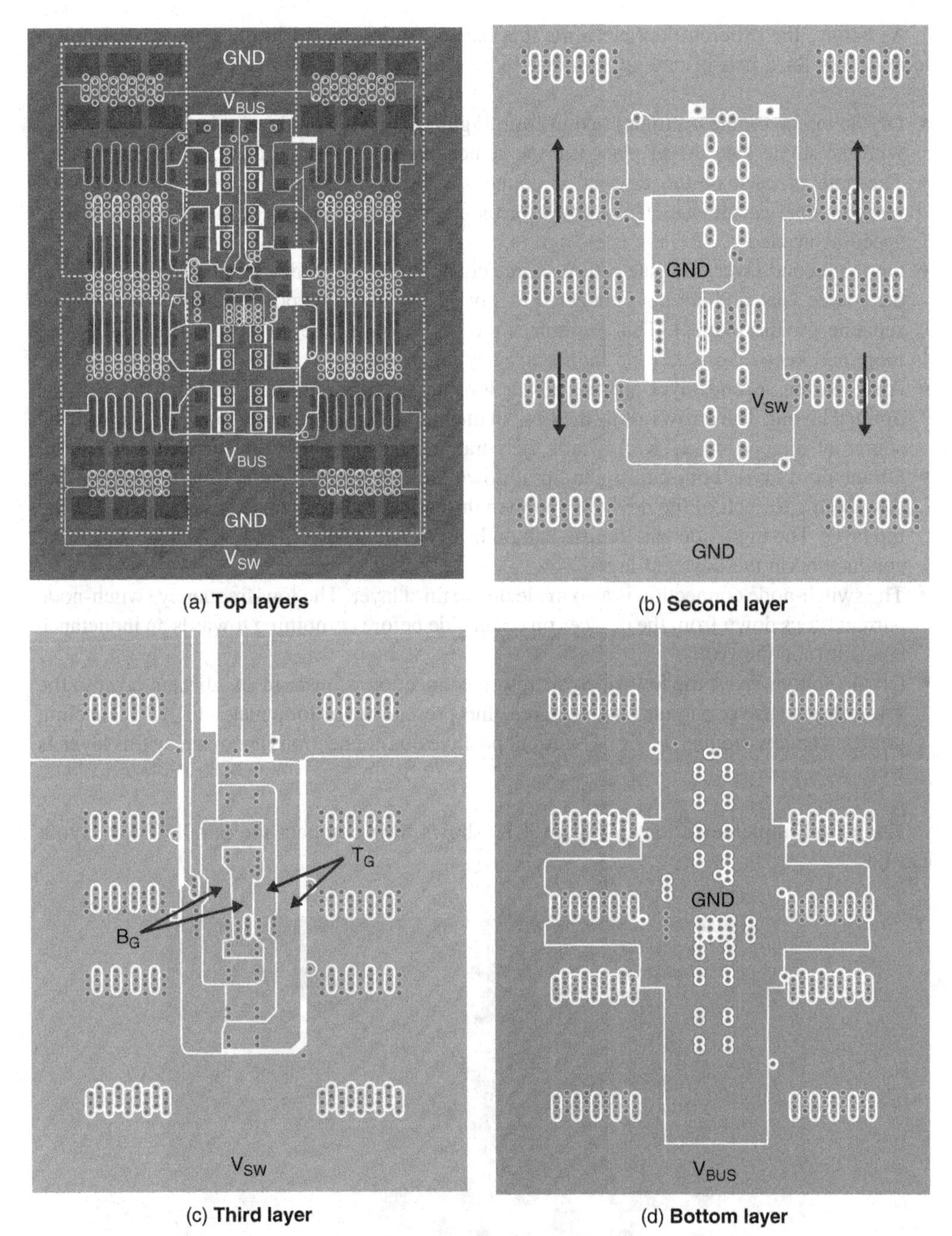

Figure 4.12 Suggested layout for a half-bridge converter with four parallel devices per switch in a parallel power loop layout

As before, the different key elements that have been integrated in the four parallel power loop design for a half-bridge are summarized as follows, with reference to Figure 4.12:

- On the top layer, the optimum layout from Figure 4.7 is mirrored in both the x- and y-axes with the single gate driver placed in the center of the layout that drives all eight devices.
- The gate of each of the devices is connected through separate pull-up and pull-down resistors, one per device. This allows for independent adjustment of the device switching speed as needed.
- On the second layer, the gate-return source connections are made directly to just one set of the source vias, and are not connected to any other part of the power circuit (note the two separate ground planes). This minimizes the CSI, isolates the gate drive from the power loop, and keeps the gate loop inductances symmetrical.
- Also, on the second layer are the power loop ground returns that form the basis of the optimal layout. The arrows show the flow of the high-frequency power loop current from the source of the bottom device towards the ceramic high-frequency bus capacitor.
- On the third layer, both pull-up and pull-down gate driver outputs are bused upwards and downwards to each of the devices, and then through the vias up to the gate resistors on the top layer. The high-side gate-return path on layer two is sandwiched between the gate drive conductors on the adjacent layers.
- The switch-node connection is also made on the third layer. The low-frequency switch-node current flows down from the devices on either side before combining towards an inductance element (not shown).
- On the bottom layer, the low-side gate return connection is made in an adjacent layer to the low-side gate driver outputs on layer three, thus preserving the loop magnetic field canceling effects. The low-frequency, or DC-current positive bus connections are made on this layer as well.

The actual implementation in Figure 4.13 shows how compact the parallel loop layout can be.

Figure 4.13 Half-bridge converter with four parallel devices per switch in a parallel power loop layout

4.6 Summary

In this chapter, layout parasitics that are important when using GaN transistors were discussed; namely the common-source inductance, the high-frequency power loop inductance, and the gate loop inductance. A number of methods to minimize these important parasitics were reviewed, starting with the most basic single transistor, through a complete half-bridge configuration and, finally, looking at the placement of multiple devices in parallel to behave as a single device for both a single-switching element and a half-bridge application.

The next step is to understand the actual behavior of these devices in-circuit. This requires not simply measuring circuit elements accurately, but also good electrical and thermal modeling approximations of aspects that cannot be measured directly. These will be discussed in Chapter 5.

References

1. Infineon, "ThinPAK 8X8 New High Voltage SMD-Package," Version 1.0, April 2010. Available from http://www.infineon.com/dgdl/Infineon+ThinPAK+8x8.pdf?folderId=db3a304314dca38901152836c5a412ab&fileId=db3a304327b897500127f6946a286519.
2. Zhou, L., Wu., Y.F., and Mishra, U. (2013) True-bridgeless totem-pole PFC based on GaN HEMTs," *PCIM Europe 2013*, pp. 1017–1022.
3. Efficient Power Conversion Corporation, "eGaN FETs in high performance DC-DC conversion," *EDN Innovation Awards Conference and Awards*, Shanghai, China, 2011, p. 28. Available from http://epc-co.com/epc/documents/presentations/EDN_Innovation_Conference_120111.pdf.
4. Direct Energy, Inc., "The destructive effects of Kelvin leaded packages in high speed, high frequency operation," Fort Collins, Colorado, Tech Note 9200-0002-1,1998. Available from www.directedenergy.com/index.php?option=com_joomdoc&task=document.download&path=ixysrf%2Fapplication-notes%2Fthe-destructive-effects-of-kelvin-leaded-packages-in-high-speed-high-frequency-operation.
5. Krausse, G.J. "DE-Series fast power MOSFET, an introduction," Directed Energy, Inc., Fort Collins, Colorado, Tech Note 9300-002 Rev 3, 2002. Available from www.directedenergy.com/index.php?option=com_joomdoc&task=document.download&path=ixysrf%2Fapplication-notes%2Fde-series-fast-power-mosfet.
6. Reusch, D. and Strydom, J. (16–21 March 2013) "Understanding the effect of PCB layout on circuit performance in a high frequency gallium nitride based point of load converter," OT *Twenty-Eighth Annual IEEE Applied Power Electronics Conference and Exposition (APEC)*, Long Beach, CA, pp. 649–655.

5

Modeling and Measurement of GaN Transistors

5.1 Introduction

The previous chapter focused on the layout parasitics that are important when using GaN transistors, and showed methods of minimizing these parasitics for layouts with various levels of complexity. In this chapter, the focus will be on how best to understand and predict the actual in-circuit behavior of the GaN transistors once the layout has been completed. Although measurement and modeling are very different, they complement each other when attempting to better understand real-world behavior. The initial discussion will focus on the electrical and thermal modeling of GaN transistors, and conclude with discussion of the requirements and limitations when directly measuring in-circuit behavior.

5.2 Electrical Modeling

Accurate electrical modeling of in-circuit GaN transistor behavior can be challenging. Apart from the active device characteristics, there is an additional need to model the high-frequency parasitic elements, such as layout inductance and skin and proximity effects. Most of these parasitics are package-dependent and cannot be modeled readily within a generic GaN transistor model. Also, cascode structures require not just an accurate model of both the active elements, but also a high-frequency model including all parasitic interconnect elements between the two devices.

Although enhancement-mode devices are made to operate similarly to silicon MOSFETs, they cannot be readily modeled with traditional physics-based MOSFET models, such as BSIM3 [1], as the physics of the GaN transistor is significantly different. A widely available model [2] for enhancement-mode GaN transistors and the methodology used in its development will be discussed in the following section.

5.2.1 Basic Modeling

SPICE models use an equivalent circuit for a device, consisting of various building blocks such as current sources, resistors, capacitors, and inductors to emulate the actual in-circuit device

GaN Transistors for Efficient Power Conversion, Second Edition.
Alex Lidow, Johan Strydom, Michael de Rooij, and David Reusch.

Companion Website: http://www.wiley.com/go/gan_transistors

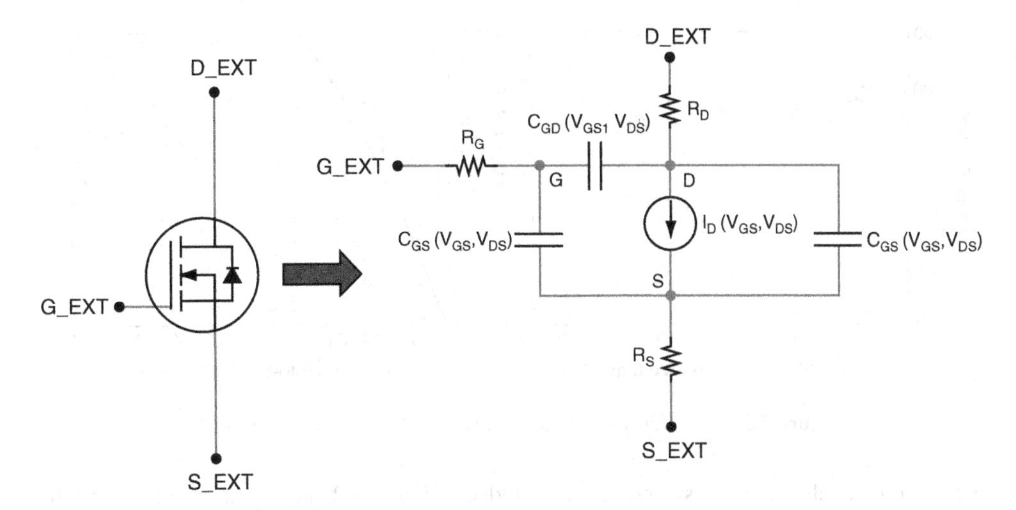

Figure 5.1 Equivalent circuit implemented by the GaN transistor model in [2]

behavior. The cross section of the device structure in reference [2] and shown in Figure 1.16, will be the basis for the model discussed throughout this section.

The basic equivalent circuit for an enhancement mode GaN transistor is shown in Figure 5.1. The main components are: a voltage-controlled current source I_D, capacitors, C_{GD}, C_{GS}, and C_{DS}, and the termination resistors, R_S, R_D, and R_G. The DC characteristics of the device depend on the voltage-controlled current source and the equivalent circuit resistances, while the AC characteristics also depend on the parasitic capacitors that vary with the bias conditions of the device. Thus, the equivalent circuit components are as follows:

1. Drain current (I_D) is a non-linear function of voltages at internal nodes D, G, and S.
   For $V_D > V_S$: $I_D > 0$ and for $V_D < V_S$: $I_D < 0$
2. Gate-source capacitance (C_{GS}) is a non-linear function of voltages at internal nodes D, G, and S.
3. Gate-drain capacitance (C_{GD}) is a non-linear function of voltages at internal nodes D, G, and S.
4. Output capacitance (C_{OSS}) is a non-linear function of voltages at internal nodes D and S.
5. Drain termination resistance (R_D) is a constant resistance that depends on the device and package termination resistances.
6. Source termination resistance (R_S) is a constant resistance that depends on the device and package termination resistances.
7. Gate termination resistance (R_G) is a constant resistance that depends on the device and package termination resistances.

The DC current–voltage models for the GaN transistor can be similar to MOSFET models [1]. In the current model, the non-linear current response is the result of saturation current that is gate-to-source voltage-dependent, with a drain-to-source voltage-dependent shaping function. For the GaN transistor model, various parameters were obtained by fitting the output curves from a large number of devices. The temperature dependence of the

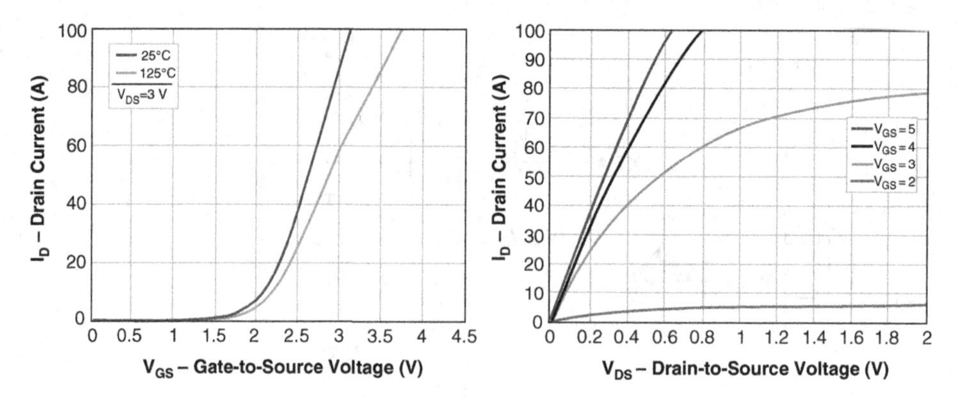

Figure 5.2 Transfer and output curves for EPC2001 device model

current–voltage characteristics were also included. Figure 5.2 shows an example of the device's transfer and output characteristics.

Unlike the silicon power MOSFET, enhancement-mode GaN transistors do not have a conventional body diode. As discussed in Chapter 1, the channel under the gate can be enhanced in either the forward or reverse direction, which gives rise to an electrical characteristic similar to the MOSFET body diode. The SPICE model accounts for this by having two parallel current sources connected in opposing orientation. Figure 5.3 shows the forward versus reverse output curves of an EPC2001 part.

The three capacitances in the equivalent circuit shown in Figure 5.1 are comprised of several elements that depend on the underlying device geometry, and vary based on the physics of the

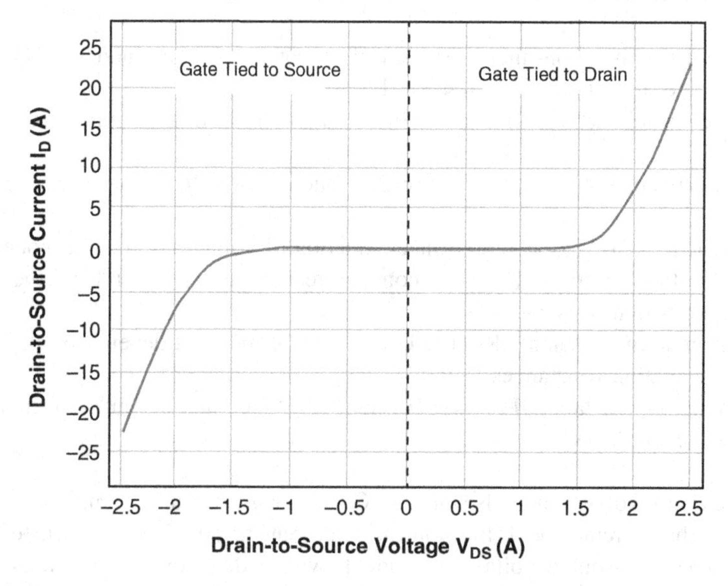

Figure 5.3 Forward and reverse current as function of drain-to-source voltage showing the symmetric nature of the device

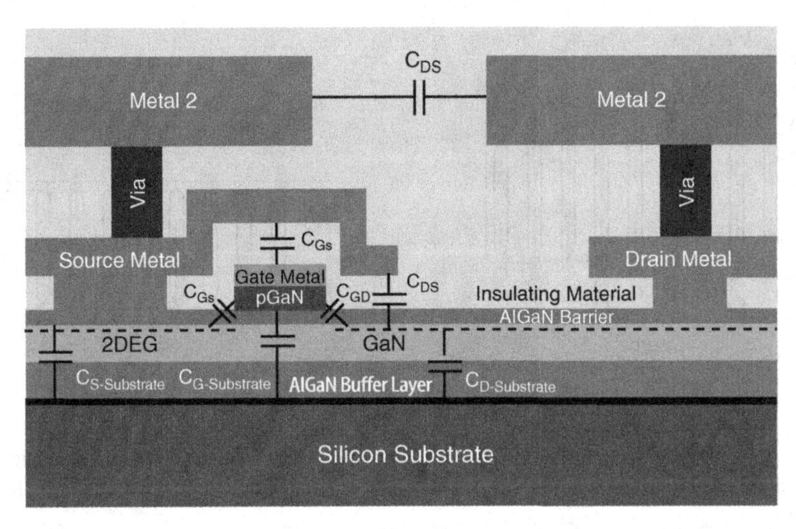

Figure 5.4 Schematic cross section of a device showing locations of various capacitances. Some of the metal layers are omitted for simplicity

device. These elements consist of both the constant metal-to-metal capacitance and the voltage-dependent capacitances associated with the 2DEG in the channel layer; and are shown in Figure 5.4. For modeling purposes, the substrate is shorted to the source and the various substrate capacitances are assigned to the source terminal. All these separate parasitic capacitances are modeled through two equivalent lumped-capacitance components for each of the three equivalent circuit capacitors: one constant capacitor and one non-linear voltage-dependent capacitor, thus yielding six equivalent capacitors overall. The non-linear equivalent capacitors are modeled using semi-empirical fits to measured values. Rather than using high-order polynomials, the models employ sum of sigmoid (or Fermi) functions. These functions provide a close fit to the data, and have better stability and convergence properties during simulation.

5.2.2 Limitations of Basic Modeling

The equivalent circuit models only account for the capacitive elements of the device's frequency-dependent behavior. In many cases, silicon power MOSFET models also include packaging inductance to accurately represent the high-frequency terminal characteristics of the packaged device. These inductances can be significant with respect to the overall non-device layout inductances, and thus cannot be omitted. As a result, and in most cases, these equivalent MOSFET circuits will give representative modeling results, even when layout parasitics are omitted.

For the LGA enhancement-mode GaN transistor with virtually no packaging inductance, layout inductance is dominant, and the way in which the layout interacts with the resultant packaging inductance is shown in Figure 5.5. Since this cannot be taken into account on a device-model level, parasitic inductance needs to be added as an additional component for accurate system modeling.

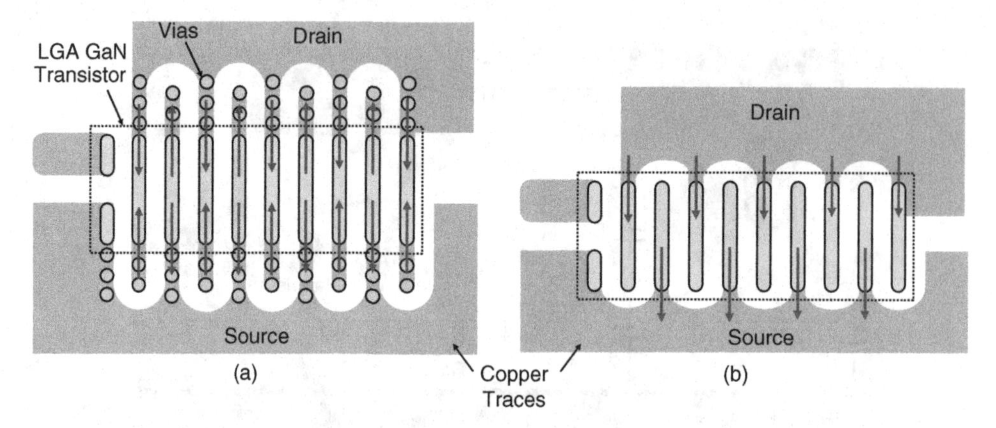

Figure 5.5 Impact of device termination on packaging and layout inductance through changes in PCB layout (a) dual-sided termination with flux cancellation (b) single-sided termination with no flux cancellation

A simulation model example with parasitic layout inductance included is shown in Figure 5.6. This model can be used to match the switch-node waveform of a GaN transistor-based buck converter to the experimental results of a simple power stage [3].

The simulation data can then be compared to the experimentally measured waveform shown in Figure 5.7. There is good correlation between the results, and in particular the ringing frequency, which is dependent on the parasitic inductances and non-linear device capacitance.

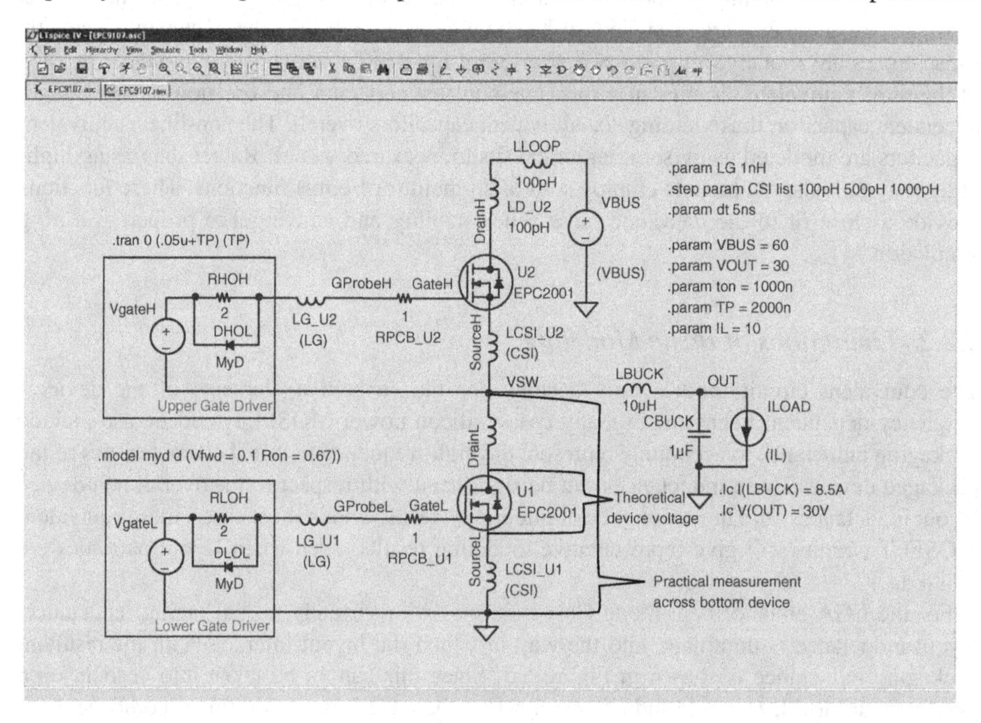

Figure 5.6 LTSPICE schematic of a buck converter power stage with layout parasitics included

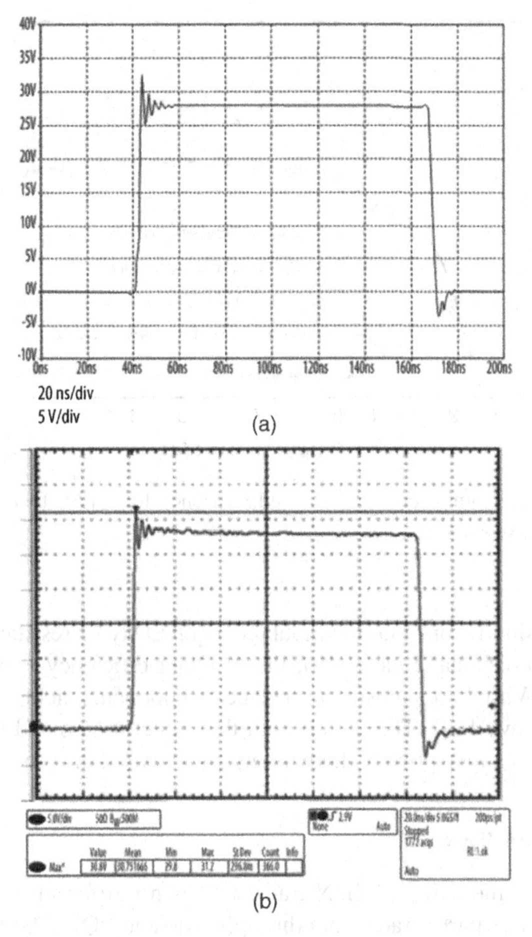

Figure 5.7 Switch node voltage versus time obtained for a 28 V to 3.3 V buck at 15 A. (a) simulation data (b) measured data [3]

From this simulation, the in-circuit power loop inductance can be estimated to be about 400 pH. The voltage overshoot damping rate is dependent on the skin and proximity effect losses at the ringing frequency. To approximate this, a parallel damping resistance of 1 Ω is placed across each of the parasitic inductance elements. The choice of this resistance value will depend on the frequency of oscillation, as SPICE is incapable of directly modeling this frequency-dependent resistance component.

5.2.3 Limitations of Circuit Modeling

The circuit modeling results will only be as good as the accuracy and complexity of the circuit model used. To best illustrate this point, simulations with varying levels of complexity were made, and the efficiency results for each simulation were calculated versus load current and compared against an actual experimental result as shown in Figure 5.8. The simulated converter efficiency was calculated by measuring the steady-state input and output power, and integrating these over a sufficiently large number of switching cycles.

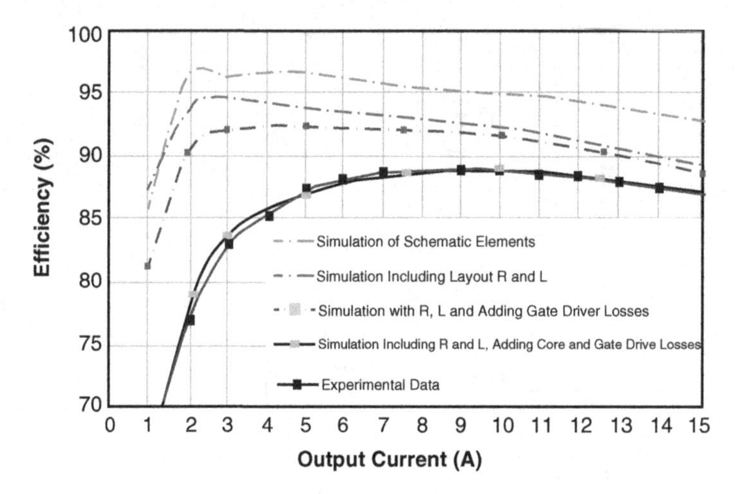

Figure 5.8 Comparison of simulated and measured efficiency for a 12 V to 1.2 V buck converter for varying levels of model complexity

First, by building a simulation circuit without the required layout resistances and inductances (purely the equivalent schematic elements), the resultant efficiency is over 5% higher than experimental results. With these parasitic impedance components included, this error drops to about 2% higher at heavy load. Then, with both the inductor core and the gate drive losses added to the simulated input power, there is excellent correlation between simulated and experimental results.

5.3 Thermal Modeling

In general, the thermal modeling of GaN transistors is no different than that of MOSFET devices. For high-voltage parts, traditional through-hole and PQFN-type packages typically are used. These plastic over-molded packages are thermally insulated on the one side, with the main direction of heat removal down through the case of the device. For such devices, heat flow is mainly one-dimensional as described in Chapter 2. For package-less devices, such as the LGA GaN transistors, the devices are flipped upside down and the back surface – the "case" – of the device is mounted away from the PCB. This arrangement is thermally similar to other "flipped" devices, such as wafer-level chip-scale packages and DirectFET® MOSFETs [4], and has two distinct thermal heat dissipation paths, as shown in Figure 5.9 and described travelling:

1. down through the LGA pads, solder bars and into the PCB
2. up through the back of the die (case). If a heatsink is used, the heat passes through a thermal interface material and through the top-side heatsink. Without a heatsink, a small amount of heat can still flow through the top of the die, but only by means of radiation and convection.

The electrically equivalent thermal resistance model for each of these two mounting approaches with a single device is shown in Figure 5.10. For the electrically equivalent circuits, the heat flux generated within the part is modeled through an equivalent power-loss

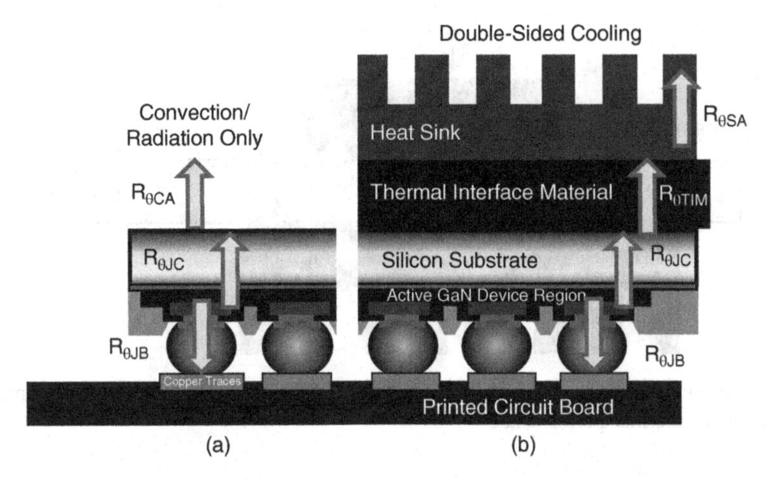

Figure 5.9 Thermal diagram of board-mounted LGA GaN transistor showing thermal paths (a) without and (b) with heatsink

current source. In both approaches, the two thermal paths are in parallel, effectively lowering the overall thermal resistance to ambient and improving the power-handling capability. Plastic over-molded packaged devices can also be modeled, using Figure 5.10(a), for large case-to-ambient temperature deltas where radiation and convection heat flow become significant.

5.3.1 Improving Thermal Performance

For traditional PQFN-packaged devices, the thermal resistance down through the PCB becomes the limiting factor, as this is the main heat flow path. This can be improved through

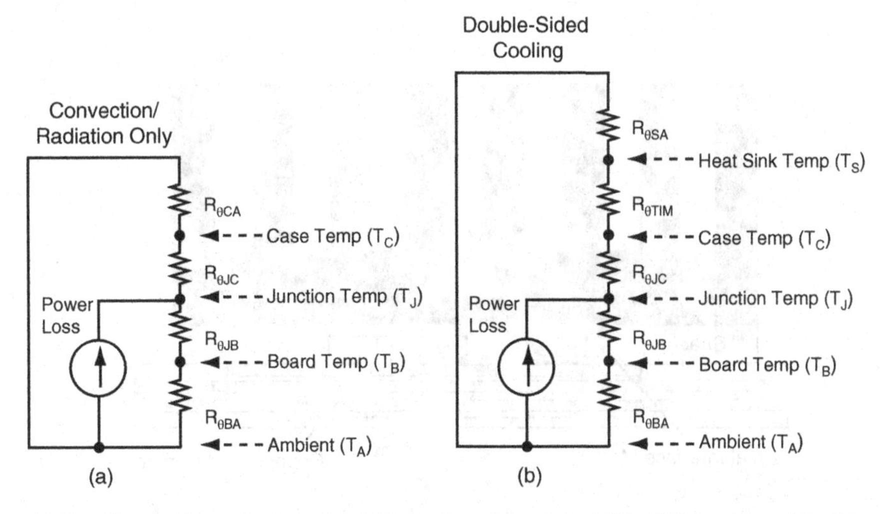

Figure 5.10 Electrically equivalent circuit thermal models of the LGA GaN transistor (a) without and (b) with heatsink

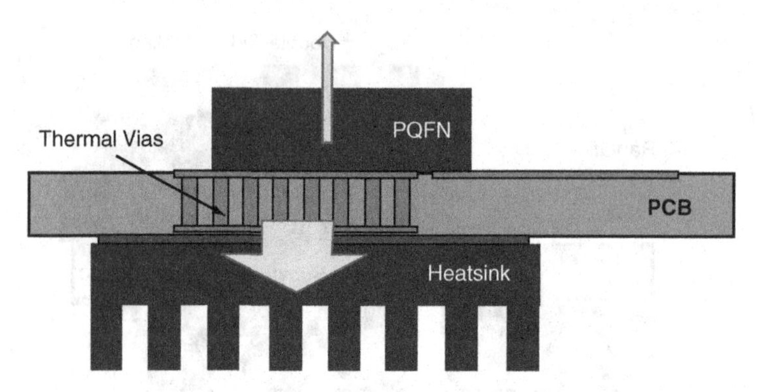

Figure 5.11 Thermal diagram of board-mounted PQFN device with thermal vias and opposite side heatsink to improve thermal performance

the addition of multiple thermal vias, allowing improved heat flow through the PCB [5]. On the opposite side, a heatsink typically is added to reduce PCB-to-ambient thermal resistance as shown in Figure 5.11.

For the LGA GaN transistor, thermal resistance improvements to both thermal heat paths can also be made, as shown in Figure 5.12. First, the PCB-to-ambient thermal resistance can be reduced through the addition of thermal vias (similar to the PQFN case), but with the thermal vias connecting to internal and external copper layers to spread the heat laterally. Second, the thermal interface material (TIM) resistance can be reduced:

1. by reducing the thickness of the device-to-heatsink interface: typically some form of spacer between heatsink and devices is required together with a TIM, as any mounting force applied to the heatsink cannot be transferred to the device for fear of cracking. This spacer determines the minimum distance between the heatsink and the GaN devices.

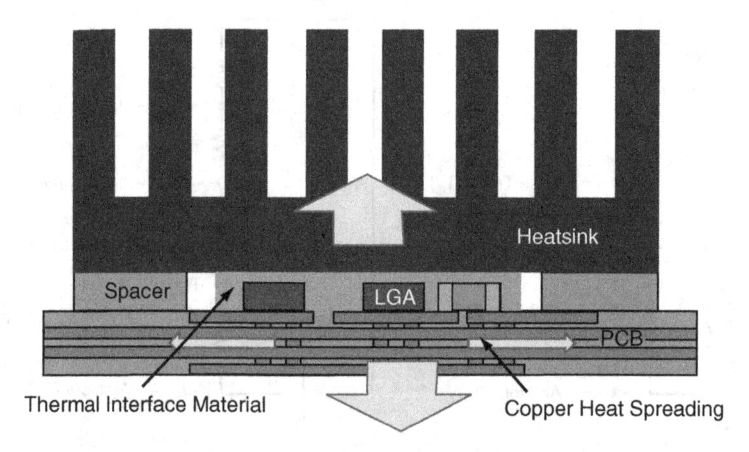

Figure 5.12 Thermal diagram of board-mounted LGA GaN transistors with dual-sided cooling showing top-side heatsink and thermal vias with heat spreading through PCB

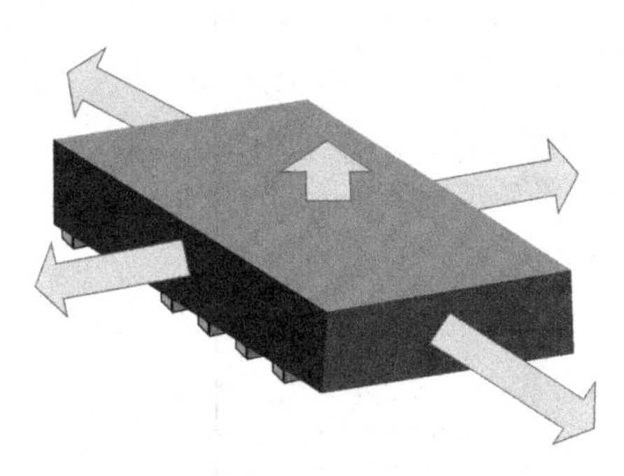

Perimeter of die adds additional surface area

Part Number	Die Area (mm²)	Perimeter Area (mm²)
EPC2001 EPC2015	6.70	7.86
EPC2007 EPC2014	1.85	3.82
EPC2010	5.80	7.10
EPC2012	1.57	3.60

Figure 5.13 Diagram of an LGA GaN transistor showing the die surface area and area of die perimeter

2. by placing the thermal interface material on all sides of the devices and not just the top (case): this reduces thermal resistance, as the surface area of the device perimeter side walls is larger than the top surface, as shown in Figure 5.13.

Further improvements are possible by utilizing dual-sided heatsinking, forced-air cooling, and through the use of exotic PCB materials, such as direct bonded copper (DBC) [6], or insulated metal substrate (IMS). Regardless of the thermal solution used, it is possible to model the overall thermal system using an electrically equivalent circuit thermal model. Some values for the device thermal resistances ($R_{\theta JC}$, $R_{\theta JB}$) are given in the datasheet; some resistances are configuration-specific, such as the PCB-to-ambient thermal resistance $R_{\theta BA}$. Values for the thermal resistance of the heatsink and thermal interface material can be extracted from their datasheets, whereas the thermal resistance for convection and radiation are configuration- and size-dependent, as well as temperature- and orientation-dependent. Furthermore, considering multiple devices and heat sources within a configuration increases the overall complexity of the model.

5.3.2 Modeling of Multiple Die

Due to the complexity of multiple heat sources, it is necessary to determine the components of thermal resistance beforehand. One method for estimating thermal resistance requires initial temperature calibration of the heat-generating transistors. This is done by utilizing a known relationship between junction temperature and on-resistance. Placing a device in a controlled thermal chamber and monitoring the $R_{DS(on)}$ versus temperature yields such a relationship, as shown in Figure 5.14. With the exact relationship between temperature and on-resistance for each device known, the system can be heated such that the in-circuit on-resistance can be monitored.

Consider an example of two devices mounted on the same board without heatsink cooling as shown in Figure 5.15. With their on-resistance as a function of temperature known, these devices can be heated by passing a known current through both devices and measuring their in-circuit on-resistance. The board temperature, as close as possible to each device, also can be measured using a thermal infrared camera. The results of these measurements are shown in Figure 5.16. If there were no heat flowing through the top (case) of the devices, then the

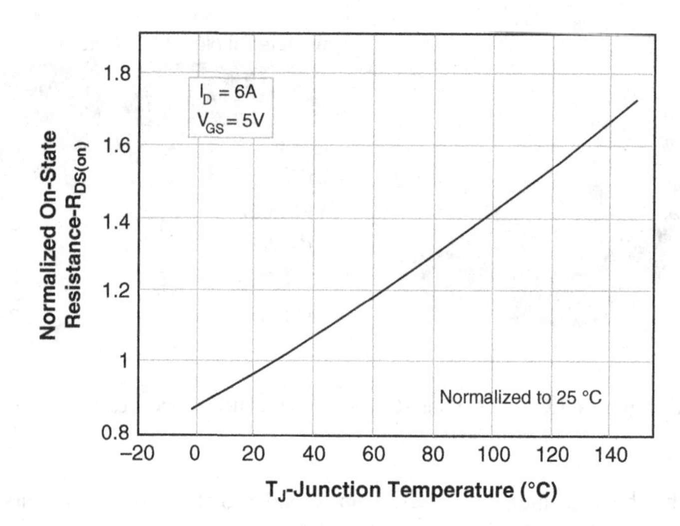

Figure 5.14 Normalized on-resistance vs. temperature for an LGA GaN transistor [7]

estimated junction temperature, based on the measured PCB temperature and device thermal resistance, would match that of the measured junction temperature. Figure 5.16, however, shows that this is not the case, and that the amount of heat flowing through the top is increasing with temperature, as expected for a radiating device.

With the junction, board, and ambient temperatures known, as well as the device thermal resistances and power losses, an electrically equivalent circuit thermal model can be constructed by first calculating the amount of heat flux down into the PCB from each device. The resultant thermal model with estimated thermal resistance values is shown in Figure 5.17 (values were chosen to yield a symmetrical system). The total resistance split between the spreading resistances ($R_{\theta SP1}$ and $R_{\theta SP2}$) and the PCB-to-ambient resistance ($R_{\theta BA}$) is determined by the temperature difference between the measured PCB temperatures at the two devices. This example also shows that a significant fraction of the total losses are dissipated through the case, even though no heatsink was attached. From the results in Figure 5.17, the following determinations can be made:

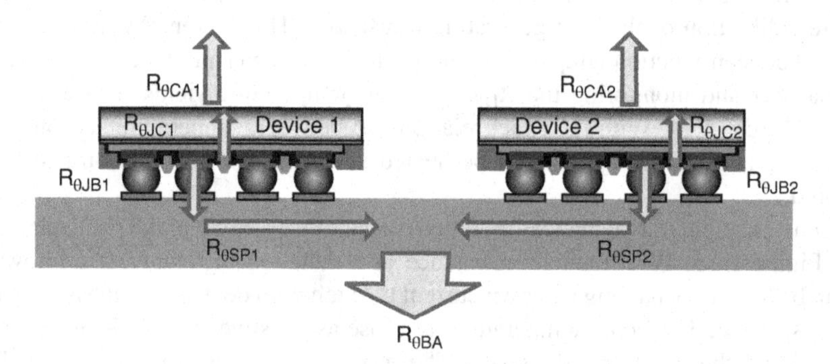

Figure 5.15 Thermal diagram of two board-mounted LGA GaN transistors without heatsink

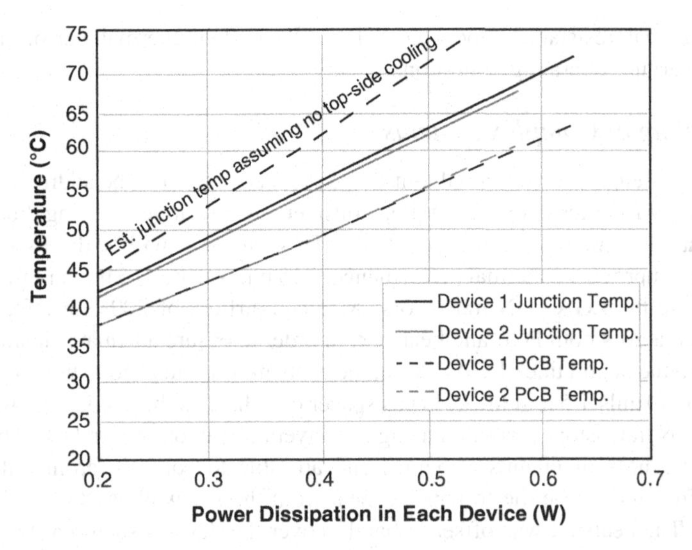

Figure 5.16 Measured device junction and PCB temperature vs. power dissipation for the two LGA GaN transistors [7] mounted on a PCB [8]

1. The thermal resistance of the top surface (case) of the LGA GaN transistor to the ambient air with no airflow ($R_{\theta BA1}$ and $R_{\theta BA2}$) is about 120°C/W, at around 70°C, but varies with case temperature.
2. The thermal spreading resistance between the two transistors mounted next to each other on the PCB ($R_{\theta SP1}$ and $R_{\theta SP2}$) is about 10°C/W.
3. The thermal resistance through the PCB-to-ambient ($R_{\theta BA}$) is about 60°C/W in still air.
4. The effective junction-to-ambient thermal resistances for each of the devices is between 69°C/W and 74°C/W, due to the thermal interaction of the two devices. With only one device dissipating heat, the effective thermal resistance decreases to about 50°C/W.

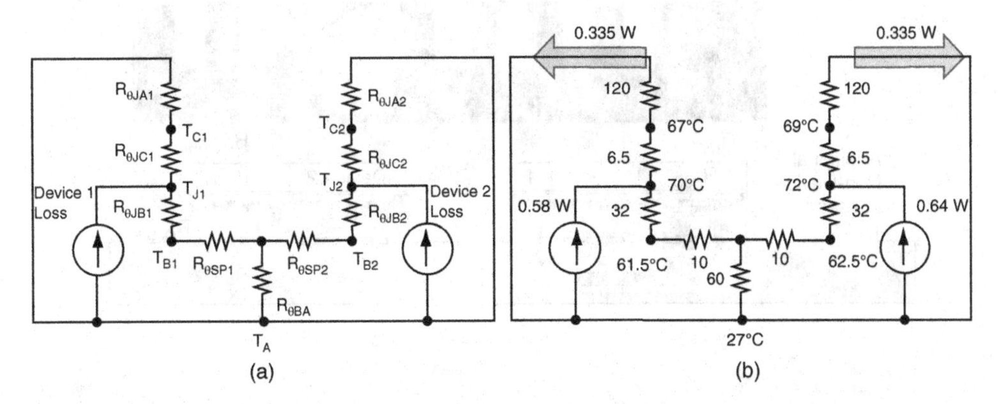

Figure 5.17 Electrically equivalent circuit thermal model of the LGA GaN transistor PCB with two devices, no heatsink, and only convection cooling: (a) parameters (b) resultant values

With the thermal resistance model complete, the system thermal performance can be estimated for varying operating conditions.

5.3.3 Modeling of Complex Systems

More complex systems with additional non-device-related losses can be addressed in a similar fashion as for two devices. For example, consider an envelope tracking buck converter with the thermal diagram shown in Figure 5.18. (This converter will be discussed in detail in Chapter 10.) To improve the thermal performance, a 15 mm square, 9.5 mm tall finned heatsink was added above the LGA GaN transistors. A forced airflow of 200 linear feet per minute (LFM) was used across both PCB and heatsink. In order to ensure adequate clearance between the heatsink and the 30 mil thick LGA device, the heatsink was attached to the board using Gap Pad® GP 1500 (60 mils/1.5 mm thick) [9] as spacer over half the heatsink area, while the area covering the GaN transistors was filled using two layers of Sarcon 30x-m [10]. The two layers have a total thickness of 60 mils (1.5 mm) and are able to conform around the die when compressed. This allows the die to conduct heat from the sidewalls as well as from the top surface (case). The heatsink was offset to barely cover the devices such that the temperature of the PCB directly adjacent to the devices could be measured using an infrared camera. A complete discussion on the component loss breakdown is given in [11].

The electrically equivalent circuit thermal model for this example is shown in Figure 5.19(a), with the resultant thermal resistance values and temperatures for full-load operation shown in Figure 5.19(b). Thermal resistance values for the heatsink and devices were taken from their datasheets, while the TIM thermal resistance was estimated based on the die contact area and material thickness. For the PCB spreading resistance, the values from Figure 5.17 were used, since the board layout was almost identical. With the loss breakdown known, the PCB-to-ambient resistance was adjusted until good correlation with the measured PCB temperature was achieved. This resulted in a PCB-to-ambient thermal resistance (not including spreading resistance) of about 5°C/W for this forced-air cooling condition, and is significantly lower that for the still-air example.

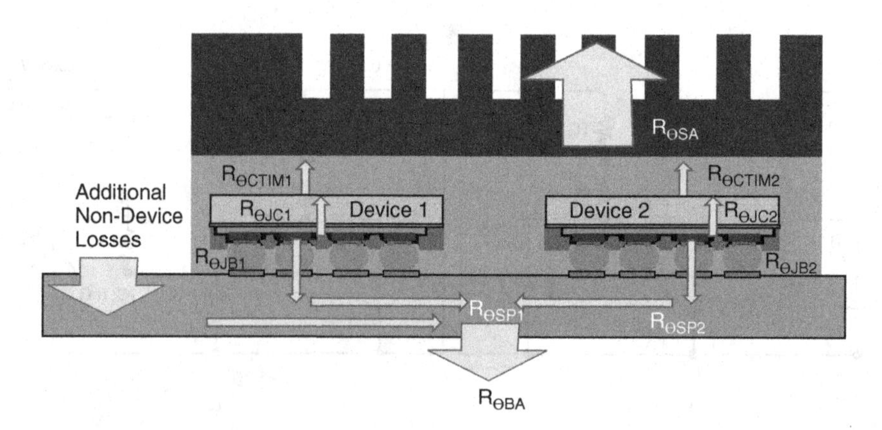

Figure 5.18 Thermal diagram of two board-mounted LGA GaN transistors with heatsink cooling on a PCB with additional non-device losses

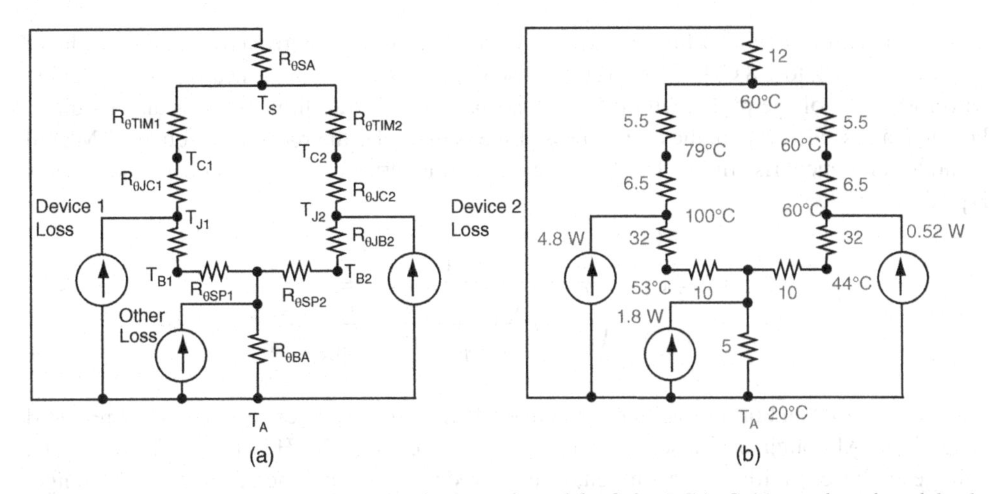

Figure 5.19 Electrically equivalent circuit thermal model of the LGA GaN transistor-based buck converter with heatsink and forced-air cooling: (a) parameters (b) resultant values at full load

5.4 Measuring GaN Transistor Performance

The increase in switching speed offered by GaN transistors with their accompanying faster di/dt and dv/dt, requires a proportional increase in the bandwidth of the measurement equipment used. First-generation GaN transistors were generating speeds close to the limit of the present oscilloscope and probe technology, making it likely that the GaN power transistor revolution will be accompanied by a corresponding step change in measurement technology. To determine the extent of these requirements, each different measurement will be evaluated in turn.

5.4.1 Voltage Measurement Requirements

To determine the requirements for voltage measurement of the high dv/dt switching waveforms, it is necessary to relate the fidelity requirements of switching waveforms with the bandwidth (BW) requirements of the oscilloscope and accompanying probes. The relationship between rise time and bandwidth is typically given by Equation 5.1, and is based on a number of assumptions [12].

$$t_{rise(10\%-90\%)} \approx \frac{0.35}{BW_{-3dB}} \tag{5.1}$$

The resultant rise time from Equation 5.1 assumes that measuring an infinitely fast rising edge will yield the given measured rise time. This does not mean that a waveform with this actual rise time will be represented accurately on an oscilloscope. To yield an accurate representation of the waveform rise time, a rule of thumb is that a bandwidth between three and five times that of Equation 5.1 is required [12].

$$BW_{-3dB} \approx \frac{k \cdot 0.35}{t_{rise(10\%-90\%)}}, \text{ k is between 3 and 5} \tag{5.2}$$

Thus, for a waveform with a 1 ns rise time, Equation 5.2 would require an overall bandwidth of between 1.1 GHz to 1.8 GHz. However, the oscilloscope alone cannot measure the switching waveforms. A voltage probe, with its own associated bandwidth limitations, is also required. For such a cascade system, the effective rise time is given by the root-mean-square (RMS) of separate component rise times. This can be expressed in terms of component bandwidths as in Equation 5.3 [13]:

$$\mathrm{BW}_{-3\mathrm{dB}} = \frac{1}{\sqrt{\frac{1}{\mathrm{BW}^2_{-3\mathrm{dB,Scope}}} + \frac{1}{\mathrm{BW}^2_{-3\mathrm{dB,Probe}}}}} \tag{5.3}$$

The oscilloscope and probe each need to have a bandwidth in excess of the overall required bandwidth. Although oscilloscope bandwidths in excess of 2 GHz are readily available, voltage probe capabilities are limited, and they decrease with increasing voltage range. Table 5.1 shows voltage probes suitable for GaN transistor measurement and their related voltage and rise-time capability. Higher-bandwidth passive and active probes are available, but these are limited to a maximum voltage of about 15 V, which is too low for most GaN transistor applications.

At these high frequencies and voltages, having sufficient rise-time capability alone isn't enough to make useful voltage measurements. It is also important to place the probe as close to the required measurement point as possible while generating an inductive loop that is as small as possible. A practical example to achieve this is shown in Figure 5.20. The probe loop needs to be minimized for two reasons:

1. The high impedance of the probe, coupled with the fast-changing currents (and associated magnetic fields), will induce noise in the probe, through electromagnetic induction. This can be mitigated by limiting the loop area and choosing the loop orientation to avoid flux coupling.
2. The probe impedance in series with the loop inductance forms an LC resonant tank that not only reduces bandwidth, but also generates oscillations not present in the actual measurement [14], as the probe gain becomes more than unity at resonance. The impact of a given loop inductance is more pronounced in a probe with higher capacitance. In short, if the waveform being measured has a fast enough rise time to contain within its Fourier spectrum the probe loop resonant frequency, then the probe itself will oscillate.

Table 5.1 Voltage probes and their associated voltage, bandwidth, and minimum rise time

	Attenuation	Input impedance	Bandwidth	Maximum voltage	Rise time (Equation 5.2)
High-end passive probe	10:1	10 MΩ/ ~10 pF	500 MHz	300 V	2.1–3.5 ns
LeCroy PP065	100:1	5 kΩ/2 pF	1 GHz	30 V	1.1–1.8 ns
Tektronix TPP1000	10:1	10 MΩ/<4 pF	1 GHz	300 V	1.1–1.8 ns
Tektronix TPP0850	50:1	40 MΩ/1.8 pF	800 MHz	2500 V	1.3–2.2 ns

*Assuming oscilloscope bandwidth high enough not to impact performance.

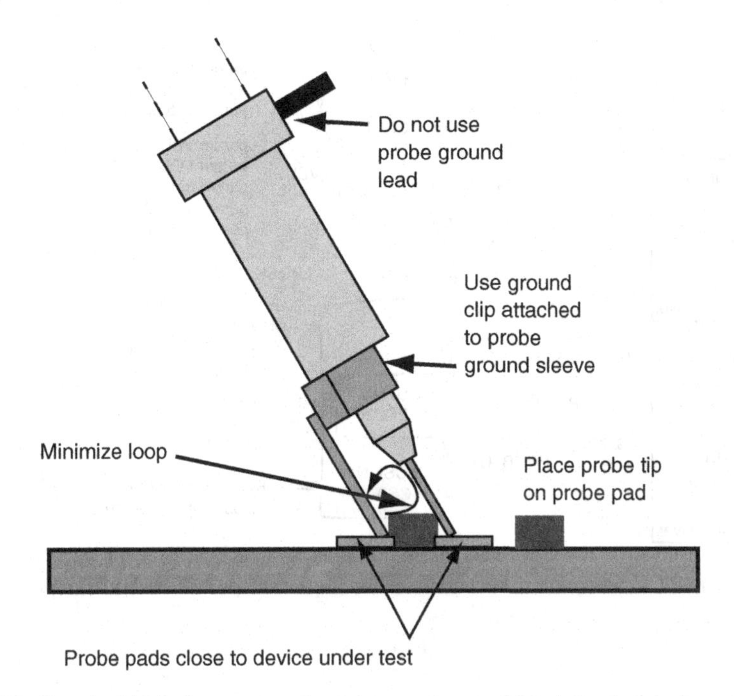

Figure 5.20 Practical high-frequency voltage measurement with minimal probe loop inductance

Figure 5.20 shows that the probe needs to be placed as close to the device under test as possible. To show the impact of probe position requires a simulation. Using the available SPICE models for the LGA GaN transistors, a buck converter simulation with layout parasitics was created as shown in Figure 5.21. The switch-node-to-ground voltage waveform (an estimation of a practical voltage measurement), with the actual voltage across the active device, is shown in Figure 5.22. It clearly shows that the "measured" voltage has much smaller ringing than is actually seen across the active device. In practice, the ringing may be higher or lower, depending on the parasitic inductance and the direction of the changing current. Secondly, there is a measured initial bump in the voltage rise time that is due to the induced voltage drop across the parasitic inductance within the measurement loop. From the active device's drain current – also shown in Figure 5.22 – the bump coincides with the current rising in the device. Thus, although this added bump adds uncertainty to the voltage rise time and shape, it indirectly does add information about the current rise time.

5.4.2 Current Measurement Requirement

As with voltage measurement, the current measurement bandwidth is determined by the rise-time requirement. For a typical rise time of 2 ns (as shown in the simulation results in Figure 5.22), a practical current measurement bandwidth (using Equation 5.2) of more than 500 MHz is required. This is well beyond the capability for traditional current probes, such as Hall element [15] and Rogowski coils [16,17], which have bandwidths of 50 MHz or less. This leaves the option of using current-sense resistors or coaxial current shunts, some of which have

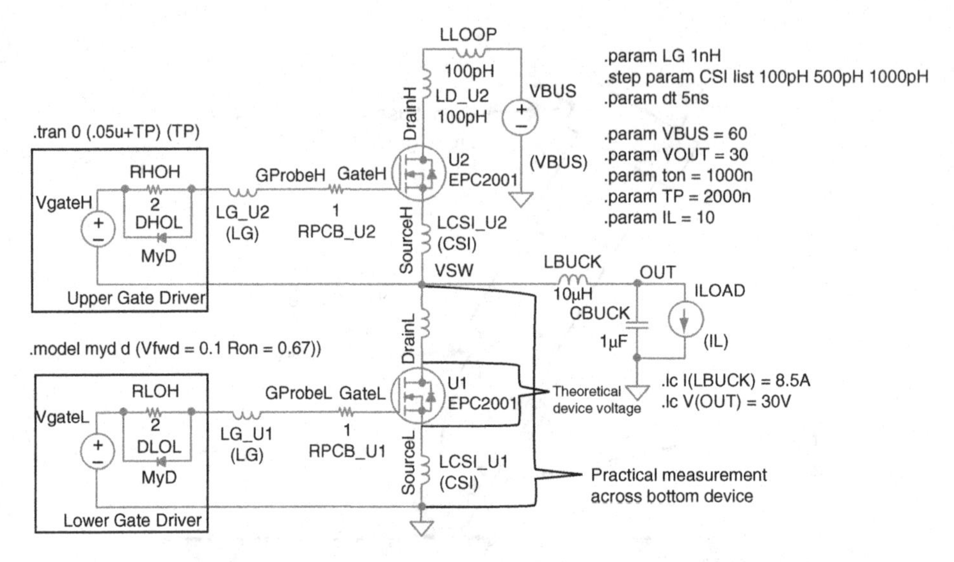

Figure 5.21 LTSPICE simulation created to show impact of probe location on practical voltage measurement

bandwidths up to 2 GHz [18]. Sufficient bandwidth, however, is only one of the criteria for accurate current measurement.

In most cases, the bandwidth of the sensing resistor or shunt is limited by the corner frequency made between the resistance and the parasitic series inductance. For a given series inductance, the bandwidth can be improved by increasing the value of the sense resistance, but this comes at the cost of increased voltage drop and power loss. This effect can be seen for a selection of coaxial shunts in Table 5.2. Thus, to measure a 20 A current pulse similar to the

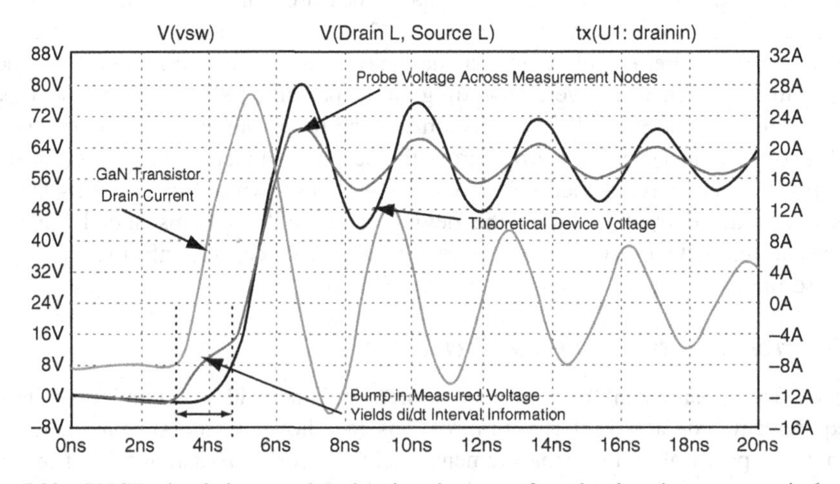

Figure 5.22 SPICE simulation results showing impact of probe location on practical voltage measurement

Table 5.2 Selection of high bandwidth coaxial shunts [18]

Model	Resistance (mΩ)	Bandwidth (MHz)	Voltage drop at 20 A (V)	Rise time (ns)
SDN-414-01	10	400	0.2	1
SDN-414-025	25	1200	0.5	0.3
SDN-414-05	50	2000	1.0	0.18
SDN-414-10	100	2000	2.0	0.18

one shown in Figure 5.22, the bandwidth requirement implies that the 10 mΩ shunt could not be used, and a minimum shunt voltage drop of 0.5 V would have to be induced. This may be excessive for a low-voltage application.

Furthermore, the main advantage of a coaxial current shunt compared with sense resistors is the reduction of parasitic inductance between the measurement nodes, increasing measurement bandwidth. Due to their size and shape, coaxial shunts add a larger amount of inductance to the overall circuit, and any significant increase in power loop inductance will adversely affect switching operation. One method is to place a large number of shunt resistors in parallel to reduce the overall inductance [19], but this still requires a significant reduction in switching speed to achieve meaningful current waveform measurements.

In summary, with present technology, it is practically impossible to measure the GaN transistor current in circuit without adversely affecting its dynamic performance.

5.5 Summary

In this chapter, the basic techniques for modeling and measuring GaN transistors in high-performance power conversion circuits were discussed. The next chapter will explore how the superior properties of GaN transistors yield significant performance improvements in hard-switching applications.

References

1. Liu, W., Jin, X., Xi, X. *et al.* (2005) "BSIM3v3.3 MOSFET Model, User's manual," Dept. of Electrical Engineering and Computer Sciences, University of California, Berkeley. Available from http://www-device.eecs.berkeley.edu/~bsim/Files/BSIM3/ftpv330/Mod_doc/b3v33manu.tar.
2. Efficient Power Conversion Corporation, Appl. Note AN005, "Circuit simulation using EPC device models," Available from http://epc-co.com/epc/documents/product-training/Circuit_Simulations_Using_Device_Models.pdf.
3. Efficient Power Conversion Corporation, "Demonstration board EPC9107 quick start guide," Available from http://epc-co.com/epc/Products/DemoBoards/EPC9107.aspx.
4. International Rectifier, Appl. Note, AN-1059, "DirectFET® technology thermal model and rating calculator," Sep. 2010, Available from http://www.irf.com/technical-info/appnotes/an-1059.pdf.
5. Infineon, Appl. Note AN 2012-04, "ThinPAK 8X8 New High Voltage SMD-Package," version 1.0, April 2010, Available from http://www.infineon.com/dgdl/Infineon+ThinPAK+8x8.pdf?folderId=db3a304314dca3890115-2836c5a412ab&fileId=db3a304327b897500127f6946a286519.
6. Reusch, D. "High frequency, high power density integrated point of load and bus converters," Ph.D. dissertation, Virginia Tech, Blacksburg, VA, 2012. Available from http://scholar.lib.vt.edu/theses/available/etd-04162012-151740/.
7. Efficient Power Conversion Corporation, "EPC2007 – Enhancement-mode Power Transistor," EPC2007 datasheet, Sep. 2011 [Revised July 2013]. Available from http://epc-co.com/epc/documents/datasheets/EPC2007_datasheet.pdf.

8. Efficient Power Conversion Corporation, "Demonstration board EPC9006 quick start guide," Available from http://epc-co.com/epc/Products/DemoBoards/EPC9006.aspx.
9. Bergquist Company, "Gap pad GP1500 thermally conductive, un-reinforced gap filling material," Gap Pad® 1500 datasheet PDS_GP_1500_12.08. Available from http://www.bergquistcompany.com/pdfs/dataSheets/PDS_GP_1500_12.08_E.pdf.
10. Fuji Polymer Industries Co. Ltd., Fujipoly®, "Sarcon® XR-m Highly Thermal Conductive and Non-Flammable Silicone Gel Sheets," datasheet, [14 Jul. 2009]. Available from http://www.fujipoly.com/usa/assets/files/2010_data_sheets/090930_Sarcon%20XR-m%20technical%20info.pdf.
11. Strydom, J., de Rooij, M., and Lidow, A. "Gallium nitride transistor packaging advances and thermal modeling," *EDN China*, Sep. 2012. Available from http://epc-co.com/epc/documents/product-training/Gallium%20Nitride%20Transistor%20Packaging%20Advances.pdf.
12. Tektronix Corporation, "Understanding oscilloscope bandwidth, rise time and signal fidelity," Technical brief no. 55W-18024-0, Available from http://www.electron.frba.utn.edu.ar/~jcecconi/Bibliografia/06%20-%20Osciloscopios%20de%20Almacenamiento%20Digital/Understanding_Oscilloscope_BW_RiseT_And_Signal_Fidelity.pdf.
13. SDE Consulting, T.J. Sobering, Technote 2, "Bandwidth and rise time," May 1999 [Revised Nov. 18, 2002] http://www.k-state.edu/ksuedl/publications/Technote%202%20-%20Bandwidth%20and%20Risetime.pdf.
14. Teledyne LeCroy, "Passive probe ground lead effects," June 2013. Available from http://teledynelecroy.com/doc/passive-probe-ground-lead-effects#pdf.
15. Teledyne Leroy, "Current probes," Available from http://teledynelecroy.com/probes/probeseries.aspx?mseries=426.
16. Power Electronic Measurement, "CWT-current probes," 2013. Available from http://www.pemuk.com/products/cwt-current-probe.aspx.
17. Athena Energy Corp, "200 A/50 MHz Rogowski Coil Current probe, Athena Energy Corp current probe," datasheet. Available from http://ecbiz122.inmotionhosting.com/~athena18/wp-content/uploads/2012/03/Rogowski_Specifications.pdf.
18. T & M Research Products, "SDN series co-axial current shunts," datasheet. Available from http://www.tandmresearch.com/.
19. Danilovic, M. *et al.* (Sept. 2011) "Evaluation of the switching characteristics of a gallium-nitride transistor," *Energy Conversion Congress and Exposition*, ECCE 2011, Phoenix, AZ, pp. 2681–2688.

6

Hard-Switching Topologies

6.1 Introduction

In hard-switching converters, the transistors are turned on and off rapidly, while there is voltage across – and current through – the drain and source of the device. These switching transitions lead to significant power losses during the switching event. The main metrics of any converter performance are: (a) efficiency, where higher is better, (b) size, where smaller is better, and (c) cost, where lower is better. Efficiency can be increased through improvements in the switching (dynamic) and conduction (static) characteristics of the devices, thereby allowing higher switching frequencies to be used. This, in turn, leads to a size reduction, which also can lead to lower cost. In this chapter, hard-switching topologies will be reviewed and we will look at how the superior properties of GaN transistors yield significant performance improvements.

Table 6.1 provides definitions for terms used in hard-switching loss analysis discussions in this chapter.

6.2 Hard-Switching Loss Analysis

To drive the frequency higher in a hard-switching converter, power devices must have very low dynamic losses. The dominant component of these losses is the hard-switching "event" where, in a turn-on switching transition, current flows through the device before the voltage across that device commutates to zero as shown in Figure 6.1(a) [1,2]. The reverse sequence occurs when the device turns off, as shown in Figure 6.1(b).

The transitions of current and voltage within the device are not the only loss contributors as there are other directly and indirectly related components. These additional factors are: output capacitance losses (P_{OSS}), gate charge losses (P_G), reverse conduction losses (P_{SD}), and reverse recovery losses (P_{RR}).

The output capacitance losses (P_{OSS}) are associated with the output capacitance of the device. The charging or discharging of this capacitor requires energy of which half is dissipated in the resistance in the current path of the capacitor in normal, positive current operation of the converter.

Gate charge losses (P_G) are similar to the output capacitance losses in that the energy required to charge the gate-to-source capacitance is dissipated in the resistance of the current path of the capacitance.

GaN Transistors for Efficient Power Conversion, Second Edition.
Alex Lidow, Johan Strydom, Michael de Rooij, and David Reusch.

Companion Website: http://www.wiley.com/go/gan_transistors

Table 6.1 Definitions used in hard-switching loss analysis

Definitions of terms used in hard-switching loss analysis	
V_{th}	Gate threshold voltage (at Q_{GS1})
V_{DR}	Gate driver on-state output voltage
V_{pl}	Gate plateau voltage (at Q_{GD})
$V_{pl(op)}$	Plateau voltage at the operating condition current
t_{CR}	Rise time for current in the transistor
t_{CF}	Fall time for current in the transistor
t_{VR}	Rise time for voltage across the transistor
t_{VF}	Fall time for voltage across the transistor
t_{SD}	Reverse diode conduction time
t_{ZVS}	Zero voltage switching transition time
t_{eff}	Effective dead-time between transistor switching
P_{on}	Power losses due to the turn-on switching transition
P_{off}	Power losses due to the turn-off switching transition
P_{COSS}	Power losses due to output charge Q_{OSS}
P_G	Power losses due to gate charge Q_G
P_{SD}	Power losses due to the forward drop of the body diode
P_{RR}	Power losses due to the reverse recovery charge of the body diode (MOSFET and Cascode transistors only)
P_{SW}	Total power losses during switching transitions
$P_{Dynamic}$	Power loss due to dynamic loss components in the transistor
$P_{Conduction}$	Power loss due to static conduction loss in the transistor
f_{SW}	Switching frequency
Q_{GS1}	Charge required to increase gate voltage from zero to the stated threshold voltage of the device.
Q_{GS2}	Charge required to increase gate voltage from the stated threshold voltage of the device to the plateau voltage (current conduction interval)
Q_{GS}	Charge required to increase gate voltage to the plateau voltage (complete current transition) $Q_{GS} = Q_{GS1} + Q_{GS2}$
Q_{GD}	Charge required into the gate to change the drain voltage down from the blocking state to near zero, at which point the device enters the linear region. (constant current region)
Q_G	Total gate charge required to drive a device from zero to rated gate voltage (fully enhanced)
$Q_{GS(op)}$	Charge required to increase gate voltage to the operating plateau voltage
$Q_{G(op)}$	Total gate charge required to drive a device from zero to rated gate voltage based on the operating conditions
g_m	Transconductance of the transistor

In most converters, an anti-parallel diode is present across the drain-to-source terminals of the device. In some cases, it is inherent to the device structure, such as in MOSFETs. In cascode GaN transistors, there is a body diode in the Si MOSFET. In the case of enhancement-mode GaN transistors, as explained in Chapter 1, there is a mechanism to conduct reverse current when the device is off, allowing operation similar to a diode. Since diode conduction is a function of the switching transient, they are included in the hard-switching loss calculation. These losses are defined as reverse conduction losses (P_{SD}).

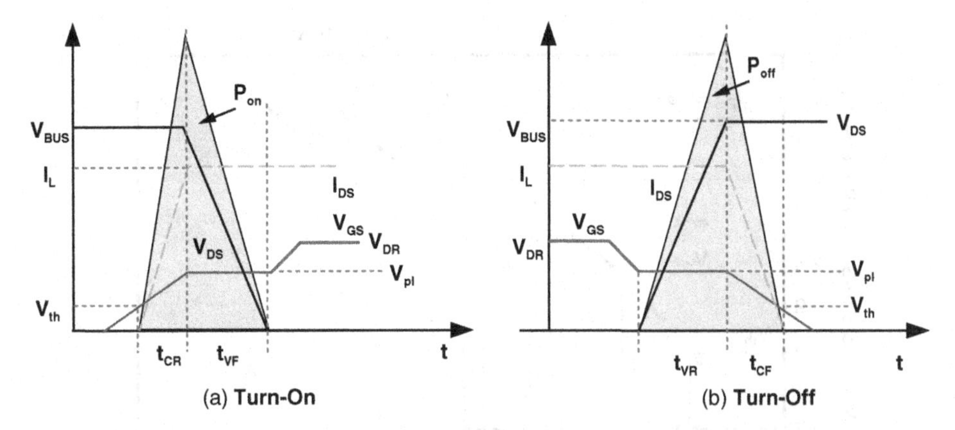

Figure 6.1 Idealized switching waveforms used for calculating switching loss (a) turn-on (b) turn-off

Reverse recovery losses (P_{RR}) are based on the amount of charge required to turn off the body diode, and are only present in MOSFET and cascode GaN transistors, because enhancement-mode GaN transistors have no reverse recovered charge.

Next, the details of each of the loss mechanisms will be presented, plus the means to determine each of the losses.

6.2.1 Switching Losses

The switching power loss can be determined graphically from Figure 6.1, by summing the voltage transition power losses (P_{Vt}) and the current transition power losses (P_{Ct}) using the following equation:

$$\begin{aligned} P_{sw} &= P_{Vt} + P_{Ct} \\ &= \tfrac{1}{2} \cdot V_{BUS} \cdot I_L \cdot (t_{xR} + t_{xF}) \cdot f_{sw} \end{aligned} \tag{6.1}$$

Where t_{xR} and t_{xF} are the switching commutation times as seen in Figure 6.1, V_{BUS} is the voltage across the device during the off-state, and I_{DS} is the on-state drain-to-source current, and where one parameter (either voltage or current) is always in transition while the other is fixed. This leads to the factor of 1/2 in Equation 6.1. The switching times (t_{xR} and t_{xF}) are not given in the circuit and need to be determined from the gate charge (Q_G) characteristics of device based on the circuit operating conditions.

GaN transistors are driven in a similar manner to MOSFETs. The gate electrodes have very high input impedance, and control of the device is accomplished by supplying or removing a certain amount of charge to/from the gate electrode. Transistor switching can be segmented into four regions, as shown in Figure 6.2: (1) the charge required to transition the drain-to-source voltage (Q_{GD}), (2) the charge required to bring the gate electrode up to device threshold (Q_{GS1}), (3) the charge required to transition the current from zero to the load current (Q_{GS2}), and (4) the incremental additional charge to overdrive the gate ($Q_G - (Q_{GS1} + Q_{GS2} + Q_{GD})$).

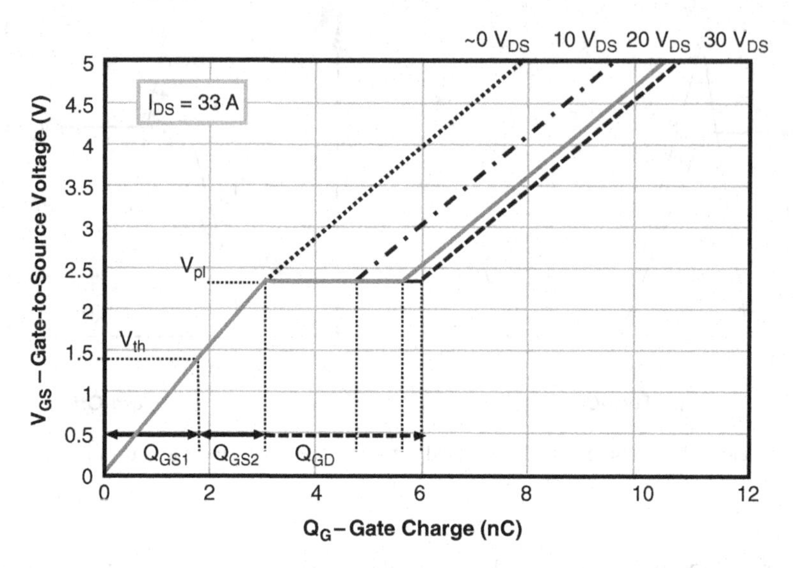

Figure 6.2 Impact of drain-source voltage and drain current on gate charge and gate voltage for EPC2015 [3]

6.2.1.1 Miller Charge (Q_{GD}): The Voltage Transition Period

The voltage commutation period of a traditional hard-switching commutation is based on the Miller charge. The Miller charge is characterized by the plateau voltage on the gate waveform and can be used to determine the switching losses during the voltage transition period. At turn-on, the period begins after the current has fully transitioned, and is complete when the drain-to-source voltage has reaches zero. The reverse process occurs for turn-off. The larger the voltage swing, the longer it will take for the transition, and the higher the losses. Figure 6.2 is a graph of the measured gate voltage as a function of gate charge, highlighting the impact of various drain-to-source voltages.

In general, the time (t) it takes to charge a capacitor to a specific charge (Q) is given by:

$$t = \frac{Q}{I} \tag{6.2}$$

Where I is the current used to charge the capacitor.

Q_{GD} for any given drain-to-source voltage can be calculated using Equation 6.3 [2] if the function of $C_{RSS}(v_{DS})$ is known.

$$Q_{GD} = \int_{0}^{V_{BUS}} C_{RSS}(v_{DS}) \cdot dv_{DS} \tag{6.3}$$

Alternatively, the $C_{RSS}(v_{DS})$ function can be captured from the graph provided in device datasheets. It can be seen from Figure 6.3 that all the devices capacitances have a non-linear relationship with drain-to-source voltage.

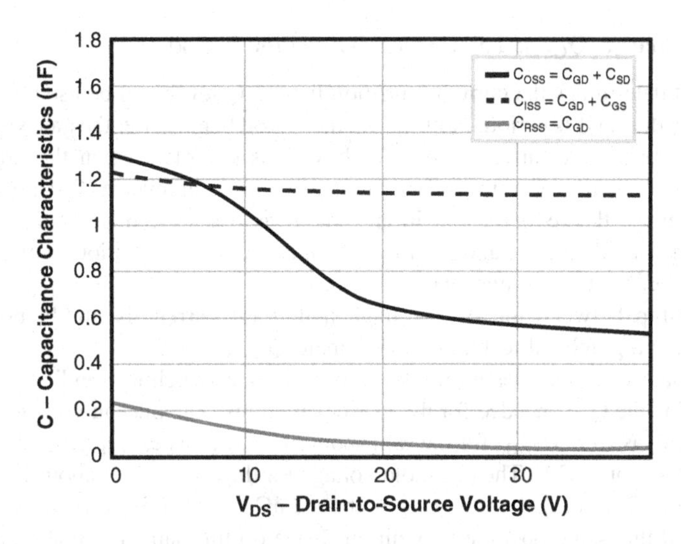

Figure 6.3 Device capacitances as a function of drain-to-source voltage for EPC2015 [3]

Substituting Equation 6.2 into Equation 6.1 for the voltage transition only, and using a linear approximation for the gate voltage and current transitions, the power losses during the voltage transition can then be approximated with Equation 6.4.

$$P_{Vt} \cong \frac{V_{BUS} \cdot I_L}{2} \cdot t_{Vx} \cdot f_{sw} = \frac{V_{BUS} \cdot I_L}{2} \cdot \frac{Q_{GD}}{I_G} \cdot f_{sw} \tag{6.4}$$

As can be seen in Equation 6.4, the gate current appears in the power loss equation. The current into the gate thus affects the transition time, and increasing the gate current will reduce this time. However, gate driver impedance and gate circuit inductance may limit this gate current and therefore the transition period.

The gate current (I_{Gvon}) during the turn-on voltage transition period (t_{VF}) can be estimated as:

$$I_{Gvon} = \frac{V_{DR} - V_{PL}}{R_{Gon}} \tag{6.5a}$$

Where V_{DR} is the gate driver on-state output voltage.

The gate current (I_{Gvoff}) during the turn-off voltage transition period (t_{VR}) can be estimated as:

$$I_{Gvoff} = \frac{V_{PL}}{R_{Goff}} \tag{6.5b}$$

The gate currents I_{Gvon} and I_{Gvoff} can be equated to I_G in Equation 6.4 to determine the voltage transition power loss.

6.2.1.2 Gate Charge (Q_{GS2}): The Current Transition Period

The charge that determines the current transition time is Q_{GS2}. It can be used to calculate the switching losses during this period. For turn-on, the period begins after the gate voltage reaches the threshold voltage and current begins to flow. It is complete when the drain-to-source voltage begins to transition. For turn-off, the sequence occurs in reverse. The larger the current swing, the longer it will take for the transition, and power losses will increase. Figure 6.4 shows a graph of the measured gate voltage as a function of gate charge for various current levels. The higher the drain current, the longer the period lasts.

The relationship between the gate voltage and drain current is highly non-linear and therefore requires a graphical technique to estimate Q_{GS2}.

The typical datasheet provides information for only one switching condition, but it can be used to determine the Q_{GS2} needed for the loss calculations. The plateau voltage in Figure 6.4 needs to be noted, as well as Q_{GS} for the same conditions. In this example, I_{DS} is 33 A and the plateau voltage is about 2.3 V. The threshold voltage for these devices is about 1.4 V, giving an estimated value of Q_{GS1} of 2 nC, and Q_{GS2} of about 1 nC at this drain current. In Figure 6.5 we have highlighted this same operating condition. If we do this same calculation at 100 A, the plateau voltage changes to 2.8 V, and Q_{GS1} is unchanged, but Q_{GS2} increases to 2 nC.

In general, Q_{GS1} can be calculated using Equation 6.6:

$$Q_{GS1} = \left(\frac{Q_{GS}}{V_{pl}}\right) \cdot V_{th} \tag{6.6}$$

Since Q_{GS} and V_{pl} both vary proportionally with current and drain-source voltage, their ratio is virtually constant. Therefore, Q_{GS1} is a fixed value regardless of operating conditions.

The $Q_{GS(op)}$, which is the Q_{GS} at the operating value of I_{DS}, can also be calculated for the operating conditions by reading off the plateau voltage $V_{pl(op)}$ from the transfer characteristic

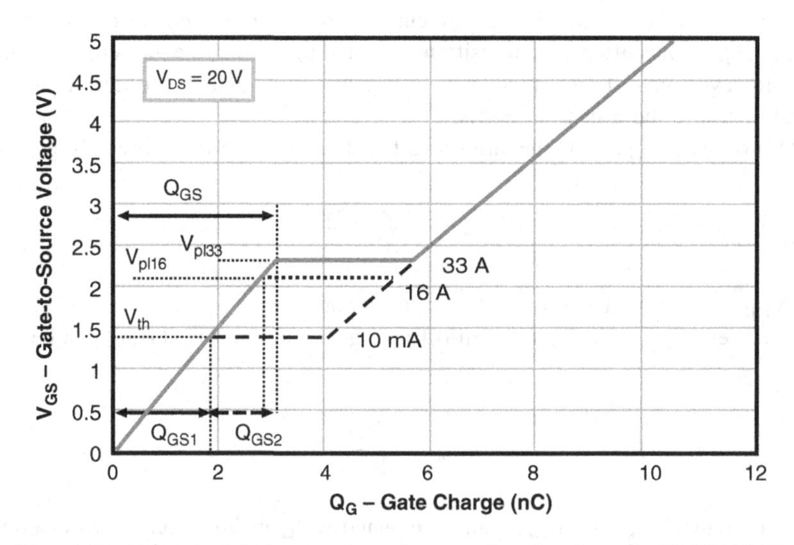

Figure 6.4 Impact of drain current on the gate plateau voltage for EPC2015 [3]

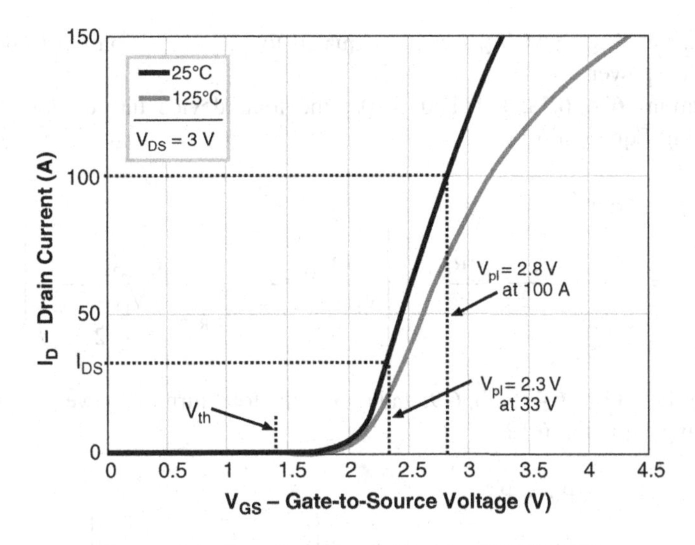

Figure 6.5 Transfer characteristic for EPC2015 [3]

graph, as shown in Figure 6.5, and using Equation 6.7:

$$Q_{GS(op)} = \left(\frac{Q_{GS}}{V_{pl}}\right) \cdot V_{pl(op)} \tag{6.7}$$

With $Q_{GS(op)}$ and Q_{GS1} determined, Q_{SG2} can be calculated using Equation 6.8:

$$Q_{GS2} = Q_{GS(op)} - Q_{GS1} \tag{6.8}$$

Substituting Equation 6.2 into Equation 6.1 for the current transition only, the losses can be approximated using Equation 6.9.

$$\begin{aligned} P_{ct} &\cong \frac{V_{BUS} \cdot I_{DS}}{2} \cdot t_{Cx} \cdot f_{sw} \\ &= \frac{V_{BUS} \cdot I_{DS}}{2} \cdot \frac{Q_{GS2}}{I_G} \cdot f_{sw} \end{aligned} \tag{6.9}$$

The gate current again appears in the power loss equation. Thus, the current into the gate affects the transition time, and increasing the gate current will reduce this time.

The gate current during the current turn-on and turn-off transition periods can be estimated as:

$$I_{Gcon} = \frac{V_{DR} - \left(\frac{V_{pl} + V_{th}}{2}\right)}{R_{Gon}} \tag{6.10a}$$

$$I_{Gcoff} = \frac{\left(\frac{V_{pl} + V_{th}}{2}\right)}{R_{Goff}} \tag{6.10b}$$

The gate currents I_{Gcon} and I_{Gcoff} can be equated to I_G in Equation 6.9 to determine the current transition power loss.

Using Equations 6.4, 6.5a, 6.9, and 6.10a, the total device turn-on loss then can be determined using Equation 6.11:

$$P_{on} = P_{Vt} + P_{Ct} = \frac{V_{BUS} \cdot I_{DS} \cdot f_{sw} \cdot R_{Gon}}{2} \cdot \left[\frac{Q_{GD}}{V_{DR} - V_{pl}} + \frac{Q_{GS2}}{V_{DR} - \left(\frac{V_{pl} + V_{th}}{2} \right)} \right] \tag{6.11}$$

Similarly using Equations 6.4, 6.5b, 6.9, and 6.10b, the total turn-off power loss (P_{off}) can be determined using Equation 6.12:

$$P_{off} = P_{Vt} + P_{Ct} = \frac{V_{BUS} \cdot I_{DS} \cdot f_{sw} \cdot R_{Goff}}{2} \cdot \left[\frac{Q_{GD}}{V_{pl}} + \frac{Q_{GS2}}{\left(\frac{V_{pl} + V_{th}}{2} \right)} \right] \tag{6.12}$$

The total switching losses can now be summarized as the sum of P_{on} and P_{off}:

$$P_{sw} = P_{on} + P_{off} \tag{6.13}$$

6.2.2 Output Capacitance (C_{OSS}) Losses

The power loss due to the output capacitance can be calculated for the specific working voltage using the Equation 6.14:

$$P_{OSS} = f_{sw} \cdot E_{OSS} = f_{sw} \cdot \int_0^{V_{BUS}} v_{DS} \cdot C_{OSS}(v_{DS}) \cdot dv_{DS} \tag{6.14}$$

The $C_{OSS}(v_{DS})$ function can be captured from the graph typically provided in the datasheet for the part being analyzed. The topology and operating conditions will determine whether P_{OSS} losses are present. As an example, self-commutation transitions have zero P_{OSS} losses, but only if the load current is sufficient to completely charge C_{OSS} to V_{BUS} or discharge to zero, and if the time to complete the self-commutation transition is much longer than the current transition time of the device itself.

6.2.3 Gate Charge (Q_G) Losses

The power loss associated with the gate charge is calculated as follows:

$$P_G = Q_G \cdot V_{DR} \cdot f_{sw} \tag{6.15}$$

The gate power losses become an important consideration at higher frequencies and at lower output power levels. It is important to note that all the gate energy is supplied during the charging phase, half of which is consumed, and during the discharge phase the remaining half of the energy is then consumed.

6.2.4 Reverse Conduction Losses (P_{SD})

Reverse or diode conduction occurs when the current in the device turning on goes through the body diode before the switch conducts the current, and is as a result of dead-time between when one device turns off and when the next device turns on. Reverse conduction follows a self-commutation transition and will only occur if the timing between the transitions is longer than needed to establish zero-voltage switching (ZVS) as shown in Figure 6.6.

The power loss due to the reverse conduction voltage through the body diode is given by the following equation:

$$P_{SD} = V_{SD} \cdot I_{DS} \cdot t_{SD} \cdot f_{sw} \tag{6.16}$$

The reverse conduction time (t_{SD}) needs to be determined from the operating conditions as it is dependent on load current, supply voltage and device parameters. To do this, a definition of effective dead-time is needed.

The time from when the gate voltage of the device reaches the turn-off plateau voltage to when the other device's load current commutates from the diode will be defined as the effective

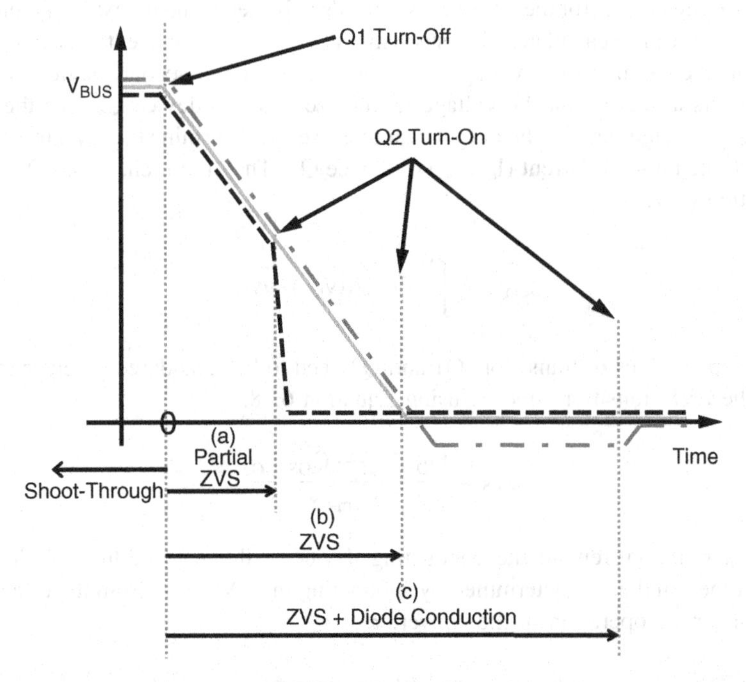

Figure 6.6 Switch-node voltage commutation with the same load current for various dead-times: (a) partial ZVS, (b) ZVS, and (c) ZVS plus diode conduction

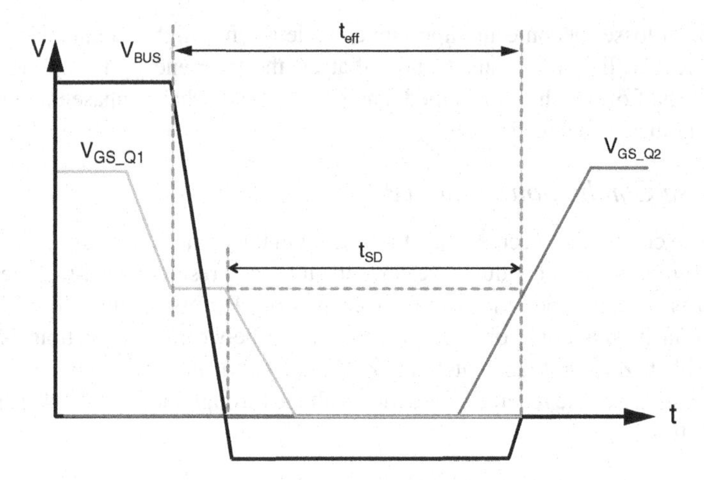

Figure 6.7 Effective dead-time definition for turn-on

dead-time (shown in Figure 6.7). Although they are related, this definition differs from the dead-time generated by the gate drive command signals. Also shown in Figure 6.7 is the definition of reverse conduction time (t_{SD}), which starts when the diode starts to conduct, and ends when the device itself turns on and conducts the current through the transistor channel.

The effective dead-time for the ZVS interval needs to be determined next, and then the diode conduction losses can be calculated. To establish ZVS, the external current must have sufficient energy to fully commutate the voltage, and the timing of the second device gate turn-on threshold occurs at the instant the voltage reaches zero across device Q2 (see the solid line labeled V_{GS_Q2} in Figure 6.7). The output charge is used to determine the ZVS transition time, together with the turn-off current ($I_{turn\text{-}off}$) of device Q1. The output charge can be calculated using Equation 6.17:

$$Q_{OSS} = \int_0^{V_{BUS}} C_{OSS}(v_{DS}) \cdot dv_{DS} \tag{6.17}$$

The output capacitances of transistors Q1 and Q2 need to be considered together to correctly determine the ZVS transition time by using Equation 6.18:

$$t_{ZVS} = \frac{Q_{OSS\_Q1} + Q_{OSS\_Q2}}{I_{Turn\text{-}off}} \tag{6.18}$$

where $I_{turn\text{-}off}$ is the current in the conducting device at the time of turn-off. The reverse conduction time can then be determined by subtracting the ZVS time from the effective dead-time, chosen for the operation of the converter.

$$t_{SD} = t_{eff} - t_{ZVS} \tag{6.19}$$

A zero result for Equation 6.19 indicates that a ZVS transition has occurred.

A positive time result for Equation 6.19 indicates that diode conduction and the associated losses can be determined using Equation 6.16.

A negative reverse conduction time means that the converter is operating in partial ZVS mode. In this case, a hard-switching event is established with associated losses, but occurs at lower voltage than the bus voltage. There is no reverse recovery loss, regardless of whether the device is a MOSFET or a GaN transistor, as no diode is conducting. The switch-node voltage when Q2 turns on (V_{PZVS}) can be determined by calculating the amount of charge transferred at the effective time relative to the total charge in the circuit using Equation 6.20:

$$V_{PZVS} = \frac{I_{Turn\text{-}off} \cdot t_{eff} \cdot V_{BUS}}{Q_{OSS\_Q1} + Q_{OSS\_Q2}} \tag{6.20}$$

In the case of partial ZVS, during the self-commutation time, the loss analysis must be treated the same as the ZVS case but at a reduced time.

6.2.5 *Reverse Recovery (Q_{RR}) Losses*

The body diode reverse recovery losses occur when the body diode transitions from the on-state to the off-state. Enhancement-mode GaN transistors, unlike standard power MOSFETs or cascode GaN devices, have no minority carriers to be stored in a junction, and therefore have no reverse recovery charge. As discussed in Chapter 2, cascode GaN transistors have a small amount of reverse recovery due to the small series-connected silicon power MOSFET.

The diode reverse recovery power loss can be calculated from the recovery charge and the bus voltage using Equation 6.21:

$$\begin{aligned} P_{RR} &= E_{RR} \cdot f_{sw} \\ &= Q_{RR} \cdot V_{BUS} \cdot f_{sw} \end{aligned} \tag{6.21}$$

The reverse recovery charge is provided in device datasheets at typical operating conditions, but it may prove inaccurate when operating conditions for the converter have large deviations from those given in the datasheet. Unfortunately there is no simple means to correctly calculate the Q_{RR}.

6.2.6 *Total Hard-Switching Losses*

The total dynamic power loss is the sum of the individual components:

$$P_{DYN} = (P_{sw} + P_{OSS} + P_G + P_{SD} + P_{RR}) \tag{6.22}$$

Owing to the GaN transistor's lower Q_{SG2} and Q_{GD}, P_{sw} is much lower than a comparable power MOSFET. The output capacitance for all types of GaN transistors is smaller than MOSFETs of comparable $R_{DS(on)}$, making P_{OSS} relatively low. Both gate drive voltages and gate charge are also lower, making P_G lower. Finally, due to the reverse current conduction mechanism, enhancement-mode GaN transistors have a higher V_{SD} when compared with the body diode forward voltage of a MOSFET, whereas cascode-connected GaN transistors have comparable forward drop. This characteristic of an enhancement-mode GaN transistor has the potential to increase the power loss P_{SD} and is influenced by the total reverse conduction time, a condition that can be controlled by the time that the rectifier switch is acting like a diode [4]. Enhancement-mode GaN transistors have zero reverse recovery, making the final term in Equation 6.22 zero (or small in the case of a cascode device).

Overall, the dynamic power losses of GaN transistors are significantly lower than power MOSFETs and enable power converters using hard-switching topologies to be more efficient and smaller. Now let's look at a simple figure of merit that can be used to estimate circuit performance and compare expected results between technologies and products within the same technology [5–8].

6.2.7 Hard-Switching Figure of Merit

Before a hard-switching figure of merit (FOM_{HS}) can be defined, we need to work through all the components that contribute to switching losses and determine which of those factors can quickly be extracted from a datasheet and are relevant enough to be compared. From Equation 6.22, it can be seen that there are five factors that need to be analyzed for inclusion in the FOM_{HS}.

The output capacitance losses are less dominant than switching losses in lower-voltage, hard-switching converters, and these can be omitted from the FOM_{HS} comparison. Gate-related losses can also be omitted as they are very small compared to other losses. Reverse conduction losses depend on operating conditions and can be mitigated by control techniques to correctly adjust for effective dead-time. Hence, reverse conduction losses will also be omitted from the FOM_{HS}. Lastly, the reverse recovery losses, although highly relevant for the comparison between different technologies, results in a complex and time-consuming process to correctly analyze. Enhancement-mode and cascode GaN transistors both have relatively low losses and so P_{RR} will also be omitted from FOM_{HS}.

The FOM_{HS} must be proportional to the power loss contribution from both the conduction and switching loss components, and is summarized in Equation 6.23 [9].

$$FOM_{HS} = (Q_{GD} + Q_{GS2}) \cdot R_{DS(on)} \tag{6.23}$$

For a given technology, a lower value of FOM_{HS} indicates lower power loss proportional to:

$$P_{Total} \propto \sqrt{FOM_{HS}} \tag{6.24}$$

where the total device power loss is the sum of the conduction losses and the dynamic losses and is given by Equation 6.25 and 6.26:

$$P_{Total} = P_{Conduction} + P_{DYN} \tag{6.25}$$

and:

$$P_{Conduction} = I_{DS\_RMS}^2 \cdot R_{DS(on)} \tag{6.26}$$

The FOM_{HS} can be plotted on a graph with $R_{DS(on)}$ on the x-axis and the charge-related terms ($Q_{GS2} + Q_{GD}$) on the y-axis as shown in Figure 6.8.

From Figure 6.8 it can be seen that 200 V GaN transistors have a similar FOM_{HS} to 40 V Si MOSFETs, and 600 V GaN transistors have a similar FOM_{HS} to 100 V MOSFETs.

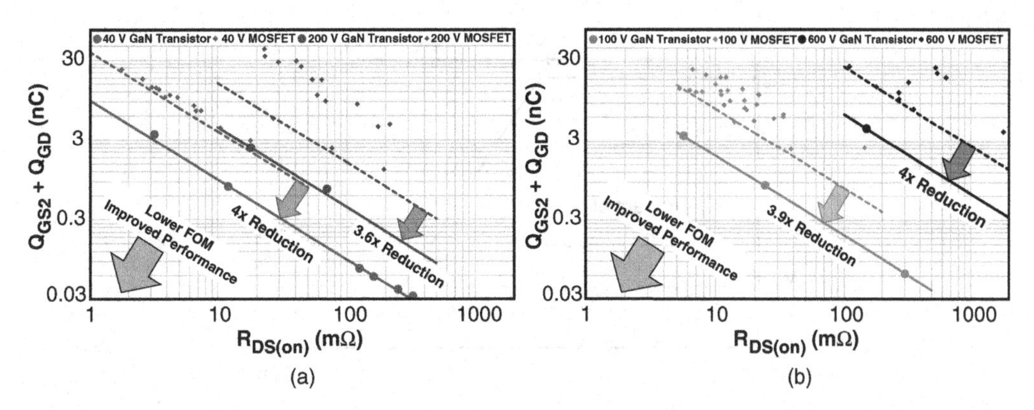

Figure 6.8 Comparison of hard-switching FOM_{HS} between GaN transistors and silicon MOSFETs at various voltages: (a) 40 V and 200 V, (b) 100 V and 600 V

6.3 External Factors Impacting Hard-Switching Losses

In the previous section, a detailed analysis of the derivation of hard-switching losses was presented. In practical applications, this picture is somewhat incomplete as there are additional factors that can further impact hard-switching losses, such as common-source inductance (CSI) (L_S) and loop inductance (L_{Loop}). These factors appear in practical circuits with physical limitations brought on by device size, device parasitics, and circuit layout parasitics. Their impact on hard-switching losses will be presented in this section.

6.3.1 Impact of Common-Source Inductance

The two main inductances, CSI and high-frequency loop inductance, are shown in Figure 6.9 for a half-bridge configuration.

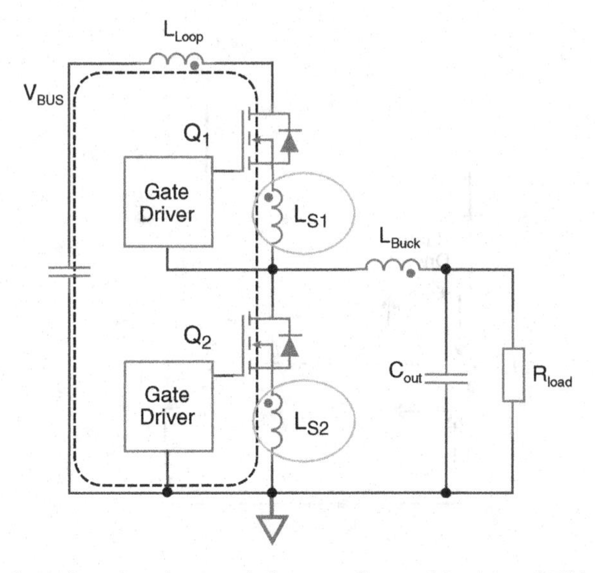

Figure 6.9 Power loop (L_{Loop}) and source inductance (L_{S1} and L_{S2}) in a half bridge configuration

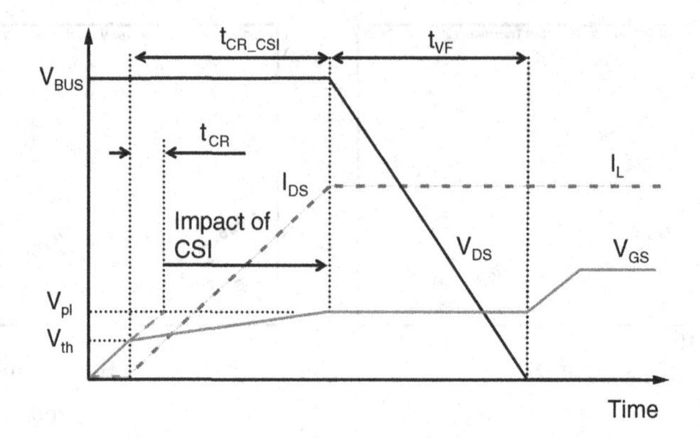

Figure 6.10 Effect of common-source inductance on the gate voltage

The impact of common source inductance on gate drive performance has been discussed in Chapter 3. In this section, its impact will be quantified for the hard-switching current transition.

During a current transition event, the voltage generated across the common source inductance will oppose the gate voltage, thereby reducing the gate current used to charge the gate capacitance. This effectively lengthens the current transition period as shown in Figure 6.10.

An analysis of the gate circuit shown in Figure 6.11 can be used to determine the amount of time by which the current transition is lengthened. Because a full analysis reveals terms with exponential and sinusoidal components, some simplifying assumptions need to be made.

The first simplifying assumption is that the voltage induced across the CSI can be regarded as a voltage source in phase with the gate voltage, and thus will only impact the magnitude of the voltage in the gate circuit. The second assumption ignores the impact of the gate circuit inductance. In Chapter 4, it was shown that this inductance contributes negligibly to circuit

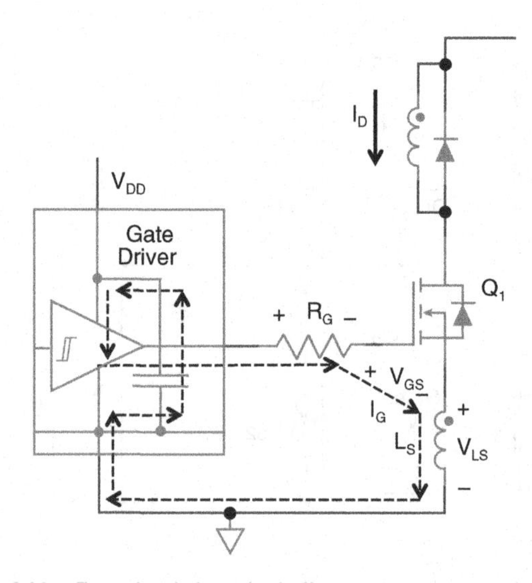

Figure 6.11 Gate circuit loop including common-source inductance

switching performance. The third assumption is that the external drain current is constant during the transition.

Neglecting the gate driver voltage drop, as it can be included as part of the gate resistance, the Kirchhoff voltage loop in the gate circuit is given by:

$$V_{DD} = V_{RG} + V_{GS} + V_{LS} \tag{6.27}$$

The gate and drain currents can then be added to yield:

$$V_{DD} = I_G \cdot R_G + V_{GS} + \frac{L_S \cdot I_{DS}}{t_{CR\_CSI}} \tag{6.28}$$

Where the gate circuit current is given by:

$$I_G = \frac{Q_{GS2}}{t_{CR\_CSI}} = \frac{C_{GS} \cdot I_{DS}}{t_{CR\_CSI} \cdot g_m} \tag{6.29}$$

Where t_{CR_CSI} is the current rise interval time, shown on Figure 6.10.

Combining Equations 6.28 and 6.29, t_{CR_CSI} can be determined:

$$t_{CR\_CSI} = \frac{I_{DS}}{V_{DD} - V_{GS}} \frac{C_{GS}}{g_m} \left(\frac{L_S \cdot g_m}{C_{GS}} + R_G \right) \tag{6.30}$$

From Equation 6.30, the equivalent common-source inductance resistance (R_{CSI}) can be extracted as:

$$R_{CSI} = \frac{L_S \cdot g_m}{C_{GS}} \tag{6.31}$$

It can be deduced from the resistance in Equation 6.31 that the impact of CSI on the gate circuit resistance is large, given that the device input capacitance is already small. The value of L_S, therefore, will need to become very small to minimize the impact of CSI. For example, 100 pH of CSI in a MOSFET circuit with C_{GS} of 2900 pF, and $g_m = 60\,S$ results in 2 Ω equivalent resistance. The same 100 pH CSI in an equivalent GaN transistor circuit with C_{GS} of only 850 pF, and $g_m = 60\,S$, results in a 7 Ω equivalent resistance.

Equation 6.31 can be used to estimate the current transition time, including the influence of the CSI. It can simply be added to the R_G term in Equations 6.11 and 6.12 for the Q_{GS2} component only.

As an example, a 1 MHz converter was evaluated with a fixed loop inductance and the CSI was varied, with the results shown in Figure 6.12. It is notable how quickly the losses increase as a function of CSI.

6.3.2 Impact of High Frequency Power-Loop Inductance on Device Losses

Another factor that impacts hard-switching losses is the high-frequency power loop inductance that impacts the commutation of voltage and current between the switching devices. This is the

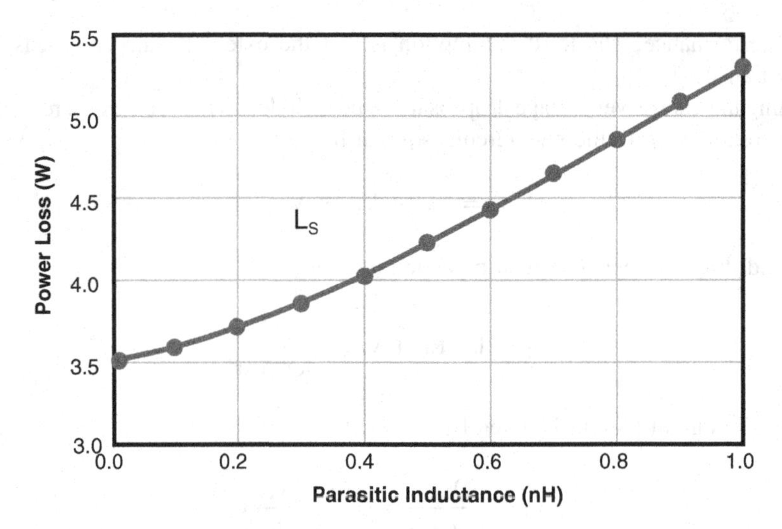

Figure 6.12 Effect of common-source inductance on power loss [24,25,26] ($V_{BUS} = 12$ V, $V_{OUT} = 1.2$ V, $I_{OUT} = 20$ A, $f_{SW} = 1$ MHz, control FET is EPC2015 [3], synchronous rectifier FET is EPC2015 [3]

inductance encompassed by the bus supply as well as the devices connected to this bus as shown in Figure 6.9. Component parasitic inductance and physical layout inductance elements all contribute to the total loop inductance.

The power loop inductance has two main negative effects on the switch during turn-off: it slows the transition and it increases the voltage across the drain and source. During device turn-on, the loop inductance reduces the device drain-to-source voltage, which decreases losses. However, the sum of the two negative effects and the positive effect has a net negative result, which means that the loop inductance will increase losses in the circuit as can be seen in Figure 6.13.

Using the circuit conditions of Figure 6.12, and by adding as little as 1 nH of CSI, losses can increase 50% over an ideal case. This is due to the negative impact of CSI on both turn-on and turn-off switching transitions. Adding 3 nH of loop inductance increases loss by 30% over the ideal case as shown in Figure 6.13. The smaller relative increase in loss is the result of the partial savings at turn-on from high frequency loop inductance.

Understanding the impact of parasitic inductance on performance, GaN transistor designers have to make the reduction of package inductance a high priority. Since all of the connections of a lateral enhancement-mode HEMT transistor are contained on the same side of the die, the die can be mounted directly to the PCB, minimizing the total inductance to the internal busing and external solder bumps. To further decrease inductance, the drain and source connections can be arranged in an interleaved land grid array, providing multiple parallel connections to the PCB from the die [10].

To illustrate the impact of loop inductance, different layouts with similar CSI and different loop inductances were compared. The impact of loop inductance on efficiency for various layout variations in a buck converter operating at 1 MHz is shown in Figure 6.14, where the CSI was kept at the minimum achievable level. An increase in the high-frequency loop inductance from around 0.4 nH to 2.9 nH decreases efficiency by over 4%.

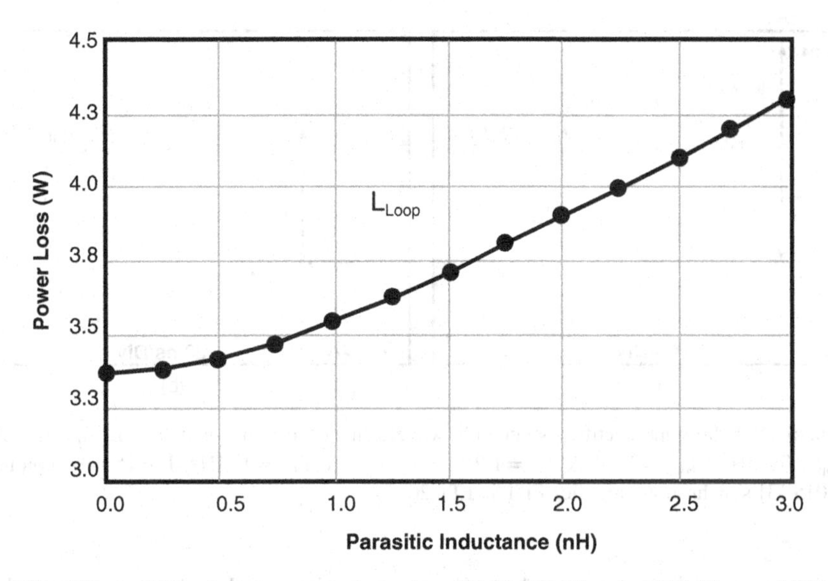

Figure 6.13 Effect of loop inductance on power loss [24,25,26] (V_{BUS} = 12 V, V_{OUT} = 1.2 V, I_{OUT} = 20 A, f_{SW} = 1 MHz, control FET is EPC2015 [3], synchronous rectifier FET is EPC2015 [3]

With these higher switching speeds, even small values of high-frequency loop inductance can increase the voltage overshoot. Decreasing this inductance therefore results in lower voltage overshoot, increased input voltage capability, and reduced electromagnetic interference (EMI). Figure 6.15 shows the drain-to-source voltage waveforms for a design with a

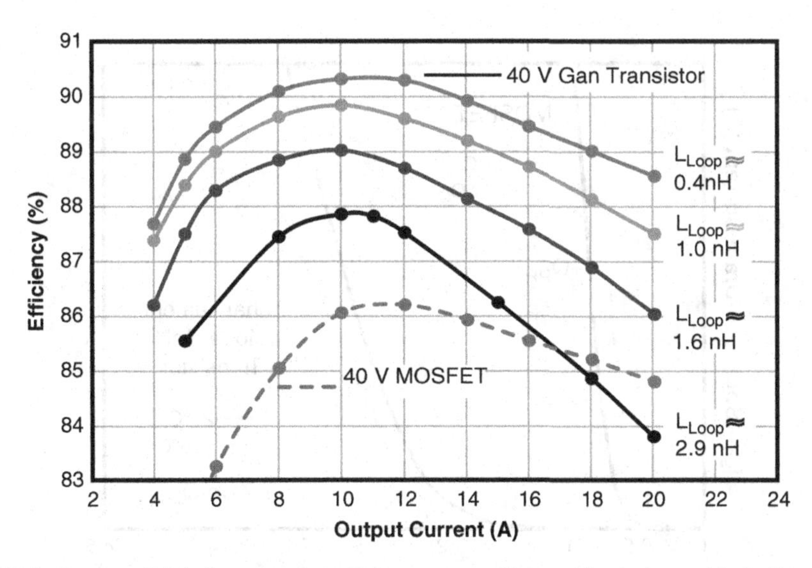

Figure 6.14 Impact of high-frequency loop inductance on efficiency for designs with similar common-source inductance (V_{BUS} = 12 V, V_{OUT} = 1.2 V, I_{OUT} = 20 A, f_{SW} = 1 MHz, control FET is EPC2015 [3] synchronous rectifier FET is EPC2015, control MOSFET is BSZ097N04LSG [11], synchronous rectifier MOSFET is BSZ040N04LSG) [12]

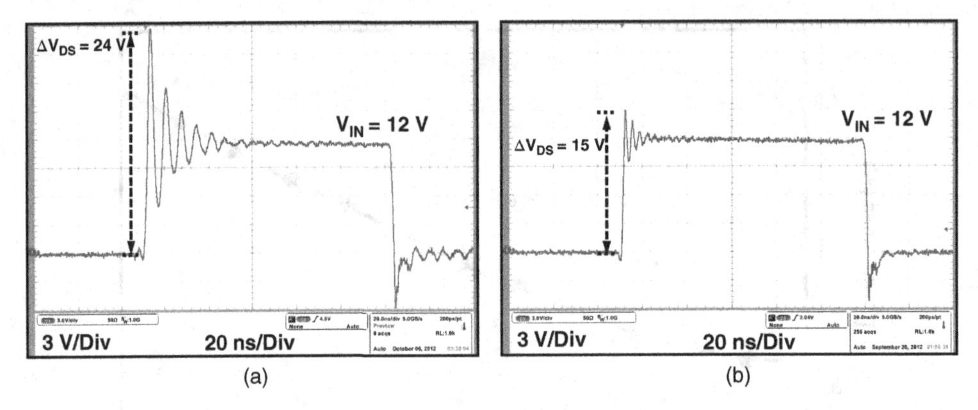

Figure 6.15 Synchronous rectifier switching waveforms of designs with (a) $L_{Loop} \approx 1.6\,nH$ and (b) $L_{Loop} \approx 0.4\,nH$ ($V_{BUS} = 12\,V$, $V_{OUT} = 1.2\,V$, $I_{OUT} = 20\,A$, $f_{SW} = 1\,MHz$, $L = 150\,nH$, control FET is EPC2015 [3] synchronous rectifier FET is EPC2015)

high-frequency loop inductance of 1.6 nH compared with 0.4 nH. The voltage overshoot is reduced from 100% of the input voltage to 25%, respectively.

6.4 Reducing Body Diode Conduction Losses in GaN Transistors

Figure 6.16 shows the forward voltage drop of both a MOSFET and a typical enhancement-mode GaN transistor. There is about 1.5 V difference between the two devices and, as the temperature increases, it can go as high as 2 V. This graph, however, does not account for

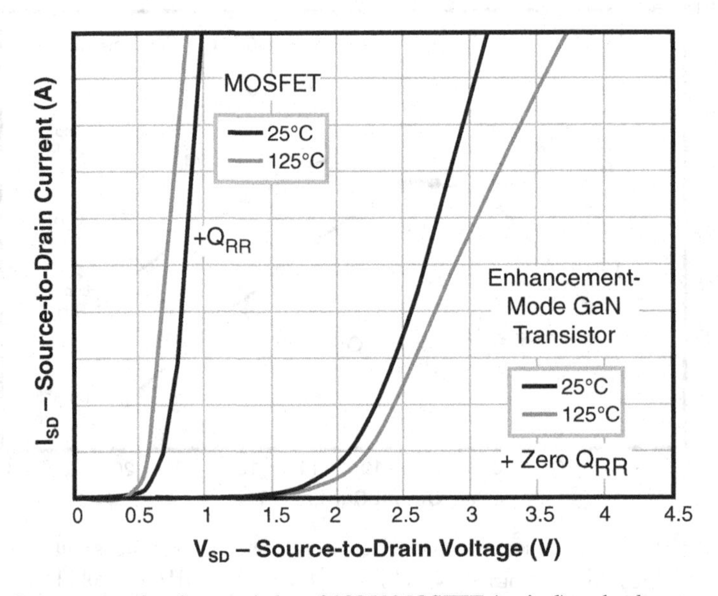

Figure 6.16 Reverse transfer characteristics of 100 V MOSFET (typical) and enhancement-mode GaN transistor (typical)

dynamic behavior, where in the case of the enhancement-mode GaN transistor, there is no reverse recovery ($Q_{RR} = 0$), in addition to lower output capacitance.

In Section 6.2.4, it was shown that optimal timing between the transistors can yield very low losses under specific conditions. Those conditions are dynamic, and depend on operating conditions such as load current and voltage, supply voltage and duty cycle. For most circuits, it is not practical to have a circuit that can actively control the dead-time to the precision needed to absolutely minimize losses. However, a simple anti-parallel Schottky diode can be connected with the GaN transistor to improve the efficiency of the body diode with less reliance on precise dead-time control.

One of the most critical requirements for the addition of the anti-parallel Schottky diode is the minimization of the connection inductance between the two devices. This comes down to three factors: the parasitic inductance between the drain and source of the GaN transistor, the parasitic inductance of the Schottky diode, and the layout inductance connecting the GaN transistor to the Schottky diode. The low parasitic inductance of the GaN transistors with LGA packages makes the addition of an external Schottky diode simple and efficient.

Using the definition of effective dead-time from Figure 6.7, the power losses associated with the diode conduction as a function of effective dead-time are shown in Figure 6.17 for a typical GaN transistor and equivalent MOSFET. It can be seen that, due to the lower forward voltage drop of the MOSFET, the losses as a function of the effective dead-time increase at a lower rate than that for a GaN transistor.

The addition of an anti-parallel Schottky diode to the GaN transistor synchronous rectifier will reduce the conduction voltage during the diode conduction period as shown in Figure 6.18.

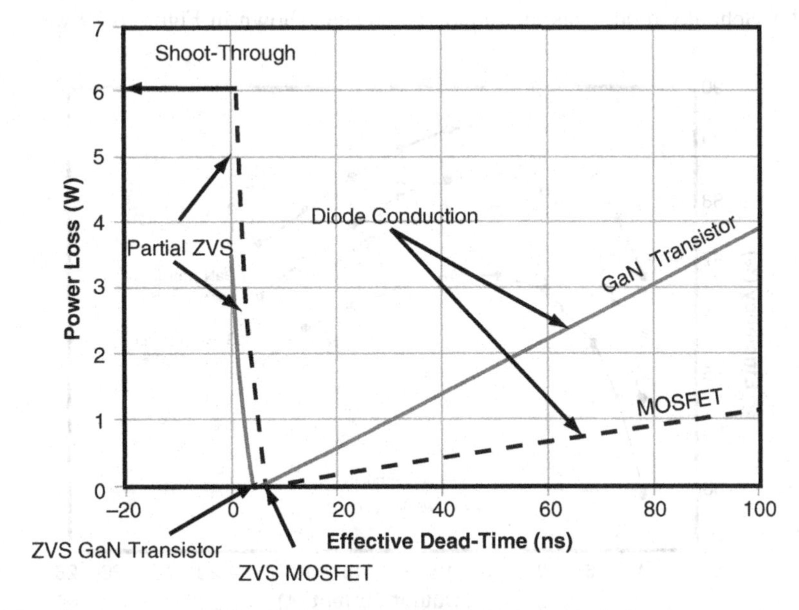

Figure 6.17 GaN transistor and MOSFET comparison of the impact of effective dead-time on power loss of the synchronous rectifier device in a buck converter operating with $V_{BUS} = 48$ V, $I_{OUT} = 16$ A, $f_{sw} = 1$ MHz

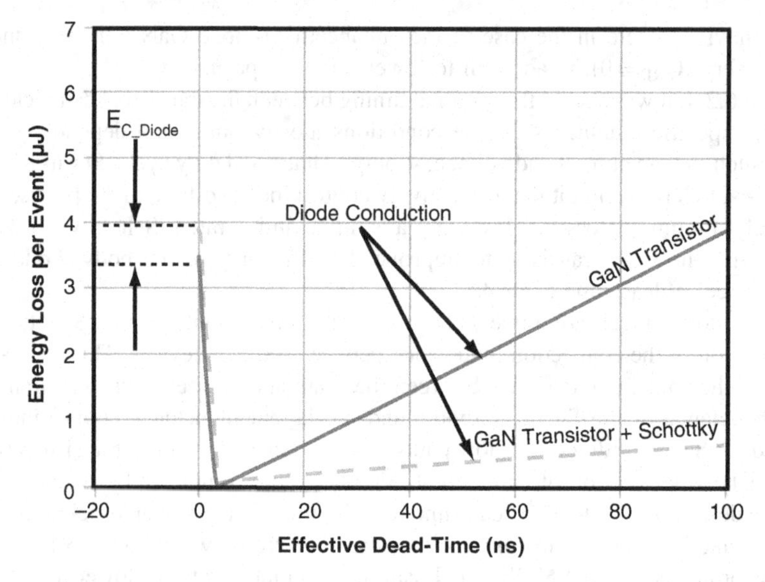

Figure 6.18 Comparison of the impact of effective dead-time on the power loss of an enhancement-mode GaN transistor-based buck converter operating with $V_{BUS} = 48$ V, $I_{OUT} = 16$ A, $f_{sw} = 1$ MHz with and without an anti-parallel Schottky diode

However, the addition of an anti-parallel Schottky diode does add some output capacitance with an associated increase in output capacitance loss.

A buck converter was tested with various effective dead-times and the effect of adding an anti-parallel Schottky diode was measured. The results shown in Figure 6.19 were obtained

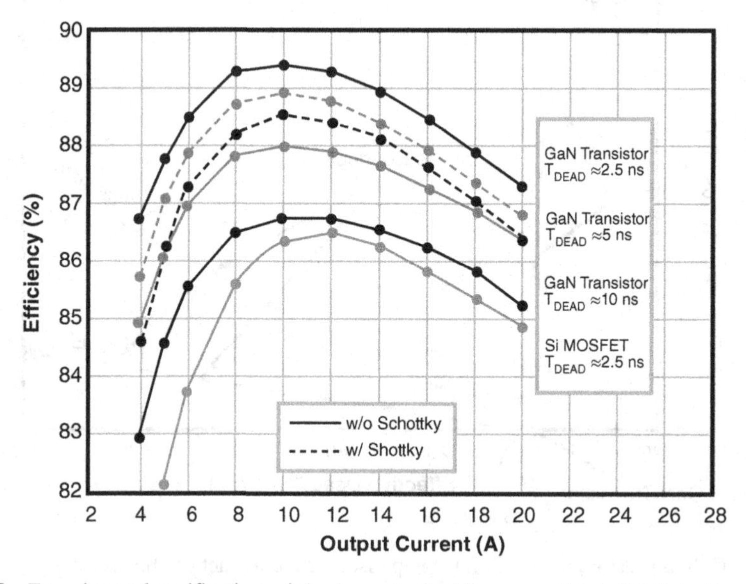

Figure 6.19 Experimental verification of the impact of adding an anti-parallel Schottky diode to an enhancement-mode transistor in a buck converter with $V_{BUS} = 12$ V, $V_{OUT} = 1.2$ V, $f_{sw} = 1$ MHz

with the converter operating with a V_{BUS} of 12 V, output voltage of 1.2 V, output current of 16 A, and operating at a switching frequency of 1 MHz.

6.5 Frequency Impact on Magnetics

Magnetic components such as transformers and inductors account for the other large contributor to power loss in switching power converters.

6.5.1 Transformers

Consider a magnetic core with a specific cross-sectional core area and specific winding window area. The core-area product is commonly used to design magnetic structures [13] and directly relates to the volume of the core. A constant core-area product results in similar losses and, consequently, converter efficiencies for a given operating frequency.

As the switching frequency is increased over a practical range for a given material, the core losses will decrease at a higher rate than the increase in frequency. This is due to the non-linearity of these losses as a function of the core flux density [14,15], an effect that can be used to an advantage in a converter using GaN transistors, compared with one using Si MOSFETs. It may be beneficial to consider magnetic materials with lower core loss density as the frequency is increased.

As an example, consider what happens when the switching frequency is increased from 300 kHz to 500 kHz. The core cross-sectional area of the 300 kHz design can be decreased to the point where there is the same flux density as the 500 kHz design. This results in a core cross-sectional area that is 60% of the 300 kHz design, as shown in Figure 6.20. Additional effects from this new core design are:

1. The magnetic core volume has decreased to approximately 60% of the original value.
2. The core losses per unit volume may have increased, but this depends upon the core material and the switching frequency.
3. The winding volume and mean length-per-turn has also been reduced to approximately 85–90% (depending on the length-to-width(l/w) ratio). This results in a lower DC winding resistance and resulting copper wire conduction losses.

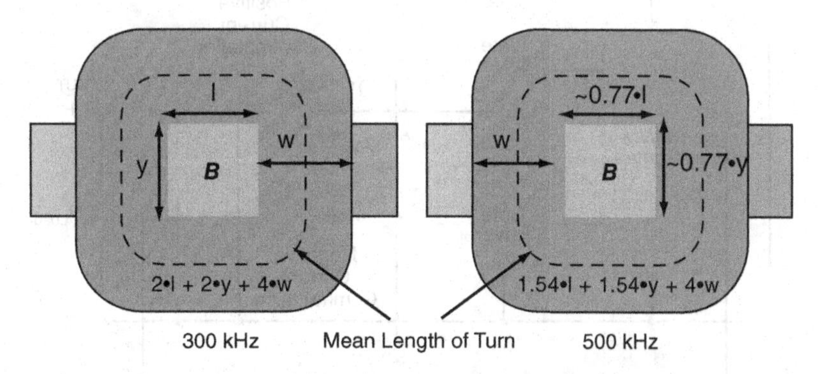

Figure 6.20 Cross-sectional view of two equivalent transformer structures (constant flux density) for different switching frequencies

4. The AC winding resistance per unit length has increased, due to reduced skin depth, and depending on the design and conductor thickness. Furthermore, the AC winding resistance change is proportional to the decrease in DC winding resistance.

Typically, (1) is greater than (2) and (3) is greater than (4) and, therefore, the transformer will yield a higher efficiency at 500 kHz than at 300 kHz. The extent by which the frequency can be increased is material-dependent; as the material is pushed beyond its intended operating frequency range, any real benefit from increasing the frequency will become negated. An alternative core material may result in higher frequency capability, but at a reduced improvement gain. In the multi-MHz frequency range, many core materials are operating at their upper limit, and, in some cases, air-core approaches may need to be investigated.

6.5.2 Inductors

In the case of inductors, the impact of the change in magnetic size is similar to the transformer, but due to a slightly different mechanism. The core material in a transformer will experience a full flux swing due to the voltage excitation but, in the case of an inductor, the current in the winding has a DC component to it. This means that the flux excitation and associated losses are proportionately lower than those of a transformer for the same frequency. However, the inductor conduction losses will be higher due to the DC component. Using the same analysis as for the transformer, a more efficient inductor will again result at higher frequencies because (1) is greater than (2) and (3) is greater than (4). The same upper frequency operating limits apply to the inductor as for the transformer.

6.6 Buck Converter Example

The analyses of the previous five sections can now be applied to an actual converter. A buck converter was chosen because it provides a simple circuit that includes a hard-switching device and a transistor acting as a synchronous rectifier, as shown in Figure 6.21. Since most of the

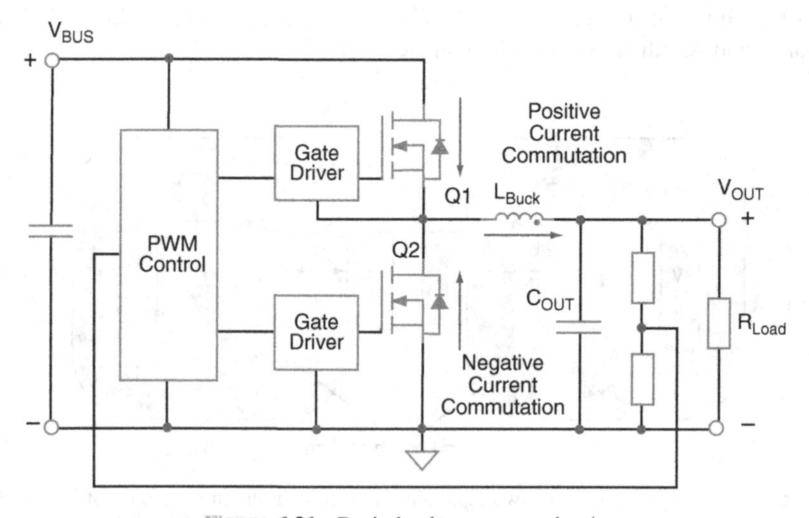

Figure 6.21 Basic buck converter circuit

dynamic losses are related to the input voltage, and to the ratio between the input voltage and output voltage, this basic rule applies: the higher the input voltage and the greater the ratio between the input and output voltage, the greater the benefits derived from using GaN transistors.

In this example, a buck converter is used, operating at 1 MHz and delivering 20 A at 1.2 V into the load with a supply of 12 V. The hard-switching loss analysis is based on the EPC2015 [3] for both the control switch (Q1) and the synchronous rectifier (Q2). The buck inductor value is 300 nH for continuous-conduction mode operation, and both devices are driven with an external turn-on gate resistance of 2 Ω, and turn-off gate resistance of 0.5 Ω. The calculated losses will be compared with experimental measurements.

The buck converter has two switching events, defined relative to the control switch, as turn-on (where the switch-node voltage will rise to the bus voltage) and turn-off (where the switch-node voltage falls to zero). Since there are two transistors involved in this design, four conditions need to be analyzed: the turn-on and turn-off events for each of the transistors, as shown in Figure 6.22. The turn-off switching transient is defined as self-commutation. The current (I_{Buck}) in the buck inductor (L_{Buck}) at the time Q1 is turned off will discharge the output capacitance, without Q2 being turned on, and hence reduces the switch-node voltage on its own. The turn-on switching transient is defined as forced commutation. Regardless of the current in the buck inductor at the time Q2 is turned off, the switch-node voltage must be forced to the bus voltage when Q1 turns on.

An analysis of the buck converter is required before calculating device losses:

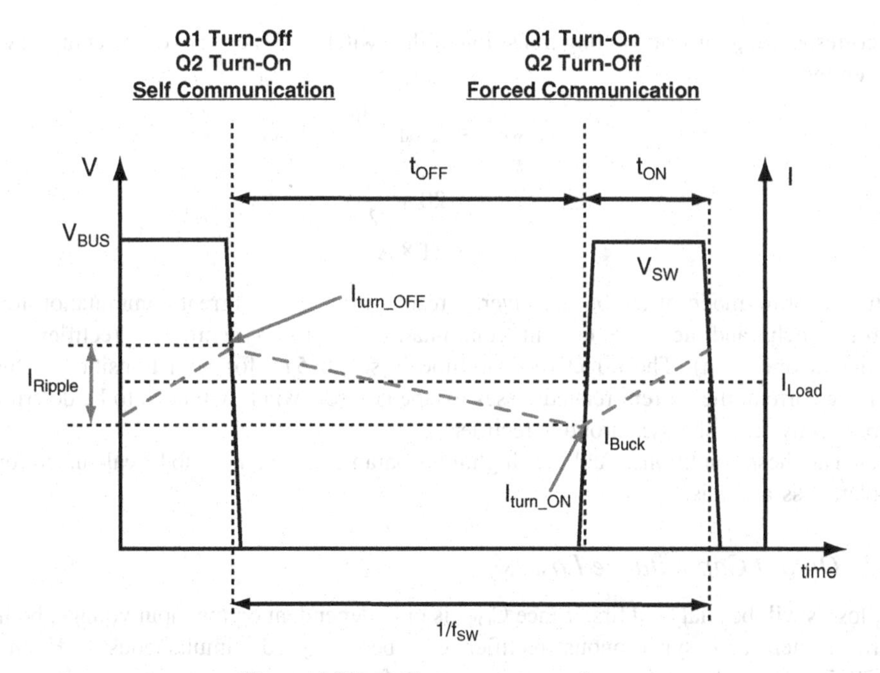

Figure 6.22 Buck converter waveforms

The control switch duty cycle (D) is given by [16,17,18]:

$$\begin{aligned} D &= \frac{V_{OUT}}{V_{Bus}} \\ &= \frac{1.2}{12} \\ &= 0.1 = 10\% \end{aligned} \tag{6.32}$$

Using this duty cycle, the peak-to-peak ripple current in the inductor can be calculated:

$$\begin{aligned} I_{Ripple} &= \frac{(V_{Bus} - V_{OUT}) \cdot D}{f_{sw} \cdot L_{Buck}} \\ &= \frac{(12 - 1.2) \cdot 0.1}{1 \cdot 10^6 \cdot 300 \cdot 10^{-9}} \\ &= 3.6\,\text{A} \end{aligned} \tag{6.33}$$

Using the load current in the inductor and Equation 6.33, the turn-on current (rising transition of the switch node) can be calculated using:

$$\begin{aligned} I_{Turn\text{-}on} &= I_{Load} - \frac{I_{Ripple}}{2} \\ &= 20 - \frac{3.6}{2} \\ &= 18.2\,\text{A} \end{aligned} \tag{6.34}$$

The corresponding turn-off (falling transition of the switch node) current of the control switch is given by:

$$\begin{aligned} I_{Turn\text{-}off} &= I_{Load} + \frac{I_{Ripple}}{2} \\ &= 20 + \frac{3.6}{2} \\ &= 21.8\,\text{A} \end{aligned} \tag{6.35}$$

The operating mode of the buck converter results in positive current commutation for the control switch, and negative current commutation for the synchronous rectifier (diode conduction operation). The effective dead-time is set to 5 ns for both transitions. Due to the ripple current, the current-related losses for the control switch will need to be determined independently from the synchronous rectifier.

Given all these conditions, Table 6.2 highlights parameters that need to be calculated for the complete loss analysis.

6.6.1 Output Capacitance Losses

P_{OSS} losses will be analyzed first. Since C_{OSS} is only dependent on the input voltage, both the control switch and synchronous rectifier can be analyzed simultaneously. From the EPC2015 [3] datasheet, the C_{OSS} as a function of drain-to-source voltage can be used to

Table 6.2 Loss analysis parameters

Transition state	Control Switch		Synchronous Rectifier	
	Turn-off	Turn-on	Turn-on	Turn-off
Commutation	Self	Forced	Diode	Diode
Effective dead-time	Fixed at 5 ns	Fixed at 5 ns	Fixed at 5 ns	Fixed at 5 ns
P_{OSS}	No	Yes	No	Induced in control FET
P_G	Yes (1/2 P_G)	Yes (1/2 P_G)	Yes (1/2 P_G)	Yes (1/2 P_G)
P_{SD}	No	No	Diode conduction	Diode conduction
P_{RR}	No	No	No	No
P_{on}	N/A	Yes	Small (from diode)	N/A
P_{off}	Yes	N/A	N/A	Small (to diode)

determine E_{OSS} as a function of drain-to-source voltage using Equation 6.14, and as plotted in Figure 6.23.

$$\begin{aligned} E_{OSS} &= \int_0^{V_{BUS}} v_{DS} \cdot C_{OSS}(v_{DS}) \cdot dv_{DS} \\ &= \int_0^{12} v_{DS} \cdot C_{OSS}(v_{DS}) \cdot dv_{DS} \\ &= 80.6\,\text{nJ} \end{aligned} \tag{6.36}$$

Since there are two devices and only one transition that creates E_{OSS} losses, the total E_{OSS} will be 161 nJ, and is dissipated in the control switch only. Therefore at 1 MHz, the P_{OSS} is 161 mW.

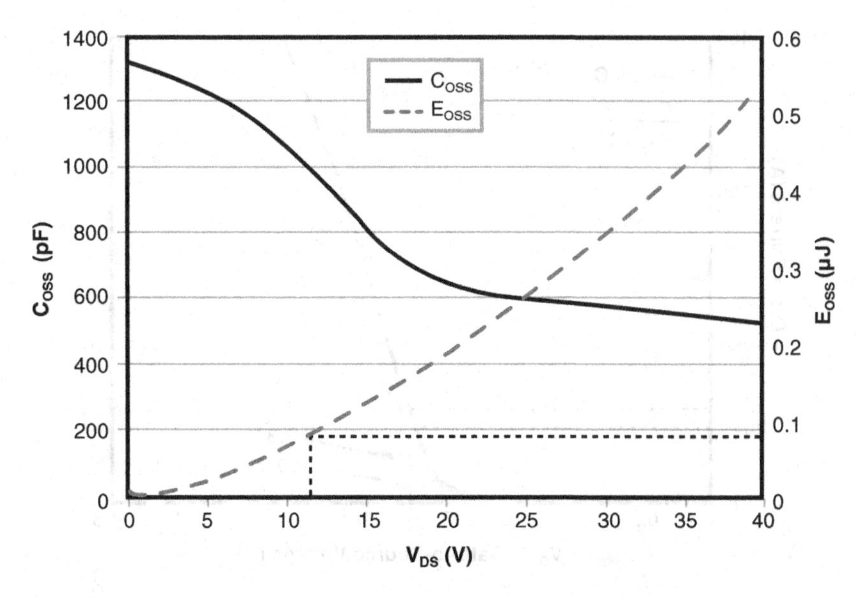

Figure 6.23 C_{OSS} and E_{OSS} as a function of drain-to-source voltage for the EPC2015 [3]

6.6.2 *Gate Losses (P_G)*

The magnitude of the gate losses is small in comparison to switching losses, but the values for Q_{GS2} and Q_{GD} need to be determined for later use in the switching loss calculations.

Both devices incur gate losses. The drain current will affect the total gate charge such that the values differ for the control switch and synchronous rectifier. In the case of the control switch, the total gate charge includes Q_{GD} as the device experiences a voltage transition. The synchronous rectifier, however, has no Q_{GD} component as it always switches from diode conduction.

Before Q_{GS} can be determined for the specific operating condition, the value of the plateau voltage must be read from the device transfer characteristic graph (extracted from the EPC2015 [3] datasheet), and as shown in Figure 6.24, with $V_{pl} = 2.3$ V at $I_{DS} = 33$ A and $V_{pl(op)} = 2.2$ V at $I_{DS} = 20$ A.

The value for Q_{GS} can now be determined using Equation 6.7 and the EPC2015 [3] datasheet, which is $Q_{GS} = 3$ nC. The $Q_{GS(op)}$ value for the buck example then can be calculated:

$$\begin{aligned} Q_{GS(op)} &= \left(\frac{Q_{GS}}{V_{pl}}\right) \cdot V_{pl(op)} \\ &= \left(\frac{3}{2.3}\right) \cdot 2.2 \\ &= 2.87\ \text{nC} \end{aligned} \tag{6.37}$$

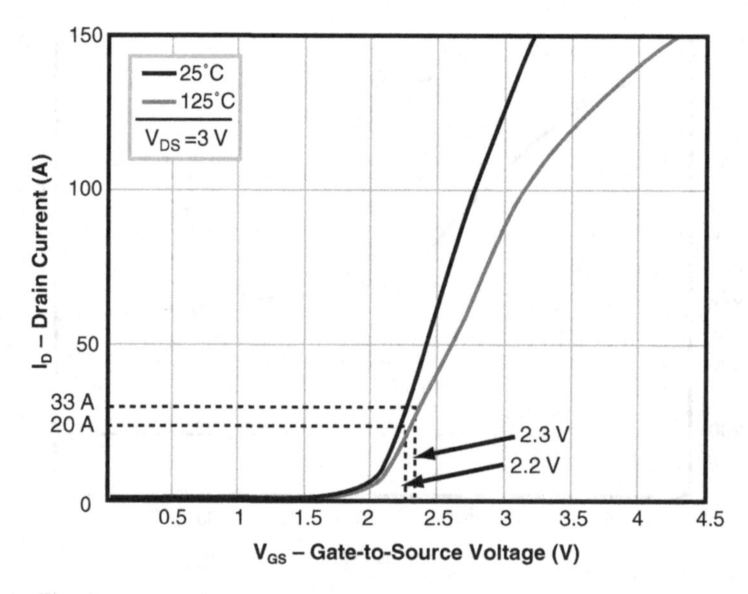

Figure 6.24 Transfer characteristic for EPC2015 [3] with V_{pl_ds} (at 33 A) and V_{pl} (at 20 A) shown

Using the same method and Equation 6.6, and the typical value given in the datasheet for V_{th} as 1.4 V, Q_{GS1} results in:

$$
\begin{aligned}
Q_{GS1} &= \left(\frac{Q_{GS}}{V_{pl}}\right) \cdot V_{th} \\
&= \left(\frac{3}{2.3}\right) \cdot 1.4 \\
&= 1.83\ \text{nC}
\end{aligned}
\tag{6.38}
$$

Q_{GS2} can then be calculated using Equation 6.8, which yields:

$$
\begin{aligned}
Q_{GS2} &= Q_{GS(op)} - Q_{GS1} \\
&= 2.87 - 1.83 \\
&= 1.04\ \text{nC}
\end{aligned}
\tag{6.39}
$$

The value of Q_{GD} cannot be linearly approximated, so in this case, it needs to be determined from the C_{RSS} capacitance. From the EPC2015 [3] datasheet, the C_{RSS} as a function of drain-to-source voltage can be used to determine Q_{GD} as a function of drain-to-source voltage using Equation 6.3. This result has been plotted in Figure 6.25, and the value of Q_{GD} at 12 V is 1.94 nC.

Knowing both Q_{GS2} and Q_{GD}, it is possible to estimate $Q_{G(op)}$. Once the device has fully turned on, the slope of the Q_G graph will always be the same, regardless of the voltage or current to which the device is switching. The slope can be determined from the datasheet for the

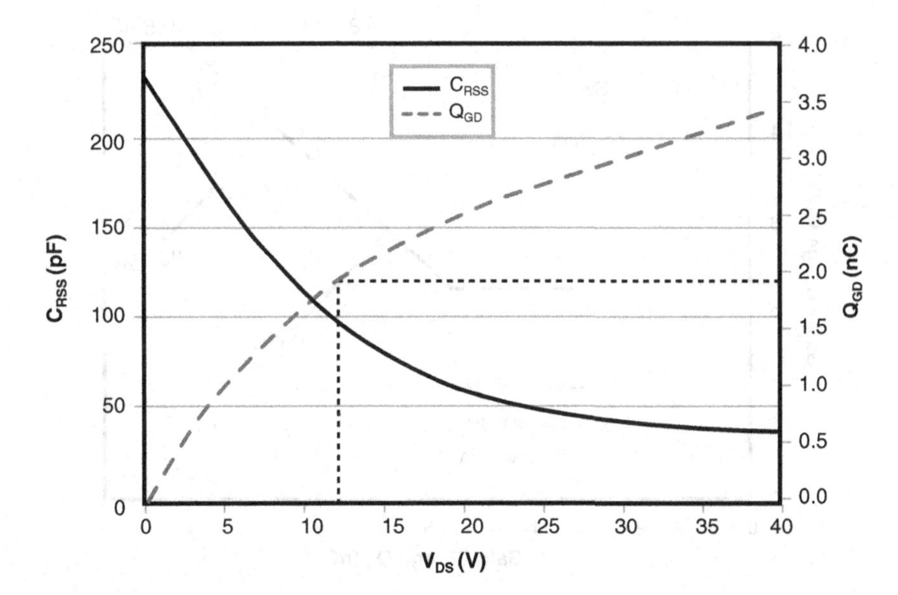

Figure 6.25 C_{RSS} and Q_{GD} as a function of drain-to-source voltage for the EPC2015 [3]

region between the plateau voltage and the final gate voltage (5 V) using Equation 6.40 for the datasheet-provided conditions:

$$\begin{aligned} m_{QGslope} &= \frac{Q_G - (Q_{GS} + Q_{GD})}{V_{DR} - V_{pl}} \\ &= \frac{10.5 - (3 + 2.5)}{5 - 2.3} \\ &= 1.85\ ^{nC}/_{V} \end{aligned} \tag{6.40}$$

Using the slope ($m_{QGslope}$) the $Q_{G(op)}$ for each of the operating conditions can be determined. First, for the control switch:

$$\begin{aligned} Q_{G(op)} &= \left(Q_{GS(op)} + Q_{GD}\right) + \left(m_{QGslope} \cdot \left(V_{DR} - V_{pl}\right)\right) \\ &= (1.04 + 1.94) + (1.85 \cdot (5 - 2.2)) \\ &= 10\,\text{nC} \end{aligned} \tag{6.41}$$

and for the synchronous rectifier:

$$\begin{aligned} Q_{G(op)} &= \left(Q_{GS(op)} + Q_{GD}\right) + \left(m_{QGslope} \cdot \left(V_{DR} - V_{pl}\right)\right) \\ &= (1.04 + 0) + (1.85 \cdot (5 - 2.2)) \\ &= 8.1\,\text{nC} \end{aligned} \tag{6.42}$$

The new calculated values for Q_{GS}, Q_{GD}, and Q_G are plotted as a function of gate-to-source voltage in Figure 6.26 together with the plot given in the datasheet. Figure 6.26 shows the gate charge difference between the control switch and synchronous rectifier.

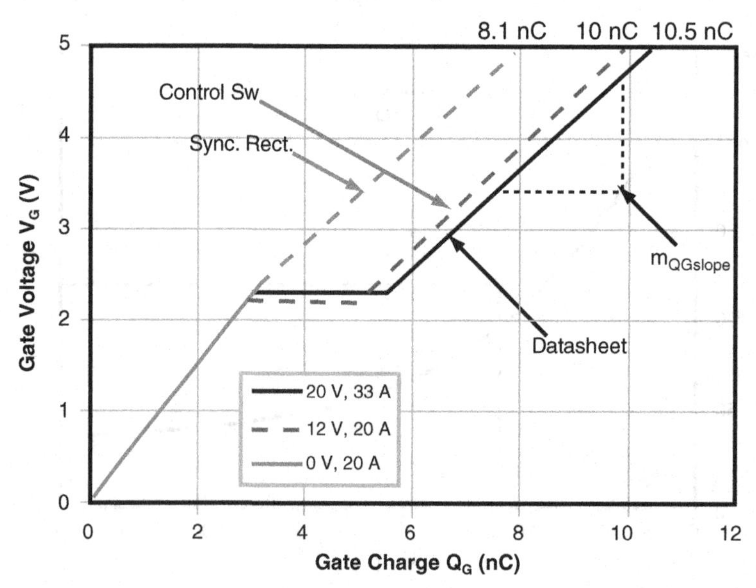

Figure 6.26 Q_G as a function of gate-to-source voltage for the EPC2015 [3] plotted with the datasheet value and re-plotted for the two cases of the example

The gate power now can be determined using Equation 6.15. For the control switch:

$$
\begin{aligned}
P_G &= Q_G \cdot V_{DR} \cdot f_{sw} \\
&= 10\,\text{nC} \cdot 5\,\text{V} \cdot 1 \cdot 10^6 \\
&= 50\,\text{mW}
\end{aligned}
\tag{6.43}
$$

and for the synchronous rectifier:

$$
\begin{aligned}
P_G &= Q_G \cdot V_{DR} \cdot f_{sw} \\
&= 8.1\,\text{nC} \cdot 5\,\text{V} \cdot 1 \cdot 10^6 \\
&= 40\,\text{mW}
\end{aligned}
\tag{6.44}
$$

6.6.3 Body Diode Conduction Losses (P_{SD})

Only the synchronous rectifier device incurs diode conduction, which is a function of the effective dead-time. This dead-time needs to be either selected or determined for each case. For this example, a fixed value of 5 ns will be selected, and the amount of diode conduction time calculated. The two cases that need to be analyzed are for turn-off (falling switching-node voltage) and turn-on (rising switch-node voltage).

6.6.3.1 Turn-Off Transient Diode Conduction Losses

Since the turn-off transition is self-commutating, the buck inductor current at the time of turn-off and the total output charge for both devices need to be determined. The turn-off current has already been calculated using Equation 6.35 and $I_{Turn\text{-}off} = 21.8\,\text{A}$. Next, the Q_{OSS} for the devices need to be determined. This can be done using Equation 6.17, the results of which are plotted in Figure 6.27. The value for Q_{OSS} at 12 V is 14 nC.

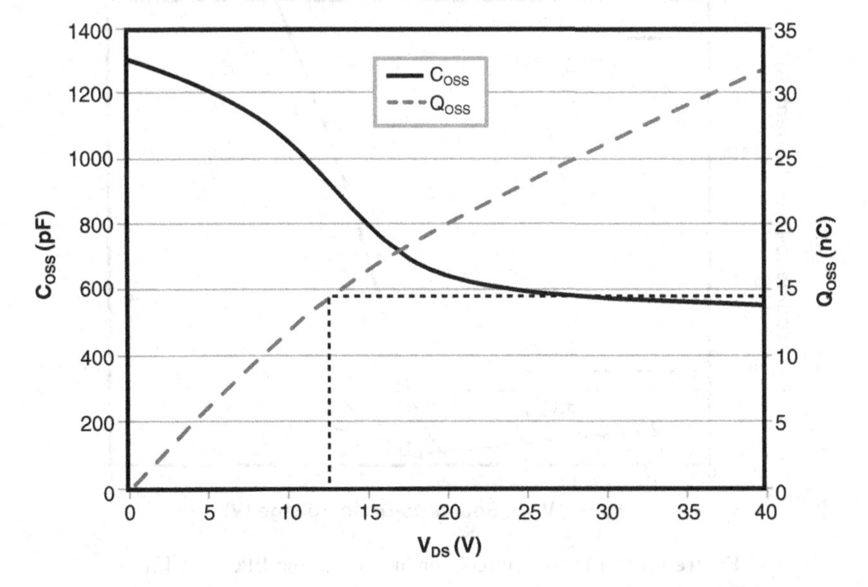

Figure 6.27 C_{OSS} and Q_{OSS} as a function of drain-to-source voltage for the EPC2015 [3]

The Q_{OSS} for the converter is used to determine the effective dead-time using Equation 6.45.

$$\begin{aligned} t_{Fall} &= \frac{Q_{OSS}}{I_{off}} \\ &= \frac{2 \cdot 14\,\text{nC}}{21.8\,\text{A}} \\ &= 1.28\,\text{ns} \end{aligned} \tag{6.45}$$

It is extremely important to include the Q_{OSS} of both devices in Equation 6.45. If both devices are the same, Q_{OSS} simply doubles. If both devices are not the same, each device's Q_{OSS} needs to be determined independently for the same voltage condition and added together, and the the total Q_{OSS} is used to calculate the fall-time. Furthermore, if the circuit includes a Schottky diode across the synchronous rectifier, it too must be included in the Q_{OSS} calculation. The fall-time is determined to be 1.28 ns. Having chosen 5 ns effective dead-time, the diode conduction time will be:

$$\begin{aligned} t_{Diode} &= t_{eff} - t_{Fall} \\ &= 5 - 1.28 \\ &= 3.72\,\text{ns} \end{aligned} \tag{6.46}$$

It is important to note that a negative result for t_{Diode} would mean that the converter is operating in the partial ZVS region, which should be avoided due to the high losses.

Next, the voltage drop across the body diode needs to be determined. Again, the datasheet is referenced for the drain current value. From Figure 6.28, it can be seen that, at 21.8 A, the voltage drop for diode conduction is 2.25 V. This value should be similar to the plateau voltage in the case of an enhancement-mode GaN transistor only.

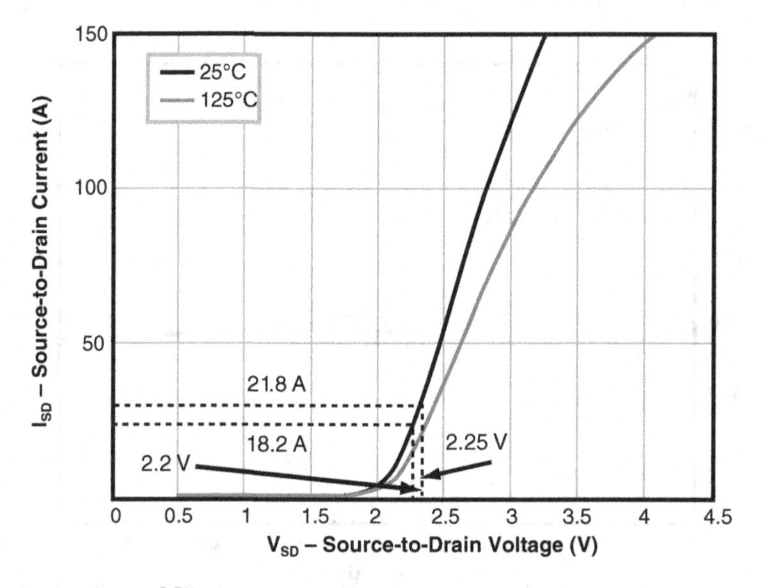

Figure 6.28 Reverse current conduction for the EPC2015 [3]

Now the turn-off reverse conduction losses can be calculated using Equation 6.16:

$$\begin{aligned} P_{SD} &= V_{SD} \cdot I_{DS} \cdot t_{SD} \cdot f_{sw} \\ &= 2.25 \cdot 21.8 \cdot 3.72 \cdot 10^{-9} \cdot 1 \cdot 10^{6} \\ &= 182\,\text{mW} \end{aligned} \tag{6.47}$$

6.6.3.2 Turn-On Transient Diode Conduction Losses

Since this transition is forced-commutating, the buck inductor current at the time of turn-on needs to be determined. In this case, the diode will conduct nearly instantaneously after the synchronous rectifier Q2 is turned off and will keep conducting until the control switch is turned on. This is due to the buck inductor's current keeping the diode in the conduction state. Thus, the diode conduction time is equal to the effective dead-time, which is 5 ns for this example. The turn-on current has already been calculated using the Equation 6.34 as 18.2 A. Using this current and the datasheet graph in Figure 6.28, the voltage drop for the diode is 2.2 V.

Now the turn-on reverse conduction losses can be calculated using Equation 6.16:

$$\begin{aligned} P_{SD} &= V_{SD} \cdot I_{DS} \cdot t_{SD} \cdot f_{sw} \\ &= 2.2 \cdot 18.2 \cdot 5 \cdot 10^{-9} \cdot 1 \cdot 10^{6} \\ &= 200\,\text{mW} \end{aligned} \tag{6.48}$$

6.6.4 Switching Losses (P_{sw})

The losses for the control switch and synchronous rectifier will be determined separately.

6.6.4.1 Control Switch Dynamic Losses

The control switch experiences both hard turn-on and turn-off losses. Since all the components have already been determined, Equations 6.11 and 6.12 can be used to determine the switching power losses.

$$\begin{aligned} P_{on} &= \frac{V_{BUS} \cdot I_{DS} \cdot f_{sw} \cdot R_{Gon}}{2} \cdot \left[\frac{Q_{GD}}{V_{DR} - V_{pl}} + \frac{Q_{GS2}}{V_{DR} - \left(\frac{V_{pl} + V_{th}}{2} \right)} \right] \\ &= \frac{12 \cdot 18.2 \cdot 10^{6} \cdot 2}{2} \cdot \left[\frac{1.94}{5 - 2.2} + \frac{1.04}{5 - \left(\frac{2.2 + 1.4}{2} \right)} \right] \\ &= 223\,\text{mW} \end{aligned} \tag{6.49}$$

$$P_{off} = \frac{V_{Bus} \cdot I_{DS} \cdot f_{sw} \cdot R_{Goff}}{2} \cdot \left[\frac{Q_{GD}}{V_{pl}} + \frac{Q_{GS2}}{\left(\frac{V_{pl} + V_{th}}{2} \right)} \right]$$
$$= \frac{12 \cdot 21.8 \cdot 10^6 \cdot 0.5}{2} \cdot \left[\frac{1.94}{2.25} + \frac{1.04}{\left(\frac{2.25 + 1.4}{2} \right)} \right] \tag{6.50}$$
$$= 94\,\text{mW}$$

6.6.4.2 Synchronous Rectifier Dynamic Losses

The synchronous rectifier's switching losses are small because it only switches to and from a diode voltage drop (see Table 6.2). Equations 6.11 and 6.12, therefore, can be used to calculate the switching power losses with V_{BUS} equated to V_{SD}.

$$P_{on} = \frac{V_{BUS} \cdot I_{DS} \cdot f_{sw} \cdot R_{Gon}}{2} \cdot \left[\frac{Q_{GD}}{V_{DR} - V_{pl}} + \frac{Q_{GS2}}{V_{DR} - \left(\frac{V_{pl} + V_{th}}{2} \right)} \right]$$
$$= \frac{2.25 \cdot 21.8 \cdot 10^6 \cdot 2}{2} \cdot \left[\frac{0}{5 - 2.25} + \frac{1.04}{5 - \left(\frac{2.25 + 1.4}{2} \right)} \right] \tag{6.51}$$
$$= 16\,\text{mW}$$

$$P_{off} = \frac{V_{BUS} \cdot I_{DS} \cdot f_{sw} \cdot R_{Goff}}{2} \cdot \left[\frac{Q_{GD}}{V_{pl}} + \frac{Q_{GS2}}{\left(\frac{V_{pl} + V_{th}}{2} \right)} \right]$$
$$= \frac{2.2 \cdot 18.2 \cdot 10^6 \cdot 0.5}{2} \cdot \left[\frac{0}{2.2} + \frac{1.04}{\left(\frac{2.2 + 1.4}{2} \right)} \right] \tag{6.52}$$
$$= 6\,\text{mW}$$

6.6.5 *Total Dynamic Losses ($P_{Dynamic}$)*

With all the dynamic loss components calculated, they can be summarized and added together to yield the total dynamic loss, which is given in Table 6.3.

6.6.6 *Conduction Losses ($P_{Conduction}$)*

All that remains to be calculated for the total hard-switching device losses are the conduction losses ($P_{Conduction}$). The conduction time for each device is based on the duty cycle. The duty

Table 6.3 Buck converter example total dynamic loss

Loss Characteristic (mW)	Control Switch		Synchronous Rectifier	
	Turn-off	Turn-on	Turn-on	Turn-off
P_{OSS}	0	161	0	0
P_G	25	25	20	20
P_{SD}	0	0	182	200
P_{RR}	0	0	0	0
P_{on}	N/A	223	16	N/A
P_{off}	94	N/A	N/A	6
Total Power Loss	528		444	

cycle (D) for the control switch already has been calculated using Equation 6.32 and is 10%. The duty cycle for the synchronous rectifier (D_{Sync}) can be determined from the control switch duty cycle using Equation 6.53:

$$\begin{aligned} D_{Sync} &= 1 - D \\ &= 1 - 0.1 \\ &= 0.9 \\ &= 90\% \end{aligned} \tag{6.53}$$

The conduction losses ($P_{Conduction}$) for the control switch can then be calculated using Equation 6.54, where the equation for the RMS of the current is given in [19]:

$$\begin{aligned} P_{Conduction} &= \left(I_{Load}^2 + \frac{I_{Ripple}^2}{12}\right) \cdot R_{DS(on)} \cdot D \\ &= \left(20^2 + \frac{3.6^2}{12}\right) \cdot 3.2 \cdot 10^{-3} \cdot 0.1 \\ &= 128\ \text{mW} \end{aligned} \tag{6.54}$$

and for the synchronous rectifier:

$$\begin{aligned} P_{Conduction} &= \left(I_{Load}^2 + \frac{I_{Ripple}^2}{12}\right) \cdot R_{DS(on)} \cdot D_{Sync} \\ &= \left(20^2 + \frac{3.6^2}{12}\right) \cdot 3.2 \cdot 10^{-3} \cdot 0.9 \\ &= 1155\ \text{mW} \end{aligned} \tag{6.55}$$

6.6.7 Total Device Hard-Switching Losses (P_{HS})

The total device hard-switching losses (P_{HS}) can be calculated by adding the conduction ($P_{Conduction}$) and dynamic (P_{DYN}) components together.

For the control switch:

$$\begin{aligned} P_{HS} &= P_{DYN} + P_{Conduction} \\ &= 0.528 + 0.128 \\ &= 656\,\text{mW} \end{aligned} \tag{6.56}$$

For the synchronous rectifier:

$$\begin{aligned} P_{HS} &= P_{DYN} + P_{Conduction} \\ &= 0.444 + 1.155 \\ &= 1599\,\text{mW} \end{aligned} \tag{6.57}$$

6.6.8 *Inductor Losses (P_L)*

The final loss component is the inductor loss (P_L). The inductor used for this example is the SLC1175-301ME [20], which has the following parameters: DC resistance (DCR) = 0.24 mΩ, ACR (at 1 MHz) = 0.24 mΩ, inductance 300 nH. The difference between DCR and ACR cannot be neglected for switching frequencies in the multi-MHz region. The manufacturer also provides a core loss calculator on its website [21], which will be utilized in this example. Alternative methods can be used to determined core losses that include referencing the manufacturers' core losses as a function of flux density and frequency.

Using the core loss calculator, the loss (P_{core}) was given as 51 mW at 1 MHz with a 3.6 A ripple current:

$$\begin{aligned} P_L &= I_{Load}^2 \cdot DCR + \frac{I_{Ripple}^2}{12} \cdot ACR + P_{Core} \\ &= 20^2 \cdot 240^{-6} + 3.6^2 \frac{3.6^2}{12} \cdot 240^{-6} + 0.051 \\ &= 147\,\text{mW} \end{aligned} \tag{6.58}$$

6.6.9 *Total Buck Converter Estimated Losses (P_{Total})*

Adding all the loss components together yields the total estimated power loss for the buck converter example:

$$\begin{aligned} P_{Total} &= P_{Control} + P_{Sync} + P_L \\ &= 656 + 1599 + 147 \\ &= 2402\,\text{mW} \end{aligned} \tag{6.59}$$

This results in an estimated efficiency for the converter of 90.9%, which excludes control circuit power.

6.6.10 *Buck Converter Loss Analysis Accounting for Common Source Inductance*

Ignoring the effect of CSI can lead to an artificially low loss prediction. Therefore, in this section, the switching losses will be recalculated to include the effect of CSI. By the very nature of CSI, it is impossible to measure without significant perturbation of the circuit.

CSI can be estimated using a commercial parametric extraction simulation program [22] that can compute the inductance from the layout and the device design. This would require knowledge of the internal design of the device that would seldom be made available. Alternatively, CSI can be estimated using circuit simulation software and using an ideal switch in the simulation. In the simulation, CSI can be added and waveforms compared with measured waveforms until enough correlation is found. For this calculation, a value of 110 pH will be used as a reasonable approximation.

Equation 6.31 can be used to determine the equivalent gate circuit resistance introduced by the CSI. This resistance will be used in conjunction with Equations 6.11 and 6.12 to determine the CSI-impacted losses. Note that CSI only impacts the current transition interval, and therefore Equations 6.11 and 6.12 need to be adjusted so that R_{CSI} is only added to R_G for the Q_{GS2} interval.

To calculate R_{CSI}, the value of C_{GS} and transconductance (g_m) at the operating conditions are required. First, the transconductance can be determined using the small signal model for a MOSFET [23] given in Equation 6.60.

$$\begin{aligned} g_m &= \frac{2 \cdot I_{DS}}{V_{pl} - V_{th}} \\ &= \frac{2 \cdot 20}{2.2 - 1.4} \\ &= 50\ \mathrm{S} \end{aligned} \tag{6.60}$$

Next C_{GS} needs to be determined. This value can be derived from Q_{GS}, which yields a time-equivalent capacitance, by reading off the values at the plateau voltage:

$$\begin{aligned} C_{GS} &= \frac{Q_{GS}}{V_{pl}} \\ &= \frac{2.87}{2.2} \\ &= 1.3\ \mathrm{nF} \end{aligned} \tag{6.61}$$

The equivalent CSI impedance (R_{CSI}) was then calculated using Equation 6.31 as:

$$\begin{aligned} R_{CSI} &= \frac{L_S \cdot g_m}{C_{GS}} \\ &= \frac{110 \cdot 10^{-12} \cdot 50}{1.3 \cdot 10^{-9}} \\ &= 4.22\ \Omega \end{aligned} \tag{6.62}$$

6.6.10.1 Control Switch Dynamic Losses Including the Effect of CSI

The control switch turn-on and turn-off losses, adjusted for CSI, can be determined by updating Equations 6.11 and 6.12 as follows:

$$
\begin{aligned}
P_{on} &= \frac{V_{Bus} \cdot I_{DS} \cdot f_{sw}}{2} \cdot \left[\frac{Q_{GD} \cdot R_{Gon}}{V_{DR} - V_{pl}} + \frac{Q_{GS2} \cdot (R_{Gon} + R_{CSI})}{V_{DR} - \left(\frac{V_{pl} + V_{th}}{2} \right)} \right] \\
&= \frac{12 \cdot 18.2 \cdot 10^6}{2} \cdot \left[\frac{1.94 \cdot 2}{5 - 2.2} + \frac{1.04 \cdot (2 + 4.22)}{5 - \left(\frac{2.2 + 1.4}{2} \right)} \right] \\
&= 373 \text{ mW}
\end{aligned}
\tag{6.63}
$$

$$
\begin{aligned}
P_{off} &= \frac{V_{Bus} \cdot I_{DS} \cdot f_{sw}}{2} \cdot \left[\frac{Q_{GD} \cdot R_{Goff}}{V_{pl}} + \frac{Q_{GS2} \cdot (R_{Goff} + R_{CSI})}{\left(\frac{V_{pl} + V_{th}}{2} \right)} \right] \\
&= \frac{12 \cdot 21.8 \cdot 10^6}{2} \cdot \left[\frac{1.94 \cdot 0.5}{2.25} + \frac{1.04 \cdot (0.5 + 4.22)}{\left(\frac{2.25 + 1.4}{2} \right)} \right] \\
&= 409 \text{ mW}
\end{aligned}
\tag{6.64}
$$

Note the degree to which turn-off losses are affected by CSI: from 94 mW to 409 mW for the control switch.

6.6.10.2 Synchronous Rectifier Dynamic Losses Including the Effect of CSI

The synchronous rectifier also experiences both hard turn-on and turn-off losses. Since all the components have already been determined, Equations 6.11 and 6.12 can be used to calculate the switching energy losses. In the case of turn-on, the rectifier switch only turns on with the diode voltage drop across it, hence V_{BUS} is equated to V_{SD} at turn-off ($I_{Turn\text{-}off}$).

$$
\begin{aligned}
P_{on} &= \frac{V_{BUS} \cdot I_{DS} \cdot f_{sw}}{2} \cdot \left[\frac{Q_{GD} \cdot R_{Gon}}{V_{DR} - V_{pl}} + \frac{Q_{GS2} \cdot (R_{Gon} + R_{CSI})}{V_{DR} - \left(\frac{V_{pl} + V_{th}}{2} \right)} \right] \\
&= \frac{2.25 \cdot 18.2 \cdot 10^6}{2} \cdot \left[\frac{0 \cdot 2}{5 - 2.25} + \frac{1.04 \cdot (2 + 4.22)}{5 - \left(\frac{2.25 + 1.4}{2} \right)} \right] \\
&= 84 \text{ mW}
\end{aligned}
\tag{6.65}
$$

$$P_{off} = \frac{V_{BUS} \cdot I_{DS} \cdot f_{sw}}{2} \cdot \left[\frac{Q_{GD} \cdot R_{Goff}}{V_{pl}} + \frac{Q_{GS2} \cdot (R_{Goff} + R_{CSI})}{\left(\frac{V_{pl} + V_{th}}{2}\right)} \right]$$
$$= \frac{2.2 \cdot 18.2 \cdot 10^6}{2} \cdot \left[\frac{0 \cdot 0.5}{2.2} + \frac{1.04 \cdot (0.5 + 4.22)}{\left(\frac{2.2 + 1.4}{2}\right)} \right] \tag{6.66}$$
$$= 55\ \text{mW}$$

Substituting these values for those in Table 6.3 and then recalculating, the total power loss increased by 582 mW to 2.95 W. The buck converter efficiency was reduced from 90.9% to 88.9%, which is 2.0% lower due to the inclusion of CSI.

6.6.11 Experimental Results for the Buck Converter

Based on the buck converter example, a practical version was designed and tested. The buck converter was operated and measured at various load currents up to the maximum rating. The dead-time was set at 5 ns. Measurements taken for the converter include the controller power and charge pump regulator for the controller IC, which accounts for as much as 152 mW of power loss in the measurements. The measured efficiency for the buck converter when operating at 12 V_{BUS}, as shown in Figure 6.29 (third trace from the top), is about 88.1% at 20 A load. This compares very well with the estimate of 88.4%. Figure 6.29 also shows the calculated efficiency results without compensating for CSI or controller consumption (top

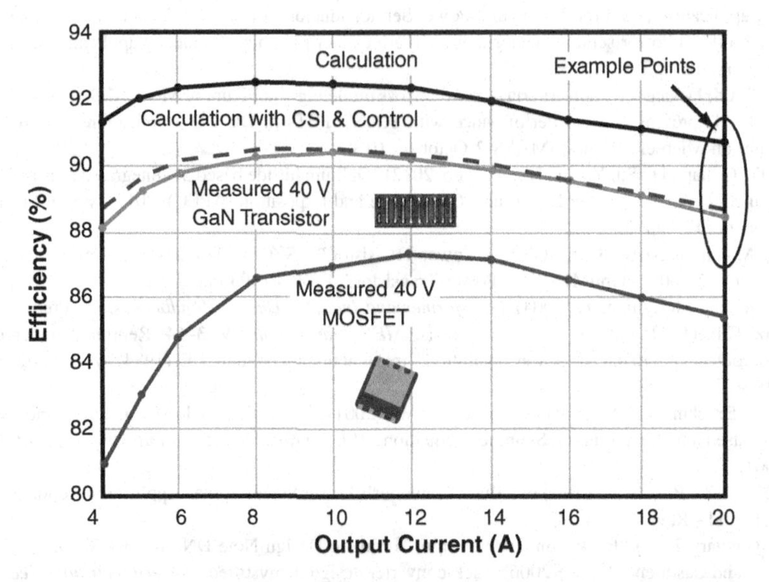

Figure 6.29 Calculated and measured efficiency results of the GaN transistor-based (EPC2015 [3]) buck converter operating at 12 V input and compared against a comparable MOSFET converter with control switch is BSC097N04LSG [11] and synchronous rectifier is BSC0400N04LSG [12]

trace), and with compensating for CSI and controller power consumption (dashed trace above measured results).

Figure 6.29 also shows the results of an equivalent MOSFET-based converter, operating under the same conditions (bottom trace). In this case, the GaN transistor-based converter efficiency exceeds that of the MOSFET converter by several percentage points.

6.7 Summary

In Chapter 6, the mechanism and key factors that contribute to hard-switching losses have been discussed. The analytical tools needed to calculate losses from the greatest contributors were developed and used in a concrete example. The calculations agreed with the actual measured results from a circuit built to these same specifications.

In the next chapter, soft-switching and resonant-switching techniques will be discussed.

References

1. Vishay (Dec. 2004) "Power MOSFET Basics: Understanding Gate Charge and Using it to Assess Switching Performance,", Appl. Note AN608.
2. On-Semiconductor (April 2012) "MOSFET Gate-Charge Origin and its Applications," Appl. Note AND9083/D.
3. Efficient Power Conversion Corporation "EPC2015 – Enhancement-mode Power Transistor," EPC2015 datasheet, March 2011 [Revised Jan. 2013]. Available from http://epc-co.com/epc/documents/datasheets/EPC2015_datasheet.pdf.
4. Strydom, Johan, "The eGaN FET-Silicon Power Shoot-Out: 1: Comparing Figure of Merit (FOM)," *Power Electronics Technology,* Sept. 1, 2010, Available from http://powerelectronics.com/power_semiconductors/power_mosfets/fom-useful-method-compare-201009/.
5. Huang, A.Q. (2004) New unipolar switching power device figures of merit. *IEEE Electron Device Letters*, **25**, 298–301.
6. Kim, I.-J.Il.-Jung., Matsumoto, S., Sakai, T., and Yachi, T. (1995) "New power device figure-of-merit for high frequency applications," in Proc. Int. Symp. Power Semiconductor Devices ICs, Yokohama, Japan, pp. 309–314.
7. Baliga, B.J. (1989) Power semiconductor device figure-of-merit for high frequency applications. *IEEE Electron Device Letters*, **10**, 455–457.
8. Ying, Y. (2008) Device selection criteria – based on loss modeling and Figure of Merit, M.S. thesis, Virginia Tech.
9. Reusch, D., "Improving System Performance with eGaN® FETs in DC-DC Applications," 46th International Symposium on Microelectronics, iMAPS 2 October 2013.
10. Reusch, D., Gilham, D., Su, Y., and Lee, F. (Feb. 2012) "Gallium nitride based 3D integrated non-isolated point of load module," in Applied Power Electronics Conference and Exposition (APEC), 2012 Twenty-Seventh Annual IEEE, Orlando, FL, pp. 38–45.
11. Infineon (March 18 2010) "OptiMOS™ 3 Power-Transistor," BSZ097N04 datasheet.
12. Infineon (Nov. 5 2009) "OptiMOS™ 3 Power-Transistor," BSZ040N04 datasheet.
13. Colonel, W. and McLyman, T. (2004) *Transformer and Inductor Design Handbook*, CRC Press.
14. Steinmetz, C.P. (1892) On the law of hysteresis. *AIEE Transactions*, **9**, 3–64. Reprinted under the title "A Steinmetz contribution to the AC power revolution," introduction by Brittain, J.E. (1984) *Proceedings of the IEEE*, **72** (2), 196–221.
15. Reinert, J., Brockmeyer, A., and De Doncker, R.W. (2001) Calculation of losses in ferro- and ferrimagnetic materials based on the modified Steinmetz equation. *IEEE Transactions on Industry Applications*, **37** (4), 1055–1061.
16. Hauke, B. (2012) "Basic Calculation of a Buck Converter's Power Stage," Texas Application Report SLVA477A – December 2011– Revised August.
17. Ejury, J. (January 2013) "Buck Converter Design," Infineon Design Note DN 2013-01 V1.0.
18. Schelle, D. and Castorena, J. (Jure 2006) Buck-converter design demystified, *Power Electronics Technology*, pp. 46–53. Available from http://powerelectronics.com/dc-dc-converters/buck-converter-design-demystified.
19. Erickson, R.W. and Maksimović, D. (January 2001) *Fundamentals of Power Electronics*, 2nd edn, Springer.
20. SLC1175-301ME datasheet http://www.coilcraft.com/pdfs/slc1175.pdf.

21. Coilcraft, "Coilcraft Core and conductor loss calculator," [Updated July 20, 2012]. Available from http://www.coilcraft.com/apps/loss/loss_1.cfm.
22. Ansys, "Ansys Q3D Extractor," Available from http://www.ansys.com/Products/Simulation+Technology/Electromagnetics/Signal+Integrity/ANSYS+Q3D+Extractor.
23. Cartwright, K.V. (2009) Derivation of the exact transconductance of a FET without calculus. *The Technology Interface Journal*, **10** (1)
24. Reusch, D. (2012) "High frequency, high power density integrated point of load and bus converters," Ph.D. dissertation, Virginia Tech, Blacksburg, VA, Available from http://scholar.lib.vt.edu/theses/available/etd-04162012-151740/.
25. Reusch, D., "eGaN® FET-Silicon Power Shoot-Out Vol. 13, Part 1: Impact Of Parasitics," *Power Electronics Technology*, March 2013, Available from http://powerelectronics.com/gan-transistors/egan-fet-silicon-power-shoot-out-vol-13-part-1-impact-parasitics#!.
26. Reusch, D., "eGaN® FET-Silicon Power Shoot-Out Vol. 13, Part 2: Optimal PCB Layout," *Power Electronics Technology,* April 2013, Available from http://powerelectronics.com/gan-transistors/egan-fet-silicon-power-shoot-out-vol-13-part-2-optimal-pcb-layout#!.

7

Resonant and Soft-Switching Converters

7.1 Introduction

The previous chapter addressed the application of GaN transistors in hard-switching power converters, and we demonstrated the benefits that GaN transistors provide – as compared to state-of-the-art silicon power MOSFETs. In this chapter, we discuss the fundamentals of resonant and soft-switching applications and evaluate the superior performance capabilities of GaN transistors over silicon MOSFETs in these applications. The chapter will conclude with a design example comparing GaN transistors and Si MOSFETs in an isolated, high-frequency 48 V intermediate bus converter (IBC) with a 12 V output, utilizing a resonant topology operating at 1.2 MHz.

7.2 Resonant and Soft-Switching Techniques

Resonant and soft-switching techniques can improve performance in converters by reducing switching-related losses compared to conventional hard-switching converters. This is accomplished by creating operating conditions where the transistor does not encounter simultaneous high voltage and high current during the switching commutation. There are many different resonant and soft-switching techniques [1–4] and the two conditions in common are zero-voltage switching (ZVS) and zero-current switching (ZCS).

7.2.1 Zero-Voltage and Zero-Current Switching

Zero-voltage switching is used to eliminate the turn-on commutation losses in a switching device. ZVS is achieved when the transistor drain-to-source voltage is reduced to zero before turning on. To reduce the drain-to-source voltage across the transistor in the majority of resonant and soft-switching topologies, the device output charge (Q_{OSS}) must be removed by conducting current from source to drain through the output capacitance (C_{OSS}) until the drain-to-source voltage reaches zero. The switching trajectory of a traditional hard-switching and ZVS, soft-switching, turn-on transition is shown in Figure 7.1(a), where the x-axis, V_{DS},

GaN Transistors for Efficient Power Conversion, Second Edition.
Alex Lidow, Johan Strydom, Michael de Rooij, and David Reusch.

Companion Website: http://www.wiley.com/go/gan_transistors

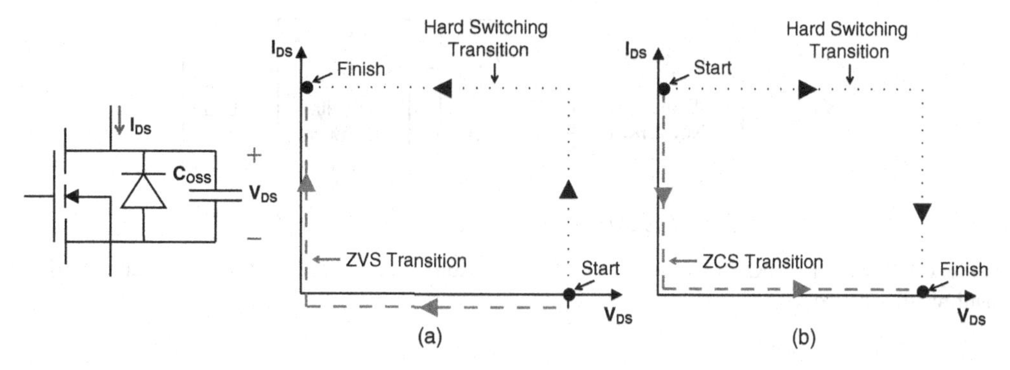

Figure 7.1 Ideal switching transition for (a) zero-voltage switching turn-on transition (b) zero-current switching turn-off transition

represents switch voltage, and the y-axis, I_{DS}, represents switch current. For the hard-switching transition, first the current rises to the load current, then the voltage falls to zero. This results in large values of voltage and current being commutated simultaneously in the transistor, generating loss as well as incurring E_{OSS} losses as discussed in Chapter 6. For the ZVS transition, before the transistor current rises to the load current, a negative current drives the drain-to-source voltage to zero, creating a soft turn-on condition. The device does not commutate high voltage and high current simultaneously, reducing the turn-on commutation losses in the device to zero, and eliminating E_{OSS} losses.

While ZVS erases turn-on commutation losses, it does not reduce the turn-off commutation losses. ZCS provides soft commutation during device turn-off, as shown in Figure 7.1(b). For the hard-switching turn-off transition, also shown in Figure 7.1(b), first the voltage rises to the bus voltage, and then the current falls to zero. This results in large values of voltage and current being commutated simultaneously in the transistor, generating losses. ZCS occurs when the transistor drain-to-source current is reduced to zero and before turning off. To achieve ZCS, the current is shaped as a sinusoidal pulse using a resonant network. When the switch current resonates to zero, the device can turn off with ZCS. During a ZCS turn-off transition, large values of voltage and current are not commutated simultaneously in the transistor, virtually eliminating the turn-off losses in the transistor. For ZCS resonant converters, the hard-switching turn-on commutation remains, and the turn-on switching transition and E_{OSS} losses are dissipated in the device.

For resonant and soft-switching topologies to operate under ZVS and ZCS, passive networks are required to shape the transistor's voltage and current. Often, the addition of passive resonant components can be avoided by utilizing the parasitics internal to the device and in-circuit elements such as package and PCB inductances. This allows the parasitics that diminished hard-switching converter performance to be used effectively to achieve soft-switching commutations in resonant and soft-switching converters. For the majority of resonant and soft-switching DC-DC power converters, zero-voltage switching is preferred to zero-current switching due to the reduction of the E_{OSS} losses, which are incurred only during the turn-on switching transition.

7.2.2 Resonant DC-DC Converters

The traditional approach for resonant converters used for DC-DC power conversion is shown in Figure 7.2. The input voltage source, V_{IN}, connects to a switching network, which outputs a

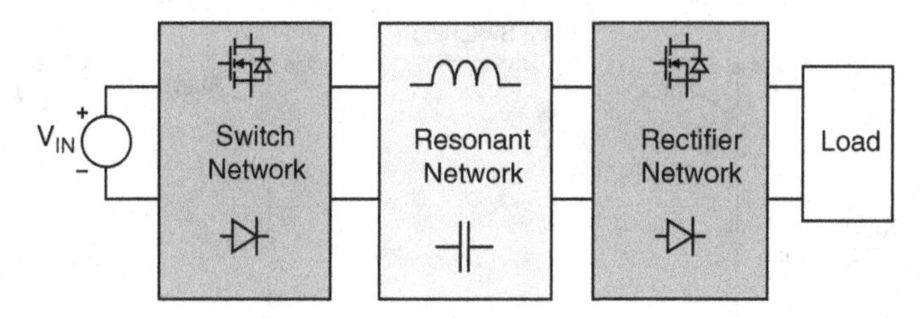

Figure 7.2 Resonant DC-DC converter block diagram consisting of switch network, resonant network, and rectifier network

pulsed waveform to the resonant network. The resonant network then shapes the voltage or current to achieve soft-switching in the switching network's power devices. Following the resonant network is the rectifier network, which rectifies and filters the voltage and current to deliver DC power to the load.

7.2.3 Resonant Network Combinations

The most basic resonant networks consist of a series network, where the load is connected in series with the resonant capacitance (C_S), as shown in Figure 7.3(a), a parallel network, where the load is connected in parallel with the resonant capacitance (C_P), as shown in Figure 7.3(b), or a combination of the series and parallel networks, also known as a series-parallel network, as shown in Figure 7.3(c). Another popular resonant network is the LLC, where the parallel capacitance (C_P) of the series parallel network is replaced with a parallel inductor (L_P), and L_R and C_S are rearranged, as shown in Figure 7.3(d). These different resonant networks can operate with ZVS or ZCS and can provide unique benefits, while having disadvantages when compared to traditional hard-switching converters [1–3,5,6]. There are also less common

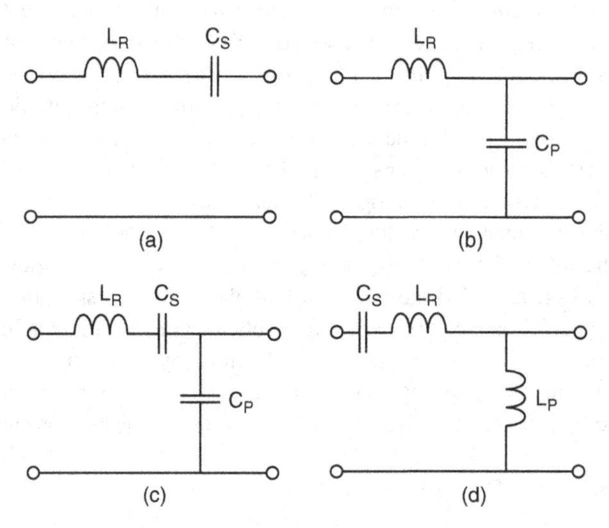

Figure 7.3 DC-DC converter resonant networks (a) series LC (b) parallel LC (c) series-parallel LCC (d) LLC

resonant network topologies as well as multi-element resonant networks that can be employed as resonant tanks [5,7,8].

7.2.4 Resonant Network Operating Principles

In this section, the basic operation of a common resonant configuration – the series-resonant network, shown in Figure 7.3(a) – will be discussed. This series-resonant network allows for either ZVS or ZCS to occur in the switching network transistors, reducing either turn-on or turn-off commutation losses. The magnitude of the impedance of the series network can be given by:

$$|Z_{SRN}| = \sqrt{R^2 + (X_L - X_C)^2} = \sqrt{R^2 + \left(\omega \cdot L_R - \frac{1}{\omega \cdot C_S}\right)^2} \tag{7.1}$$

$$X_L = \omega \cdot L_R \tag{7.2}$$

$$X_C = \frac{1}{\omega \cdot C_S} \tag{7.3}$$

where X_L and X_C are the reactances of the inductor and capacitor respectively, R is the equivalent load resistance, Z_{SRN} is the magnitude of the impedance of the series-resonant network, and ω is the angular frequency of the resonant network. The standard frequency common to DC-DC converters is given by:

$$f = \frac{\omega}{2 \cdot \pi} \tag{7.4}$$

where f is frequency measured in hertz.

The magnitude of the impedance plot for a series-resonant network is shown in Figure 7.4. The resonant frequency is the point where the network transitions from appearing capacitive to inductive. The resonant frequency provides the minimum impedance and is given by:

$$f_0 = \frac{1}{2 \cdot \pi \cdot \sqrt{L_R \cdot C_S}} \tag{7.5}$$

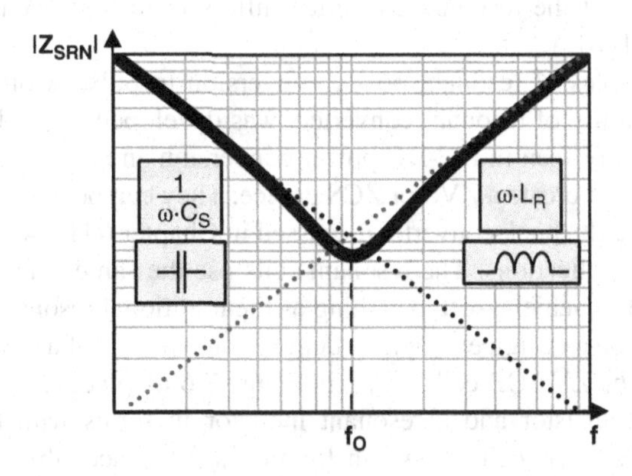

Figure 7.4 Impedance plot of series-resonant network

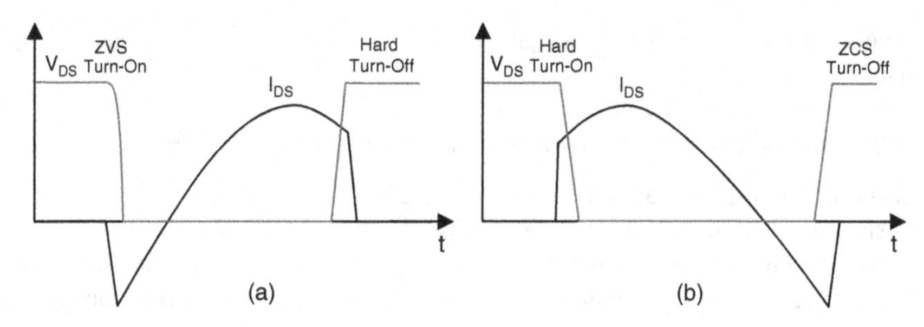

Figure 7.5 Operation of series-resonant converter transistor (a) above the resonant frequency (b) below the resonant frequency

The turn-on commutation of a transistor connected to a series-resonant network operating above the resonant frequency will switch under ZVS. Operation above the resonant frequency makes the resonant tank appear inductive, as shown in Figure 7.4. For an inductive load, the current lags the voltage, as shown in Figure 7.5(a). The inductive-resonant tank produces a negative current in the device before turn-on, discharging the transistor output charge and allowing for a soft ZVS turn-on commutation. For operation above the resonant frequency, there will be voltage and current in the device at turn-off, leading to a hard turn-off commutation.

For operation below the resonant frequency, the transistor will turn off under ZCS. For a series-resonant converter, operation below the resonant frequency makes the resonant tank appear capacitive, as shown in Figure 7.4. For a capacitive load, the current leads the voltage, as shown in Figure 7.5(b). The capacitive-resonant tank produces a negative current in the device before turn-off, allowing for a soft ZCS turn-off commutation. For operation below the resonant frequency, there will be voltage and current in the device at turn-on, leading to a hard turn-on commutation and E_{OSS} losses.

7.2.5 Resonant Switching Cells

The best performance is achieved for traditional resonant converters operating close to the resonant frequency, yet output regulation is achieved by varying the switching frequency. The further the operating frequency is moved from the resonant point to maintain regulation, the more the performance of the resonant converter suffers from higher circulating energy and component stresses [1–8].

To apply the principles of resonant power conversion to pulse width modulated (PWM) converters, another family of resonant converters was developed [1,9]. These quasi-resonant (QR) cells are commonly seen in DC-DC power conversion and combine a resonant network with a single transistor to create a ZVS or ZCS device. They can be applied to traditional non-isolated topologies like the buck converter discussed in Chapter 6 [1,3,4], as well as to various other topologies and applications. The resonant cells use the same concept of shaping of the voltage and or current to achieve soft-switching as the traditional resonant networks. (Later in this chapter there will be a design example to demonstrate the use of a QR cell in a GaN-based resonant converter.) The ZVS QR cell, shown in Figure 7.6(a), places a resonant capacitor, C_R, in parallel with the transistor and a resonant inductor in series with the capacitor-switch network, while the ZCS QR cell, shown in Figure 7.6(b), places the resonant capacitor in parallel with the series combination of the resonant inductor-switch network.

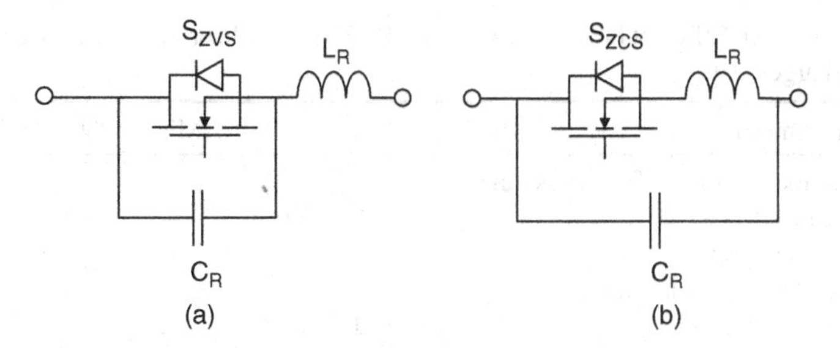

Figure 7.6 Quasi-resonant switching cells (a) ZVS (b) ZCS

7.2.6 Soft-Switching DC-DC Converters

Soft-switching converters can be seen as a hybrid between hard-switching PWM converters and frequency-controlled resonant converters. Soft-switching converters employ resonant techniques for a portion of the operating period to achieve a soft device commutation, with the remaining period operating as a PWM converter [2,3]. This allows for transistor soft commutation while reducing the higher circulating energy and device stresses associated with resonant converters, as well as offering PWM control for output regulation.

7.3 Key Device Parameters for Resonant and Soft-Switching Applications

In resonant and soft-switching applications, the switching-related losses are minimized by using techniques to achieve ZVS and ZCS. With the reduction of switching losses, the Q_{GD} and Q_{GS2} terms that dominated losses in hard-switching applications are no longer the critical device parameters determining in-circuit performance. The two device parameters key to high performance in resonant and soft-switching applications are device output charge, Q_{OSS}, and gate charge, Q_G.

7.3.1 Output Charge (Q_{OSS})

Output charge has a large impact on the performance of resonant and soft-switching converters as it directly impacts the energy required to achieve ZVS. A reduction in energy can result in reduced ZVS transition times and currents, providing both a longer power delivery period and lower RMS currents in high-frequency resonant and soft-switching converters. In a ZCS topology, the energy of the output capacitance (E_{OSS}) is dissipated when the transistor turns on in the same manner as a hard-switching commutation.

Before a ZVS transition can occur, the output capacitance must be discharged, bringing the drain-to-source voltage of the transistor to zero volts before turning on the transistor. The time required to achieve ZVS is given by:

$$t_{ZVS} = \frac{C_{OSS(TR)} \cdot V_{DS}}{I_{ZVS}} = \frac{Q_{OSS}}{I_{ZVS}} \tag{7.6}$$

where t_{ZVS} is the time required to discharge the output capacitance, $C_{OSS(TR)}$ is the time-related output capacitance, V_{DS} is the transistor drain-to-source voltage, I_{ZVS} is the soft-switching

Table 7.1 Data from an Efficient Power Conversion EPC2001 datasheet showing transistor capacitances and associated charges [10]

	Parameter	Test conditions	Min	Typical	Max	Unit
Dynamic characteristics ($T_j = 25\,°C$ unless otherwise stated)						
C_{ISS}	Input capacitance	$V_{DS} = 50V$, $V_{GS} = 0\,V$	—	850	950	pF
C_{OSS}	Output capacitance		—	450	525	
C_{RSS}	Reverse transfer capacitance		—	20	30	
Q_G	Total gate charge ($V_{GS} = 5\,V$)	$V_{DS} = 50\,V$, $I_D = 25\,A$	—	8	10	nC
Q_{GD}	Gate-to-drain charge		—	2.2	2.7	
Q_{GS}	Gate-to-source charge		—	2.3	2.8	
Q_{OSS}	Output charge	$V_{DS} = 50\,V$, $V_{GS} = 0\,V$	—	35	40	
Q_{RR}	Source-drain recovery charge		—	0	0	

All measurements were done with substrate shorted to source.

current used to discharge the transistor's output capacitance, and Q_{OSS} is the output charge of the transistor.

7.3.2 Determining Output Charge from Manufacturers' Datasheet

To properly design a ZVS transition, the values of C_{OSS} and Q_{OSS} at the proper in-circuit operating conditions are critical. Values for C_{OSS} and Q_{OSS} are generally given for a single drain-to-source operating voltage in manufacturers' datasheets as shown in Table 7.1 for a 100 V EPC2001 [10] enhancement-mode GaN transistor, and Table 7.2 for a 100 V silicon MOSFET from Infineon, the BSC060N10NS3G [11].

The single output charge and capacitance point given in manufacturers' datasheets does not provide enough information to properly design for a wide range of operating conditions. The output capacitance of both GaN transistors and MOSFETs is highly non-linear, and the output charge varies with drain-to-source voltage. Figure 7.7 shows the capacitance curves for the 100 V EPC2001 [10] and the 100 V BSC060N10NS3G [11] MOSFET, where it can be seen

Table 7.2 Data from an Infineon BSC060N10NS3G datasheet showing transistor capacitances and associated charges [11]

	Parameter	Test conditions	Min	Typical	Max	Unit
Dynamic characteristics						
C_{ISS}	Input capacitance	$V_{GS} = 0\,V$, $V_{DS} = 50\,V$,	—	3700	4900	pF
C_{OSS}	Output capacitance	$f = 1\,MHz$	—	650	860	
C_{RSS}	Reverse transfer capacitance		—	25	—	
Q_{GS}	Gate-to-source charge	$V_{DD} = 50\,V$, $I_D = 25\,A$, $V_{GS} = 0$	—	15	—	nC
Q_{GD}	Gate-to-drain charge	to 10 V	—	9	—	
Q_{SW}	Switching charge		—	13	—	
Q_G	Gate charge total		—	51	68	
$Q_{Plateau}$	Gate plateau voltage		—	4.2	—	V
Q_{OSS}	Output charge	$V_{DD} = 50\,V$, $V_{GS} = 0\,V$	—	68	91	nC

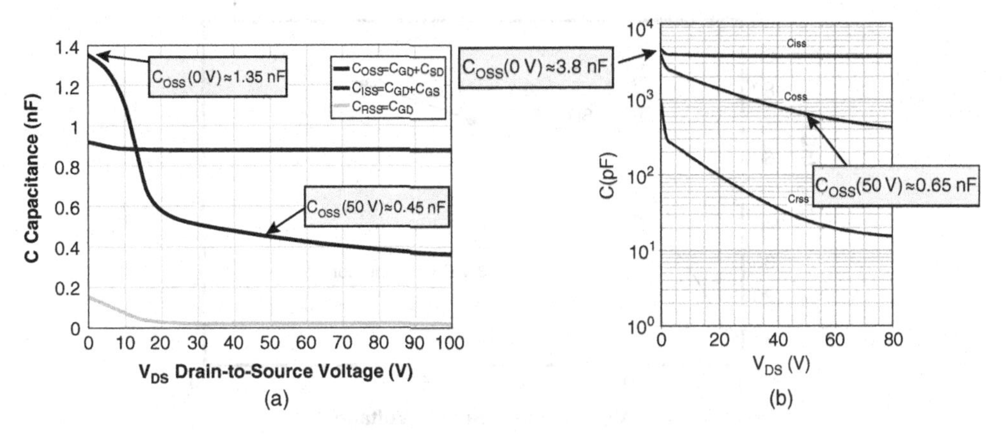

Figure 7.7 Capacitance curves of (a) EPC2001 GaN transistor [10] (b) BSC060N10NS3G Si MOSFET [11]

that the output capacitance changes by factors of three and six, respectively, for GaN transistors and Si MOSFETs from 0 V to 50 V.

The output charge and effective time-related capacitance for a transistor at any given voltage can be calculated from the manufacturers' datasheets in the same manner as the hard switching converters discussed in Chapter 6:

$$Q_{OSS}(V_{DS}) = \int_{0}^{V_{DS}} C_{OSS}(V_{DS}) \cdot dV_{DS} \tag{7.7}$$

$$C_{OSS(TR)}(V_{DS}) = \frac{Q_{OSS}(V_{DS})}{V_{DS}} \tag{7.8}$$

where V_{DS} is the drain-to-source operating voltage of the transistor.

The output charge for the EPC2001 [10] GaN transistor and BSC060N10NS3G [11] Si MOSFET are calculated and plotted in Figure 7.8 using Equation 7.7 for drain-to-source voltages varied from 0 V to the maximum voltage listed in the manufacturers' datasheets. The GaN transistor, with a similar on-resistance, offers a significant reduction in output charge over the entire voltage range.

7.3.3 Comparing Output Charge of GaN Transistors and Si MOSFETs

To compare the output charge figure of merit between GaN and MOSFET technologies in resonant and soft-switching applications, the product of the on-resistance and output charge for state-of-the-art 40 V, 100 V, 200 V, and 600 V GaN and Si MOSFETs are plotted in Figure 7.9. The GaN devices offer a significant reduction in output charge FOM, with the gains increasing as voltage increases. The reduction in FOM allows the circuit designer to reduce the transistor conduction losses, shorten the ZVS transition, or reduce the ZVS current, all of which would lower converter loss and improve efficiency. These benefits will be demonstrated experimentally in the design example later in this chapter.

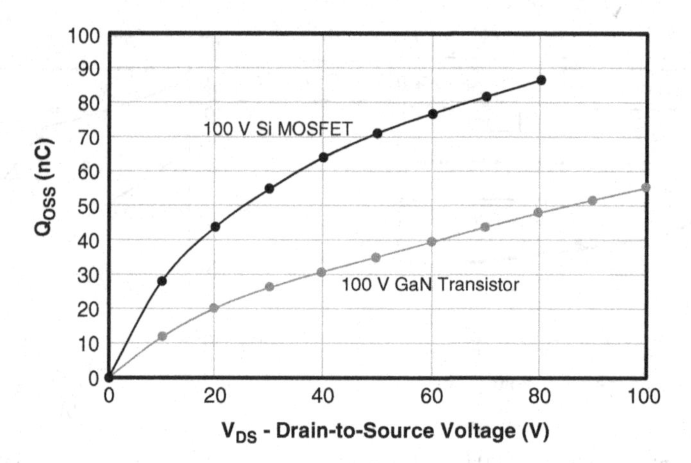

Figure 7.8 Output charge for varying drain-to-source voltages for 100 V EPC2001 GaN transistor and 100 V BSC060N10NS3G Si MOSFET

7.3.4 Gate Charge (Q_G)

The frequency capability of resonant and soft-switching topologies are also significantly impacted by the gate charge, Q_G. The gate charge is the amount of charge required to fully turn on or off the transistor. Voltage source drivers are employed for the vast majority of DC-DC converters. This voltage source appears in series with the input capacitance of the transistor and has an effective resistance equal to the sum of the gate driver circuit's internal and external resistance and the internal gate resistance of the transistor. The gate charge is dissipated each switching cycle, resulting in a gate drive loss equal to:

$$P_G = Q_G \cdot V_{DR} \cdot f_{sw} \tag{7.9}$$

where V_{DR} is the gate drive voltage and f_{sw} is the switching frequency.

Looking beyond the gate drive power loss, the gate driving speed also has a large impact on the performance of high-frequency resonant and soft-switching converters. The switching period is inversely proportional to switching frequency and, as frequencies increase, the gate rising and falling speeds can become limitations to the minimum switching time. The design example at the end of this chapter will illustrate this issue and show how GaN technology can offer superior performance in high-frequency DC-DC converters compared to Si MOSFETs.

7.3.5 Determining Gate Charge for Resonant and Soft-Switching Applications

To enable users to calculate gate charge, manufacturers supply a gate charge curve similar to Figure 7.10 for the 100 V EPC2001 [10] GaN transistor. The gate charge, Q_G, is given for a hard-switching transition and not directly applicable to resonant and soft-switching applications. The definition of charges Q_{GS1}, Q_{GS2}, Q_G, Q_{GD}, and Q_{GS} are given in Chapters 3 and 6. For a ZVS application, the voltage commutation period occurs before the device turns on, and the Miller plateau region as well as the accompanying charge, Q_{GD}, are eliminated from the

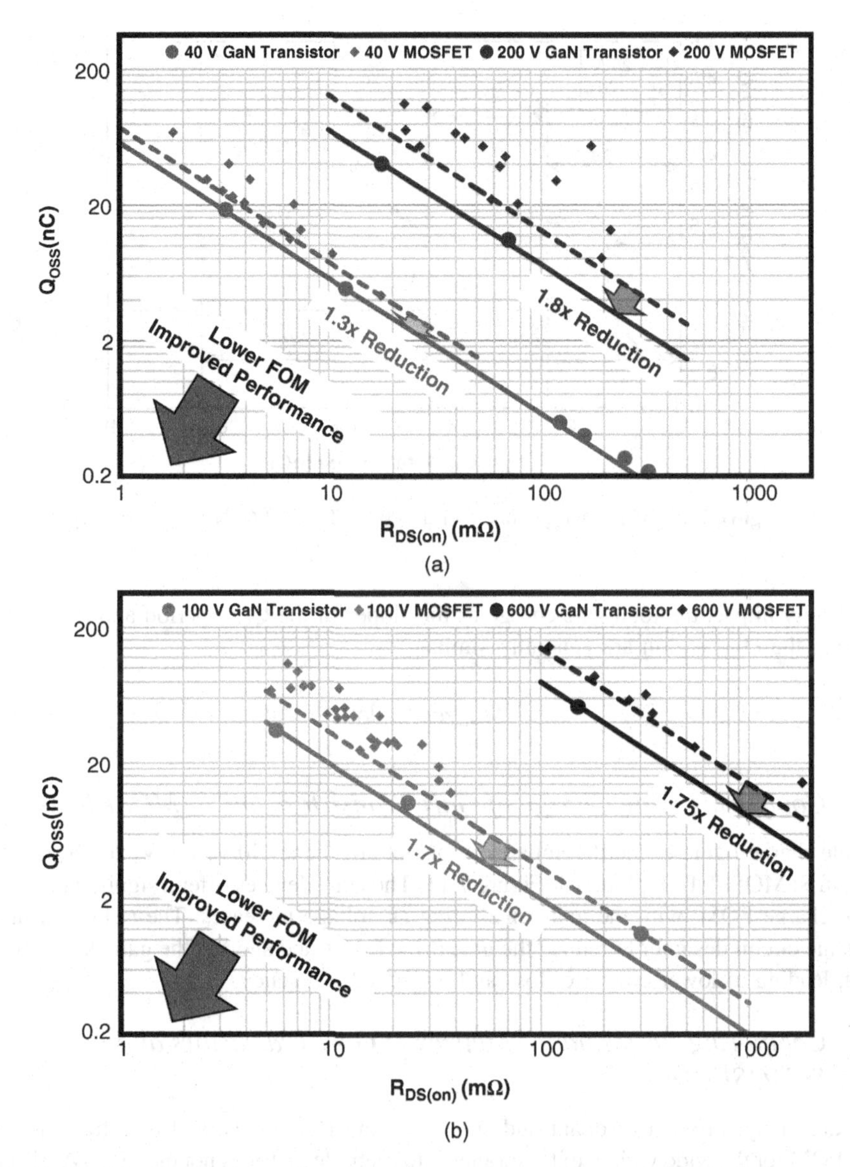

Figure 7.9 Output charge figure of merit comparison between GaN and Si devices

total gate charge. The gate charge for ZVS topology can be given by:

$$Q_{G\_ZVS} = Q_G - Q_{GD} \tag{7.10}$$

where Q_G is the gate charge of a hard-switching application, and Q_{GD} is the gate-to-drain charge.

For a ZCS transition, the current commutation period occurs before device turn-off, and the Q_{GS2} period is eliminated. While this reduces switching commutation losses, it does not

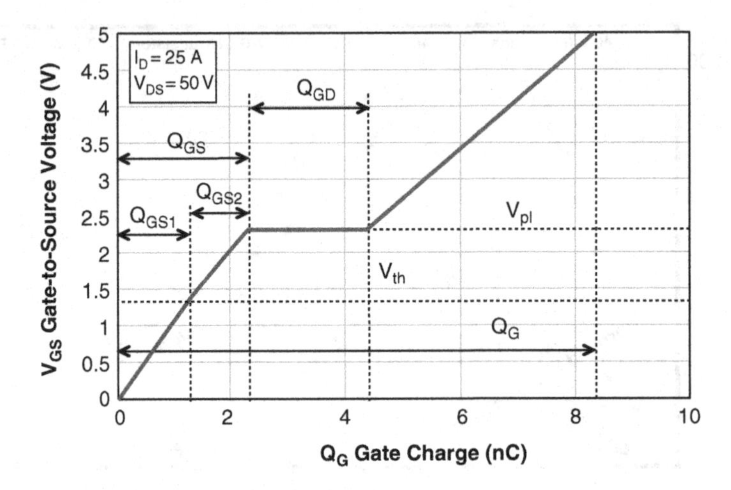

Figure 7.10 Gate charge curve for a 100 V EPC2001 GaN transistor [10]

significantly impact the total gate charge as the slopes of the Q_{GS2} period and the Q_G period following the Q_{GD} region are generally similar.

$$Q_{G\_ZCS} = Q_G \tag{7.11}$$

7.3.6 Comparing Gate Charge of GaN Transistors and Si MOSFETs

The gate charge figure of merit comparison for state-of-the-art 40 V, 100 V, 200 V, and 600 V GaN and Si MOSFETs is plotted in Figure 7.11. The GaN devices offer a significant reduction in gate charge FOM, with the gains increasing as voltage increases. The reduction in FOM allows the circuit designer to reduce the gate drive losses and shorten the gate drive transition period, leading to lower converter loss and improved efficiency.

7.3.7 Comparing Performance Metrics of GaN Transistors and Si MOSFETs

There are many different resonant and soft-switching techniques, and therefore distilling a single FOM for the wide variety of topologies into a simple metric is not practical. As discussed earlier, for resonant and soft-switching applications, the output charge, Q_{OSS}, and gate charge, Q_G, are the dominating device parameters. To allow designers to simply compare different devices to determine the technology providing the best relative in-circuit performance for resonant and soft-switching applications, a practical soft-switching FOM is:

$$FOM_{SS} = (Q_{OSS} + Q_G) \cdot R_{DS(on)} \tag{7.12}$$

The comparison of this soft-switching FOM for 40 V, 100 V, 200 V, and 600 V GaN and Si MOSFETs is plotted in Figure 7.12. The GaN technology offers a significant FOM reduction for all voltages, indicating significant performance improvements in high-frequency

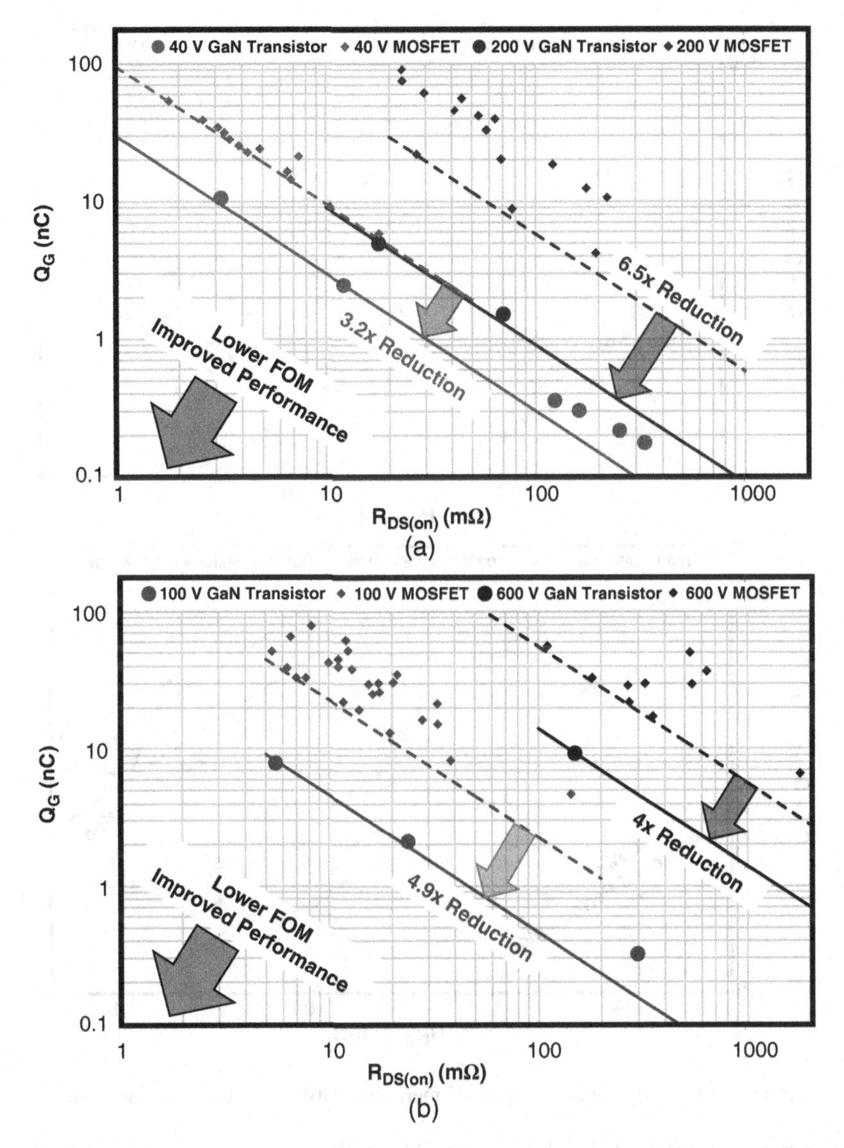

Figure 7.11 Gate charge figure of merit comparison between GaN and Si devices

soft-switching applications. The benefits of replacing an Si MOSFET with an enhancement-mode GaN transistor in a high-frequency resonant converter will be quantified next.

7.4 High-Frequency Resonant Bus Converter Example

Distributed power systems are prevalent in telecommunications, networking, and high-end server applications, and generally utilize a 48 V bus voltage adopted from the telecom industry. The traditional distributed power architecture (DPA), shown in Figure 7.13(a), employs a

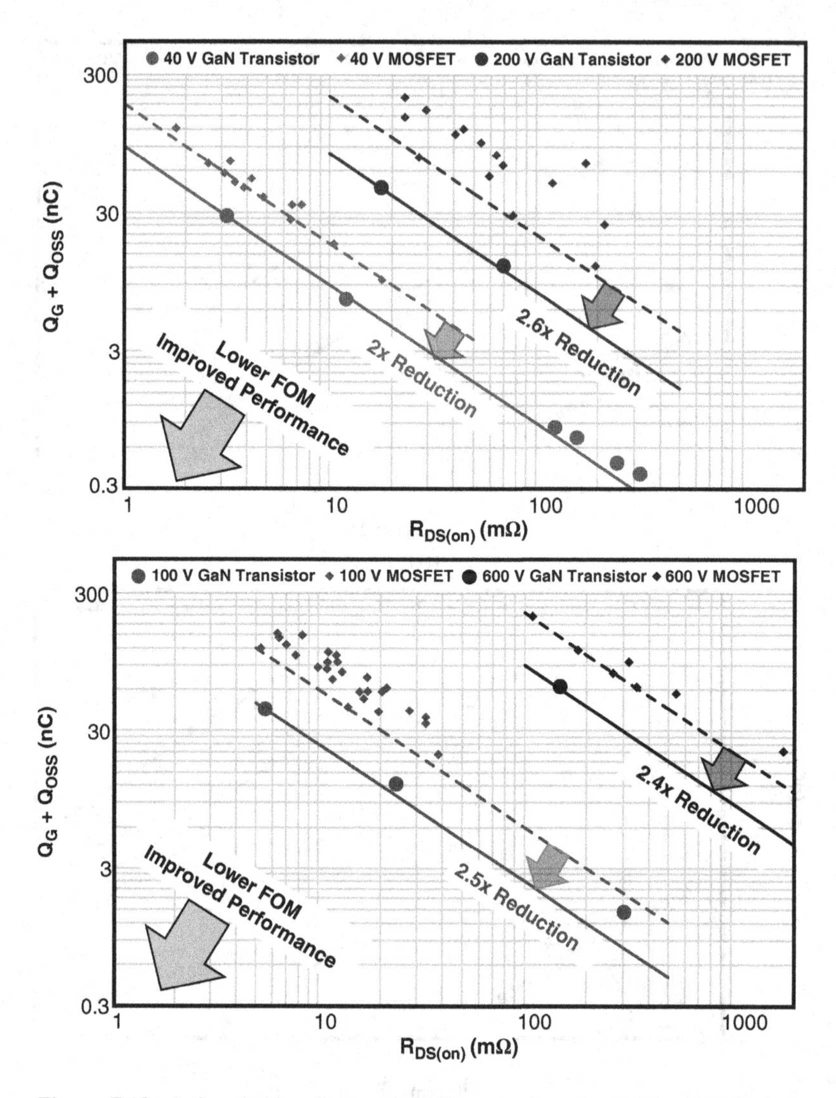

Figure 7.12 Soft-switching figure of merit comparison for GaN and Si devices

number of 48 V isolated hard-switching point-of-load (POL) converters to power the end loads. However, having a large number of regulated and isolated POLs, significantly increases the cost, volume, and complexity of the system. To simplify design and improve performance, the intermediate bus architecture (IBA) has been widely adopted [12,13]. A popular IBA approach, shown in Figure 7.13(b), employs a lower number of 48 V isolated bus converters that satisfy isolation requirements and supply an intermediate 12 V bus voltage. With the final regulation to the loads provided by smaller, more efficient, non-isolated POL buck converters, the bus converters can be operated as unregulated DC/DC transformers, improving efficiency and reducing cost.

The unregulated bus converter, also known as a DCX, or DC/DC transformer, can provide the highest efficiency by being designed to deliver power close to 100% of the operating

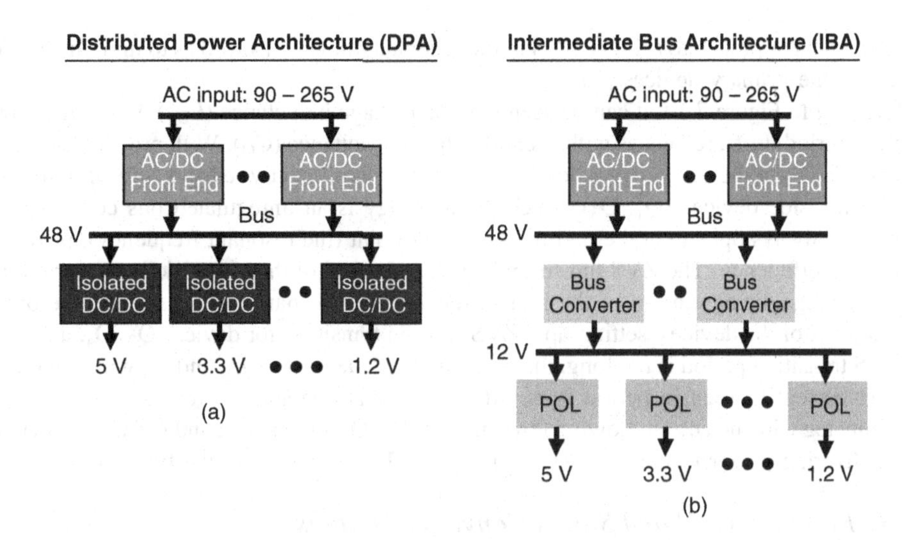

Figure 7.13 (a) Traditional distributed power architecture (b) Intermediate bus architecture

period, something not possible for a regulated converter that requires the converter to vary the power-delivery period to provide a regulated output voltage for various input voltages. The majority of today's bus converters use traditional hard-switching bridge topologies, which, due to large switching related losses, are forced to operate at lower frequencies where the bulky isolation transformer and output inductors occupy a large portion of board area. In an effort to improve power density and performance, the operating frequency can be increased through the use of resonant and soft-switching converters [14–17], shrinking passive components, and improving performance [18].

An experiment was undertaken to verify the superiority of enhancement-mode GaN transistors over Si MOSFETs in this application. The subject design was a high-frequency resonant converter, 48 V to 12 V unregulated isolated bus converter operating at a switching frequency of 1.2 MHz, and an output power of up to 400 W. The topology, shown in Figure 7.14, employs a soft-switching technique to achieve ZVS for the primary devices,

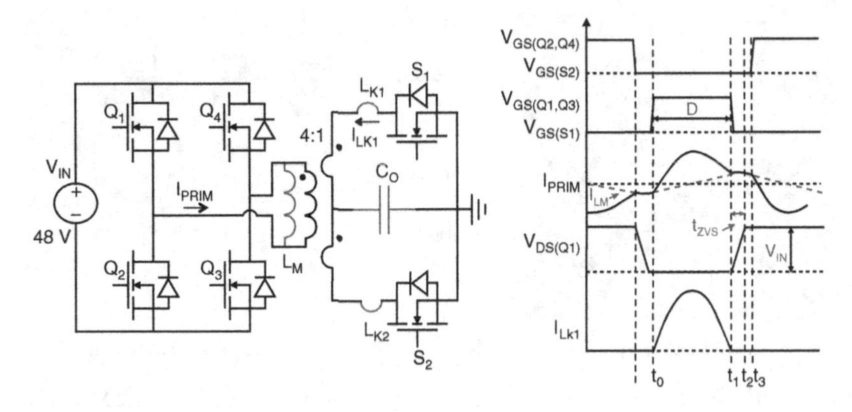

Figure 7.14 High-frequency resonant bus converter schematic and key waveforms

and a resonant approach to achieve ZCS in the secondary devices as well as to limit the turn-off current in the primary devices [14].

Referring to Figure 7.14, it can be seen that the leakage inductance (L_{K1}) during the power delivery period, t_0–t_1, resonates with a small output capacitance (C_O). With proper timing, this results in ZCS for the secondary-side device (S_1), and significantly reduces turn-off current in the primary-side devices (Q_1, Q_3). Since the topology is an unregulated bus converter, the circuit can always operate at the optimal operating point (the resonant frequency), providing the highest efficiency. The ZVS transition begins at the end of the power delivery period. For t_1–t_2, the magnetizing current of the transformer is used to charge and discharge the output capacitances of the devices, setting up a ZVS turn-on transition for devices Q_2, Q_4, and S_2. If the ZVS transition period is too long, the body diode of the devices Q_2 and Q_4 will turn on and conduct current as seen in period t_2–t_3. At time t_3, this operation is repeated for the other switching leg with the current flowing through switches Q_2, Q_4, and S_2, and leakage inductance L_{K2}, delivering power to the load while providing flux balancing in the transformer.

7.4.1 Resonant GaN and Si Bus Converter Designs

To obtain a direct comparison in performance between GaN transistors and Si MOSFETs in an isolated converter, having identical layouts and using the same topology are critical. Isolated DC-DC converter performance is heavily dependent on topology selection, PCB layout, number of PCB inner layers, copper weight of inner layers, and design of the transformer. To accurately compare the performance of GaN and Si in a high-frequency resonant bus converter application, the same circuit topology was used and a similar layout was maintained for both designs.

Two bus converters, shown in Figure 7.15, were built based on the schematic in Figure 7.14, to operate at a switching frequency of 1.2 MHz. Both PCBs were constructed with 12 layers and two-ounce copper thickness for all layers. To accurately compare only device performance, these converters both had the same transformer core material, core shape, and winding arrangement, designed from [18]. The placement of the primary-side input capacitors and secondary resonant capacitors were similar for the two designs to ensure similar parasitic inductances for the primary and secondary loops, with the only differences being those

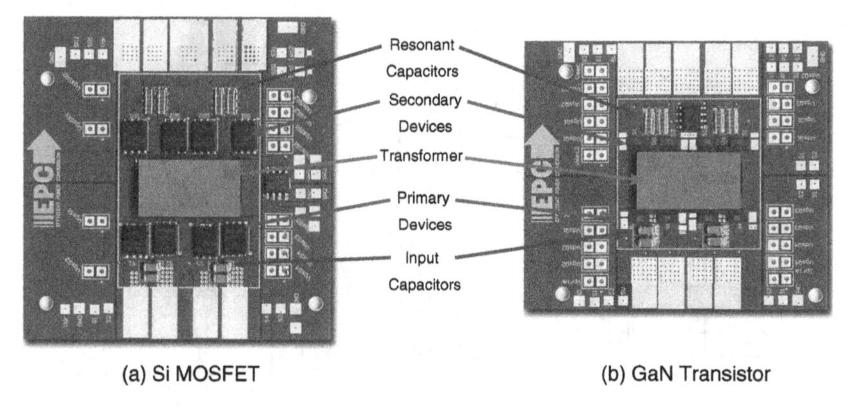

Figure 7.15 48 V to 12 V bus converters operating at a switching frequency of 1.2 MHz constructed with (a) silicon MOSFETs (b) gallium nitride transistors

Table 7.3 Device comparison between GaN and Si devices for primary devices for $V_{IN} = 48$ V, $V_{OUT} = 12$ V

	GaN transistor [10]	Si MOSFET [19]	
Voltage rating (V_{DSS})	100 V	80 V	
$R_{DS(on)}$	5.6 mΩ at 5 V	5.2 mΩ at 8 V[a]	
Q_G	5.8 nC at 5 V	25.9 nC at 8 V[a]	
Q_{GD} at V_{IN}	2.2 nC	8.1 nC[a]	
Q_{OSS} at V_{IN}	35 nC	62 nC[a]	
$Q_G \times R_{DS(on)}$	32.5 pC·Ω	134.7 pC·Ω	4.14 × reduction
$Q_{OSS} \times R_{DS(on)}$	196 pC·Ω	322.4 pC·Ω	1.64 × reduction
FOM_{SS} $(Q_{OSS} + Q_G) \times R_{DS(on)}$	228.5 pC·Ω	457.1 pC·Ω	2.00 × reduction

[a] Calculated from manufacturers datasheet curves.

introduced by the different packages of the Si MOSFETs and GaN transistors. It can be seen that by using GaN transistors with smaller size for the same on-resistance the active footprint area shrank significantly, reducing the power stage size by 30% compared to the Si MOSFET design.

7.4.2 GaN and Si Device Comparison

To obtain a direct comparison in performance between GaN transistors and Si MOSFETs in the high-frequency resonant bus converter application, GaN and Si devices with similar on-resistance were selected. Comparisons of the key parameters for the GaN transistors and Si MOSFETs are shown in Tables 7.3 and 7.4 for the primary and secondary devices respectively. The soft-switching FOM for the GaN devices ($(Q_{OSS} + Q_G) \times R_{DS(on)}$) is reduced by over a factor of two for both the primary and secondary devices, leading to proportionally shorter resonant transition periods and increased power delivery periods. The GaN transistor provides additional performance improvements in the form of reduced Miller charge, Q_{GD}, further reducing the turn-off switching losses incurred in the primary devices. As a further

Table 7.4 Device comparison between GaN and Si for secondary devices for $V_{IN} = 48$ V, $V_{OUT} = 12$ V

Parameter	GaN transistor [20]	Si MOSFET [21]	FOM ratio
Voltage rating (V_{DSS})	40 V	40 V	
$R_{DS(on)}$	3.2 mΩ at 5 V	2.9 mΩ at 5 V[a]	
Q_G	8.3 nC at 5 V	27.5 nC at 5 V[a]	
Q_{GD} at 20 V	2.2 nC	6.5 nC	
Q_{OSS} at 20 V	18.5 nC	40 nC	
$Q_G \times R_{DS(on)}$	26.6 pC·Ω	79.8 pC·Ω	3.00 × reduction
$Q_{OSS} \times R_{DS(on)}$	59.2 pC·Ω	116 pG·Ω	1.96 × reduction
FOM_{SS} $(Q_{OSS} + Q_G)$ $\times R_{DS(on)}$	85.8 pC·Ω	195.8 pC·Ω	2.28 × reduction

[a] Calculated from manufacturers' datasheet curves.

advantage, the LGA packaging of the GaN transistor has lower parasitic package inductance compared to the traditional Si MOSFET package (TDSON-8). When putting all these benefits together, efficient multi-MHz switching frequencies can be obtained through the use of advanced topologies combined with low-loss GaN transistors.

7.4.3 Zero-Voltage Switching Transition

The experimental waveforms for the ZVS transition period are shown in Figure 7.16 for the GaN and Si designs. By replacing a Si MOSFET with a GaN transistor, the ZVS transition period is reduced from 87 ns to 42 ns as a result of the reduced output charge enabled by GaN technology. Looking at the gate waveforms, it can also be seen that the gate drive speed for the GaN transistor is significantly faster than the Si MOSFET counterpart even when driven with a lower gate drive voltage, providing both a longer power delivery period and reduced gate losses. The Si MOSFET takes almost 100 ns to reach its steady-state gate voltage; this is over 10 times longer than the GaN device, and reflects the gate charge reduction enabled by GaN technologies.

The power delivery period for a half cycle is shown in Figure 7.17 for the GaN- and Si-based resonant bus converters. The effective duty cycles, D, which represent the power delivery period for each half cycle for the GaN and Si designs, were measured to be 42% and 34%, respectively. The soft-switching FOM from Equation 7.12 predicts the duty cycle gains,with a 50% reduction of the soft-switching FOM translating into a 50% reduction in the dead-time, including the ZVS transition and gate charging periods.

As the power delivery periods increase in duration, the circulating energy and resonant currents decrease, reducing the conduction losses in the resonant converter. For the resonant

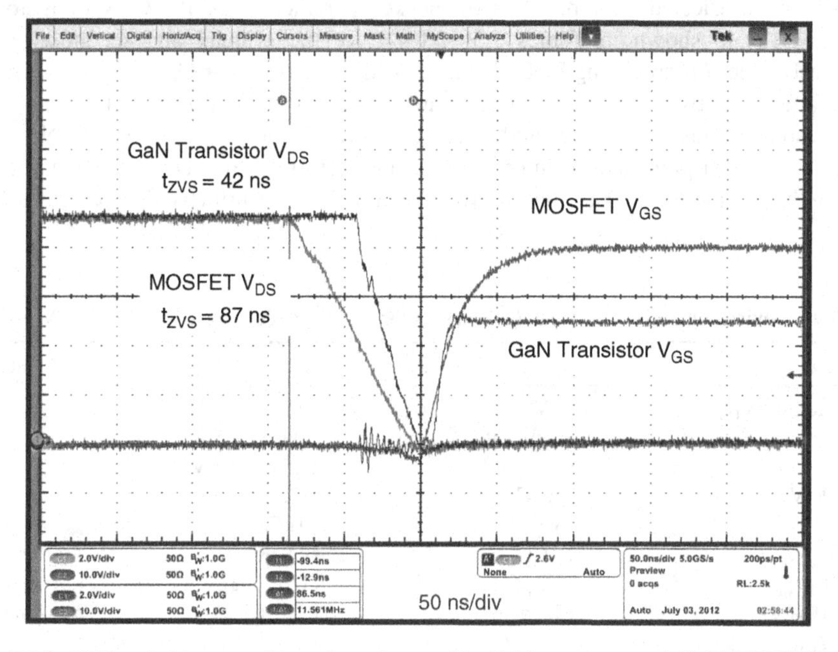

Figure 7.16 ZVS switching transitions for primary side GaN transistor and Si MOSFET designs at $f_{sw} = 1.2\,\text{MHz}$, $V_{IN} = 48\,\text{V}$, and $I_{OUT} = 26\,\text{A}$

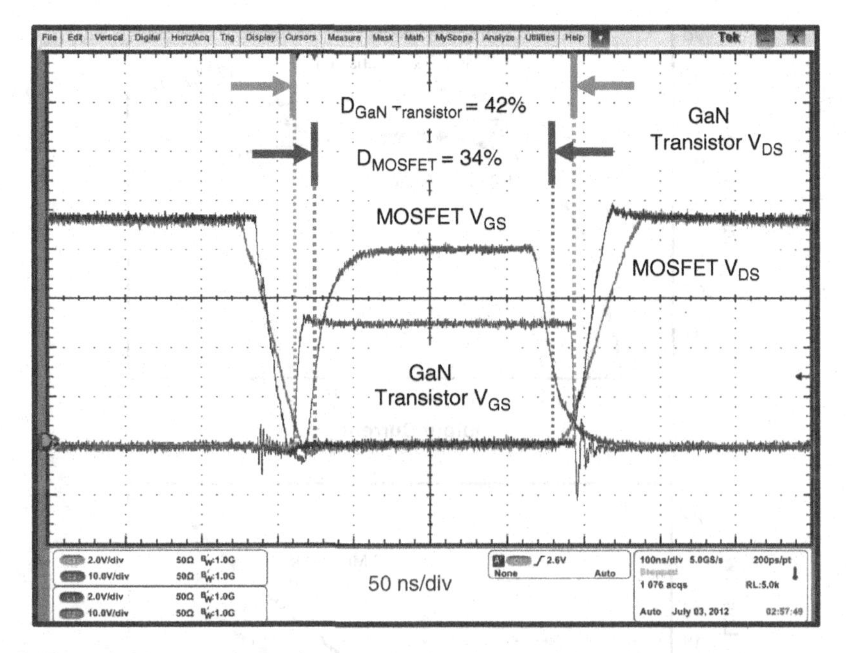

Figure 7.17 Switching waveforms showing effective duty cycle for primary side GaN transistor and Si MOSFET designs at $f_{sw} = 1.2\,MHz$, $V_{IN} = 48\,V$, and $I_{OUT} = 26\,A$

converter used in this design, the conduction losses, related to the resonant converters RMS current, I_{RES}, are inversely proportional to effective duty cycle, D, by:

$$I_{RES} \propto \frac{1}{\sqrt{D}} \tag{7.13}$$

For this design example, the increased duty cycle provided by GaN devices can reduce the conduction losses in the devices, transformer, PCB, and components by almost 20%.

7.4.4 Efficiency and Power Loss Comparison

The comparison in efficiency and power loss between the two designs operating at 1.2 MHz are shown in Figure 7.18. The GaN transistor-based converter offers a 1% improvement in peak efficiency over its Si MOSFET counterpart, resulting in about a 25% decrease in power loss. Since it is typical for bus converter designs to be thermally limited by losses under full load for a fixed converter size, the reduction in power loss translates directly into higher output power handling capability. For a design capable of dissipating 14 W, the GaN transistor converter can increase the output power capability by up to 65 W while maintaining the same total converter loss when compared to the benchmark Si MOSFET design. Assuming a 12 W maximum power loss for both designs, the output power of the GaN transistor based converter increases by 55 W, from 270 W to 325 W.

The loss breakdown for the 1.2 MHz designs at output currents of 2.5 A and 20 A is shown in Figure 7.19, which leads to the conclusion that the GaN technology improves efficiency for all

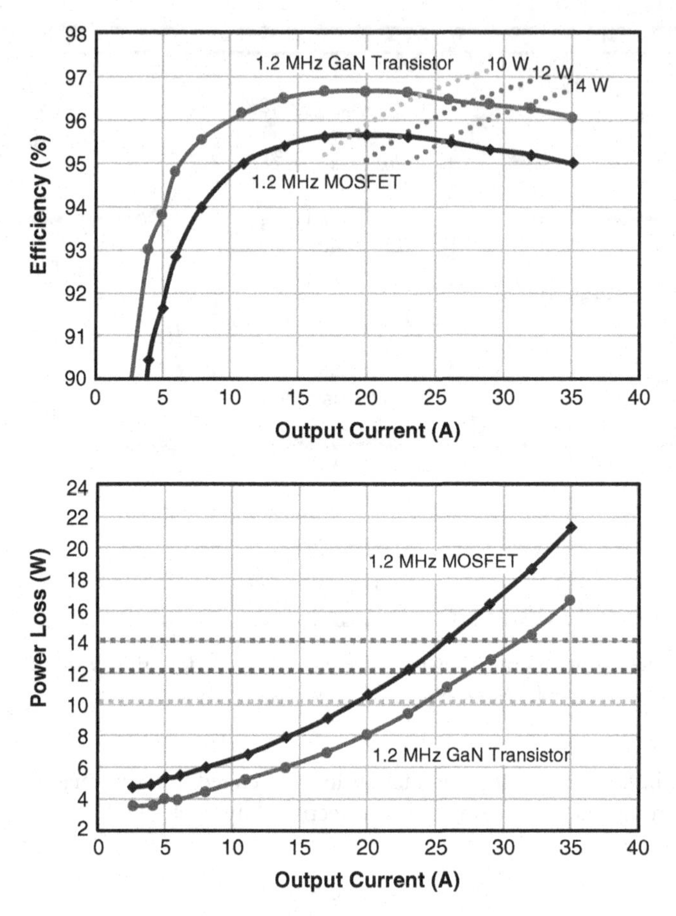

Figure 7.18 Experimental comparison between a GaN transistor and an Si MOSFET-based resonant bus converters ($V_{IN} = 48$ V, $V_{OUT} = 12$ V, $f_{sw} = 1.2$ MHz)

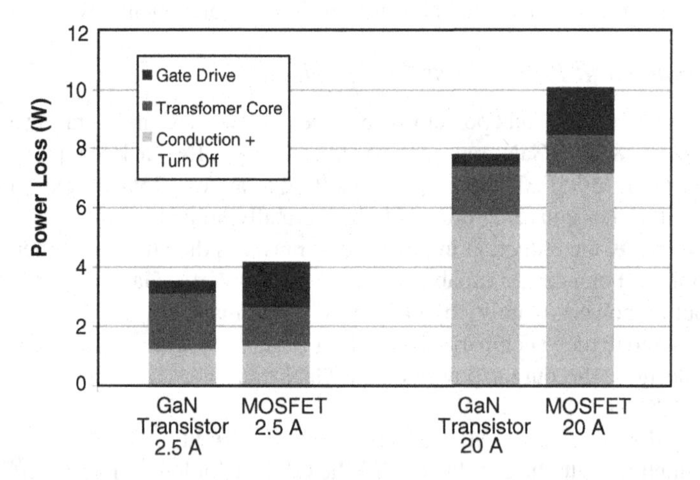

Figure 7.19 Loss breakdown of a GaN transistor and Si MOSFET-based resonant bus converter ($V_{IN} = 48$ V, $V_{OUT} = 12$ V, $f_{sw} = 1.2$ MHz)

load conditions. At lower currents, the gate driving losses dominate the transistor-related losses, and the GaN device's lower gate charge enables substantially reduced drive loss. At high currents, the conduction loss dominates total power loss, and the GaN-based converter's shorter ZVS dead-time and gate charging time provides lower conduction losses proportional to the effective duty cycle. The one area where the Si-based design offers lower loss is in the transformer core. The longer power delivery period provided by the GaN-based design increases the transformer flux density, leading to higher core loss, but the increase in transformer core loss is more than offset at lower currents by the gate drive loss savings, and at high currents by the combination of conduction and gate loss savings.

From the results at 1.2 MHz, it can be seen that the Si MOSFET converter is approaching its frequency limit because the ZVS transition time and gate charging time are becoming a significant portion of the overall period. To compare the frequency improvements possible with GaN transistors over Si MOSFETs, the converter frequency was reduced to 800 kHz for the Si MOSFET design, while increasing the eGaN transistor design to 1.6 MHz. In both cases, the core structure remained the same and was not optimized for the different operating frequencies. The efficiency and loss comparisons between the designs are shown in Figure 7.20,

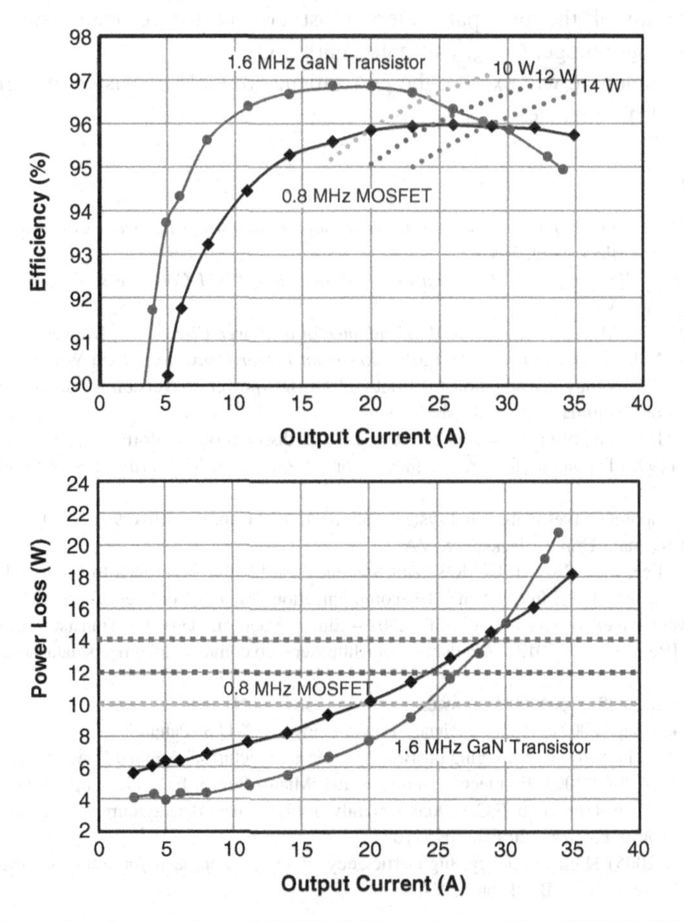

Figure 7.20 Comparison between $f_{SW} = 1.6$ MHz GaN transistor and $f_{sw} = 800$ kHz Si MOSFET-based $V_{IN} = 48$ V, $V_{OUT} = 12$ V resonant bus converter

with the GaN transistor-based design offering a 0.9% peak efficiency improvement and less power loss up to an output current of 29 A. The sharp drop-off in efficiency at currents above 20 A for the GaN transistor-based converter is a result of increased AC transformer winding losses, and an effective duty cycle reduction. Conversely, the flattening out of the efficiency for the 800 kHz Si MOSFET design was a result of reduced AC transformer winding losses and a higher effective duty cycle at the lower frequency.

7.5 Summary

It was shown in previous chapters that enhancement-mode GaN transistors have a distinct advantage over silicon MOSFETs in hard-switching applications due to reduced Q_{GD} and Q_{GS2}, both of which are critical in hard-switching applications, but have little impact in resonant and soft-switching converters. In this chapter, the benefits of GaN technology as applied to resonant and soft-switching applications have been discussed, and the in-circuit superiority of GaN transistors demonstrated in a 48 V bus converter operating at 1.2 MHz. This chapter also discussed a simple soft-switching FOM that allows designers to quickly compare device technologies for use in resonant and soft-switching applications. The soft-switching FOM is made up of the two parameters most critical for resonant and soft-switching applications: output charge, Q_{OSS}, and gate charge, Q_G.

In the next chapter we will explore the performance of GaN transistors designed for power conversion at RF frequencies.

References

1. Lee, F.C. (1989) *High-Frequency Resonant, Quasi-Resonant, and Multi-Resonant Converters*, Virginia Power Electronics Center, Blacksburg, VA.
2. Lee, F.C. (1989) *High-Frequency Resonant and Soft-Switching PWM Converters*, Virginia Power Electronics Center, Blacksburg, VA.
3. Erickson, R.W. and Maksimović, D. (2001) *Fundamental of Power Electronics*, Kluwer, Norwell, MA.
4. Kazimierczuk, M.K. and Czarkowski, D. (2011) *Resonant Power Converters*, John Wiley & Sons, NJ.
5. Yang, B. (2003) Topology investigation for front end DC/DC power conversion for distributed power system, Ph.D. dissertation, Virginia Tech, Blacksburg, VA.
6. Vorperian, V. (1984) Analysis of resonant converters, Ph.D. dissertation, California Inst. Technol., Pasadena, CA.
7. Severns, R.P. (1992) Topologies for three-element resonant converters. *IEEE Transactions, Power Electron*, **7** (1), 89–98.
8. Fu, D. (2010) Topology investigation and system optimization of resonant converters, Ph.D. dissertation, Virginia Polytechnic Inst. State Univ., Blacksburg, VA.
9. Liu, K.H. and Lee, F.C. (Nov. 1984) Resonant Switches – a Unified Approach to Improved Performances of Switching Converters, IEEE International Telecommunications Energy Conference, pp. 334–341.
10. Efficient Power Conversion Corporation, EPC2001 – Enhancement-mode Power Transistor, EPC2001 datasheet, March 2011 [Revised Jan. 2013]. Available from http://epc-co.com/epc/documents/datasheets/EPC2001_datasheet.pdf.
11. Infineon (21 Oct. 2009) OptiMOS™ Power-Transistor, BSC060N10NS3 G datasheet.
12. Schlecht, M. (11 Sep. 2007) "High Efficiency Power Converter," U.S. Patent 7,269,034.
13. White, R.V. (9–13 Feb. 2003) "Emerging on-Board power architectures," Applied Power Electronics Conference and Exposition (APEC) 2003, Eighteenth Annual IEEE, Miami Beach, FL, vol. 2, pp. 799–804.
14. Ren, Y., Xu, M., Sun, J. and Lee, F.C. (2005) A family of high power density unregulated bus converters. *IEEE Transactions, Power Electron*, **20** (5), 1045–1054.
15. Ren, Y. (April 2005) High frequency, high efficiency two-stage approach for future microprocessors, Ph.D. Dissertation, Virginia Tech, Blacksburg, VA.

16. Ren, Y., Lee, F.C. and Xu, M. (27 March 2007) Power Converters Having Capacitor Resonant With Transformer Leakage Inductance, U.S. Patent 7,196,914.
17. Vinciarelli, P. (5 Dec. 2006) Point of load sine amplitude converters and methods, U.S. Patent 7,145,786.
18. Reusch, D. (2012) High frequency, high power density integrated point of load and bus converters, Ph.D. Dissertation, Virginia Polytechnic Inst. State Univ., VA.
19. Infineon (22 Oct. 2009) OptiMOS™3 Power-Transistor, BSC057N08NS3G datasheet.
20. Efficient Power Conversion Corporation, EPC2015 – Enhancement-mode Power Transistor, EPC2015 datasheet, March 2011 [Revised Jan. 2013]. Available from http://epc-co.com/epc/documents/datasheets/EPC2015_datasheet.pdf.
21. Infineon (22 Oct. 2009) OptiMOS™3 Power-Transistor, BSCO27N04LSG datasheet.

8

RF Performance

8.1 Introduction

The main focus to this point in the book has been the switching capabilities of GaN transistors. Now, the RF capabilities of these same GaN transistors and, in particular, enhancement-mode transistors will be examined, highlighting specific RF applications that can benefit from their adoption.

High electron mobility transistors (HEMT), using GaN as a semiconductor material, are available from Cree, Nitronex, NXP, Integra, RFMD, and TriQuint, among others. These transistors are all depletion-mode, which is not as limiting for RF power amplifiers as it is for switching converters. This is because the potential of device failure upon power up, due to the short condition, can be mitigated easily in the RF design, which is not the case for a switching converter. Depletion-mode transistors require additional circuitry for gate circuit biasing, due to the negative gate voltage requirements to regulate the drain current, which is also viewed as a disadvantage in cases where enhancement-mode devices are an option.

The main alternative today for GaN RF transistors operating in the range of 500 MHz to 3 GHz is the laterally diffused metal oxide semiconductor (LDMOS) FET made using silicon. Compared to LDMOS transistors, GaN is well-suited for RF transistors for many of the same reasons as for switching applications [1–3]. GaN transistors exhibit superior RF performance over LDMOS in general, most notably with respect to power density [4], frequency range (bandwidth) [1,5], and noise figure. This leads to improvements in RF power capability [2,6] with transistors being specified over a very wide frequency range [7]. In addition, the lower input and output capacitances lead to higher impedances that allow for higher drain efficiencies and reduced impedance transformation ratios required for matching. Both these factors lead to improvements in amplifier efficiency, size reduction, and ultimately cost.

When used in pulsed-RF applications, the bias circuit will also need to be operational before main power and RF are applied, to prevent starting with very high currents that can potentially damage the circuit. Therefore, RF circuits can benefit from using enhancement-mode transistors for many of the same reasons as in switching converters.

The measurement and performance metrics used for RF transistors differ significantly from switching devices. These metrics will be presented, along with their relevance, and guidance on how to measure and use them.

GaN Transistors for Efficient Power Conversion, Second Edition.
Alex Lidow, Johan Strydom, Michael de Rooij, and David Reusch.

Companion Website: http://www.wiley.com/go/gan_transistors

Table 8.1 Definitions of terms used in RF analysis and design

V_{GSQ}	RF circuit gate voltage quiescent bias point
I_{DQ}	Drain current in the transistor at the quiescent operating point
P_{DQ}	Power losses at the quiescent operating point
P_{DC}	DC power delivered to the RF transistor
$P_{\mathbf{RFout}}$	Output RF power
η_D	Drain efficiency – the ratio of P_{RFou}/P_{DC}
s_{11}	Input port reflection coefficient: the percentage of the input incident wave that is reflected back from the input port
s_{12}	Reverse gain: the percentage of the output port incident wave that is reflected to the input port
s_{21}	Forward gain: the percentage of the input port incident wave that is reflected to the output port
s_{22}	Output port reflection coefficient: the percentage of the output incident wave that is reflected back from the output port
K	Rollett stability factor
C_S	The source-side stability circle center on a Smith chart
C_L	The load-side stability circle center on a Smith chart
C_A	Constant available gain circle center on a Smith chart
R_S	The source-side stability circle radius on a Smith chart
R_L	The load-side stability circle radius on a Smith chart
R_A	The radius of the constant available gain circle on a Smith chart
Γ_{in}	Input reflection coefficient of the transistor
Γ_{out}	Output reflection coefficient of the transistor
Γ_S	Input-side matching reflection coefficient
Γ_L	Output-side matching reflection coefficient
G_T	Transducer power gain
G_{TU}	Unilateral transducer power gain
U	Unilateral figure of merit
g_u	Normalized unilateral transducer gain
G_A	Available gain
G_{MSG}	Maximum stable gain of the transistor
X	Matching network series reactance
B	Matching network shunt susceptance

8.2 Differences Between RF and Switching Transistors

The main difference between switching transistors and RF transistors is that RF transistors are designed to work best in the linear region of the transfer characteristic to maximize RF power gain and minimize RF signal distortion. Switching devices are designed to work best in the on-state and transition states to minimize losses [8–14]. For performance comparison, RF devices have different evaluation metrics [1] from switching devices, among which are power gain, linearity or 1 dB compression [8], and drain efficiency. The power gain determines how much the transistor amplifies the input signal with power. The 1 dB compression point determines the maximum output power that the transistor can deliver without distorting the signal. The drain efficiency determines how efficient the transistor is at amplifying.

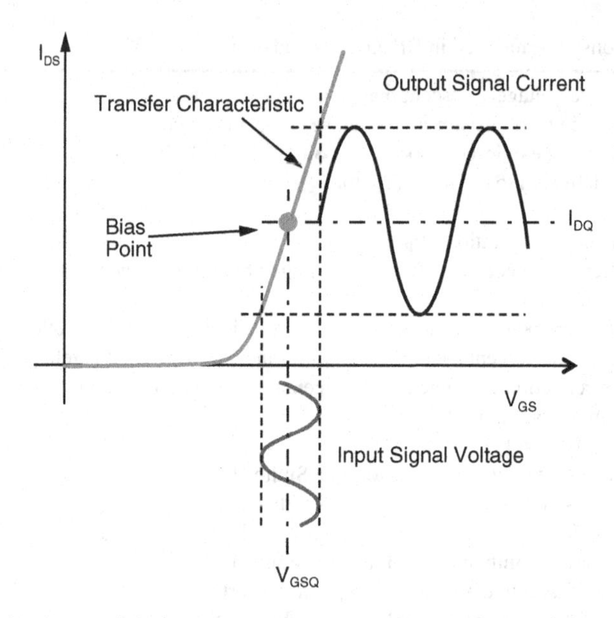

Figure 8.1 Transfer characteristic for an enhancement-mode transistor showing the bias point and input voltage signal with corresponding output current

Typically, RF devices are biased to an operating point onto which the RF signal is superimposed. To bias the transistor, a drain-to-source current is established such that the voltage remains at the supply voltage. Figure 8.1 illustrates this concept, using the transistor transfer characteristic. The bias voltage and current are named quiescent points and denoted as V_{GSQ} and I_{DQ} respectively, and typically are provided in datasheets for baseline performance reporting.

The bias point will be associated with losses ($P_{DQ} = I_{DQ} \cdot V_{Supply}$), and as such, RF devices have a higher operating power (P_{DQ}) loss ratio with respect to power delivered (P_{RFout}), compared to switching devices. The ratio P_{RFout}/P_{DC} is referred to as drain efficiency (η_D) where P_{RFout} is the output RF power and P_{DC} is the DC power delivered to the transistor. When employing an RF FET as a Class A amplifier, this power loss ratio reaches a theoretical maximum of 50%, but as discussed in earlier chapters, a switching converter can have efficiencies as high as 98%. This means that the thermal dissipation of an RF device is significantly higher than equivalently sized switching devices, and necessitates that RF transistors have the ability to efficiently dissipate heat to the environment. Figure 8.2 shows the difference between a packaged RF FET (left) and a chip-scale switching FET (right), both with similar drain voltage and current ratings.

Another key difference between RF devices and switching devices is in how they are characterized. RF transistors are characterized as part of a transmission line in terms of incident and reflected waves, whereas switching devices are characterized in terms of energy commutation. This opens up many new terms and definitions unfamiliar to many power circuit designers. The common metric used to characterize RF devices are the s-parameters, which represent a measure of incident, reflection, and transmission of electromagnetic waves. S-parameters will also be used in this chapter to characterize the enhancement-mode GaN transistor originally designed for switching converter applications.

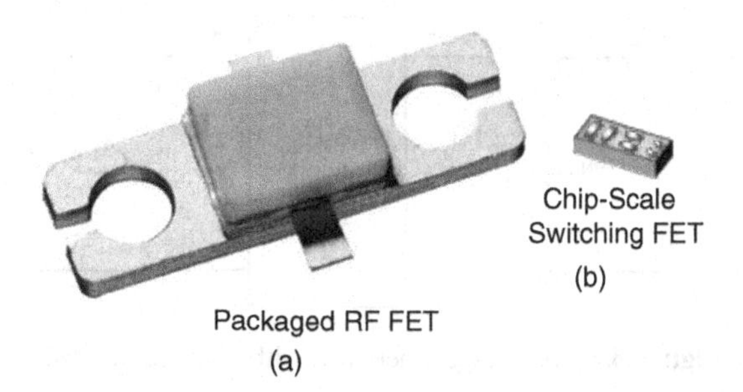

Figure 8.2 (a) An RF-packaged FET [15] and (b) the equivalent chip-scale switching FET [19], with similar voltage and current rating

8.3 RF Basics

Before delving into RF transistor measurement and analysis, some basics need to be reviewed. In this chapter, all discussions will be limited to two-port networks, as this is sufficient to describe a transistor. Figure 8.3 shows the basic diagram for a two-port network showing the incident waves, denoted a_1 and a_2, and reflected waves denoted b_1 and b_2. Any incident wave at a port can be reflected to either of the ports. For example, an incident wave on port 1 (a_1) can reflect from port 1 (b_1) and/or port 2 (b_2).

An s-parameter is defined as the ratio between reflected (b_1 and b_2) and incident (a_1 and a_2) waves and is given by Equation 8.1.

$$s_{nm} = \frac{b_n}{a_m} \tag{8.1}$$

And s_{nm} is a complex number with the general form:

$$s_{nm} = \mathcal{Re}(s_{nm}) + i \cdot \mathcal{Jm}(s_{nm}) \tag{8.2}$$

In this chapter, port 1 will be designated as the input port, or gate, and port 2 as the output port, or drain.

From the s-parameters, it is possible to derive useful characteristics of the two-port network such as impedance, gain, and isolation. (The methodology is described in detail in [8].)

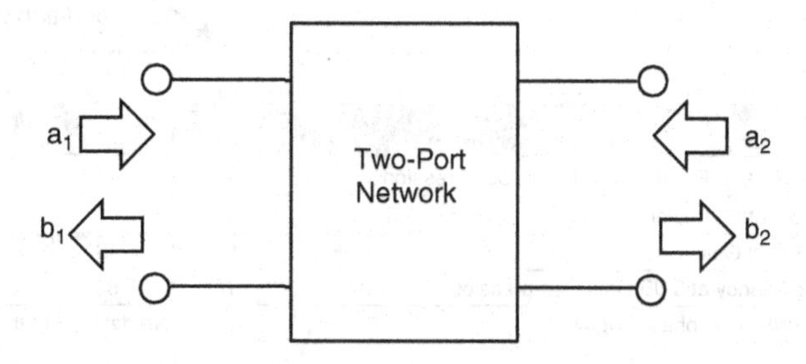

Figure 8.3 Two-port network with (a) incident waves and (b) reflected waves

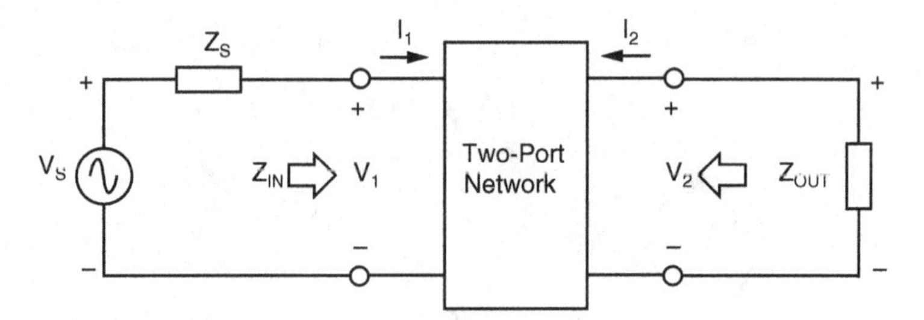

Figure 8.4 A two-port network with source and load

Figure 8.4 shows the two-port network connected to a load and source, and gives the input and output impedance for those ports.

The Smith chart is a useful tool [8], which simplifies the conversion of s-parameters into impedances, and will be extensively used in this chapter.

8.4 RF Transistor Metrics

The main metric for RF transistor performance evaluation is RF power gain. Maximum power gain is defined by the limit of linear performance for a transistor. Table 8.2 shows data from the Nitronex NPT1012 datasheet [7] showing the key RF metrics for the depletion-mode GaN-on-silicon HEMT under certain operating conditions.

RF power gain is a measure of how much power is increased or decreased by when incident on a port. Mathematically it is expressed as:

$$G = \frac{P_{out}}{P_{in}} \tag{8.3}$$

Gain can also be expressed logarithmically, with units of decibel (dB) as follows:

$$G(dB) = 10 \cdot \log\left(\frac{P_{out}}{P_{in}}\right) \tag{8.4}$$

Table 8.2 Data from Nitronex NPT1012 datasheet [7] showing the Key RF metrics for the transistor

Operating Conditions

RF Specifications (CW, 3000 MHz): V_{DS} = 28V, I_{DQ} = 225mA, T_C = 25°C, Measured in Nitronex Test Fixture

Symbol	Parameter	Min	Typ	Max	Units
P_{3dB}	Average Output Power at 3 dB Gain Compression	43	44	-	dBm
P_{1dB}	Average Output Power at 1 dB Gain Compression	-	43	-	dBm
G_{SS}	Small Signal Gain	12	13	-	dB
η	Drain Efficiency at 3 dB Gain Compression	57	65	-	%
VSWR	10:1 VSWR at all phase angles	No damage to the device			

Key Metric

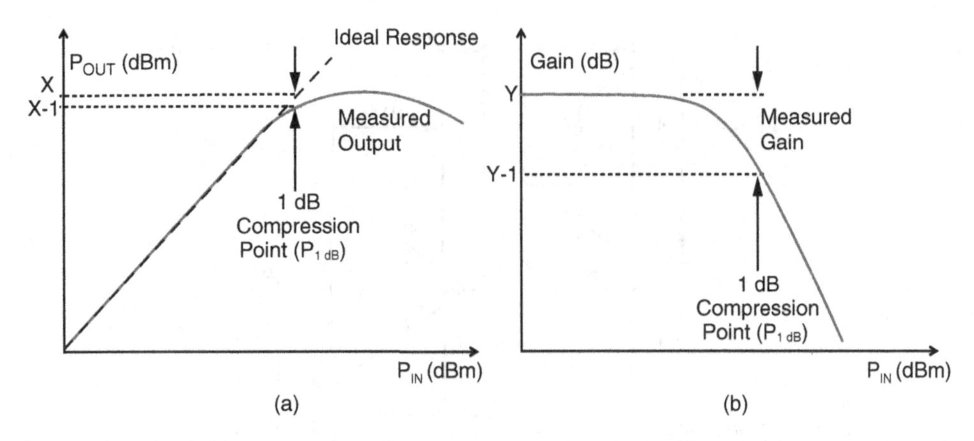

Figure 8.5 Graphical representation of the definition of linearity based on (a) the 1 dB compression point and (b), linearity on the gain graph

Using the definition of gain, linearity can be defined for the amplifier as having a fixed gain value and is characterized by the linear relationship between the input power and the output power. At its limit, there is a loss in gain due to amplification saturation. Linearity is also known as linear dynamic range with its limit being the 1 dB compression point [8]. For an RF transistor, the gain will be a constant as a function of input power until it exceeds a specific value. The 1 dB compression point is defined as the point when the measured amplifier output power (dB) deviates by 1 dB from the ideal predicted power, and is shown in Figure 8.5. Figure 8.5(a) is the output power as a function of the input power, and Figure 8.5(b) shows the same result for gain as a function of input power. Above a specific input power level, the gain will begin to decrease and the output power will no longer be a linear function of the input power.

Typically, the linearity of a transistor is provided in a datasheet at a specific frequency and bias condition. Power gain is given over a frequency range, based on a specific bias setting. Small-signal s-parameters can be used to design a Class A RF power amplifier, but large-signal s-parameters are needed to predict power performance.

8.4.1 Determining the High-Frequency Characteristics of RF FETs

To determine the RF characteristics of a transistor requires a few steps. The procedure starts with s-parameter measurements of the device itself. The s-parameters are used to test for stability and to identify if the device is unilateral (negligible reverse gain) or bilateral (reverse gain is high enough to affect stability). Once the stability criteria have been determined, an amplifier can be designed, typically a Class A or Class AB.

Small-signal s-parameters are measured using a vector network analyzer (VNA) with the transistor under specific bias conditions. Several measurements may be required to determine the bias conditions that yield the highest performance metrics.

To measure the RF characteristics requires that the device be mounted to a test fixture. The test fixture will need to be calibrated using a set of standards to obtain the device-only s-parameters. The thru-reflect-line (TRL) [16] and short-open-load-thru (SOLT) methods are the most popular methods of calibration. The calibration process and accuracy between the two methods is well documented in [17,18].

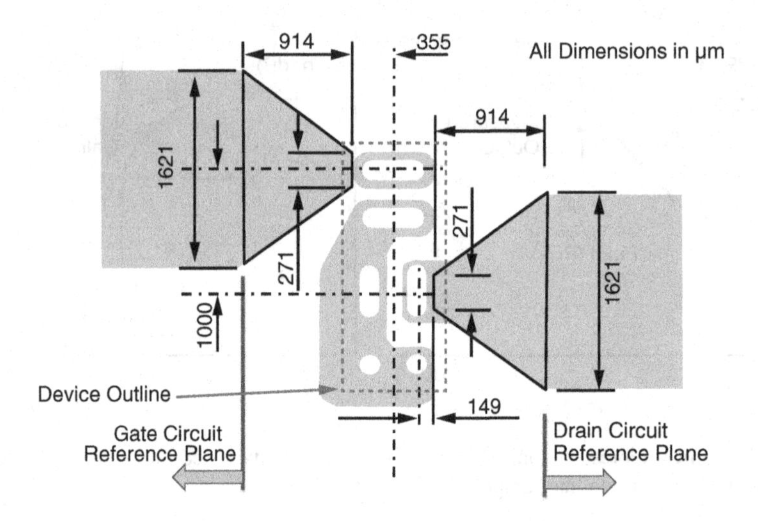

Figure 8.6 Reference plane design for RF connection to the EPC8009 [19] FET

Figure 8.6 shows a reference plane design suitable for the EPC8009 [19] enhancement-mode switching FET that can be used to test its RF characteristics. Due to the small size of this device and its connections, a reducing taper microstrip is required to make the connection. The taper's interface to the 50 Ω microstrip transmission line in this example is based on a 30 mil thick, one-once copper, Rogers 4350 substrate [20].

8.4.2 Pulse Testing for Thermal Considerations

It has already been mentioned that RF devices have a high ratio of power dissipation to power output, and therefore require substantial cooling. In the case where power dissipation exceeds the capability of the device, pulsed-mode testing can be employed.

Most RF amplifiers are operated with a continuous RF signal and bias. This is known as continuous wave (CW) operation. Pulse testing is used to lower the average power dissipation in the device. The bias pulse and pulsed RF signal used in pulse testing are shown in Figure 8.7.

It is important to maintain the drain bias current stable during the on-state of the pulse as changes can cause inaccurate measurements. This is difficult to achieve, particularly for Class A amplifiers, as the drain current will increase in response to an increase in RF power without any means to differentiate between what is the bias component and what is the RF

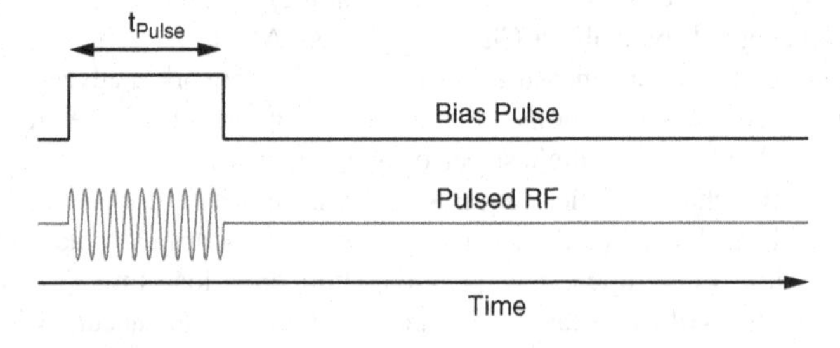

Figure 8.7 Bias pulse and pulsed RF signal for pulsed RF testing

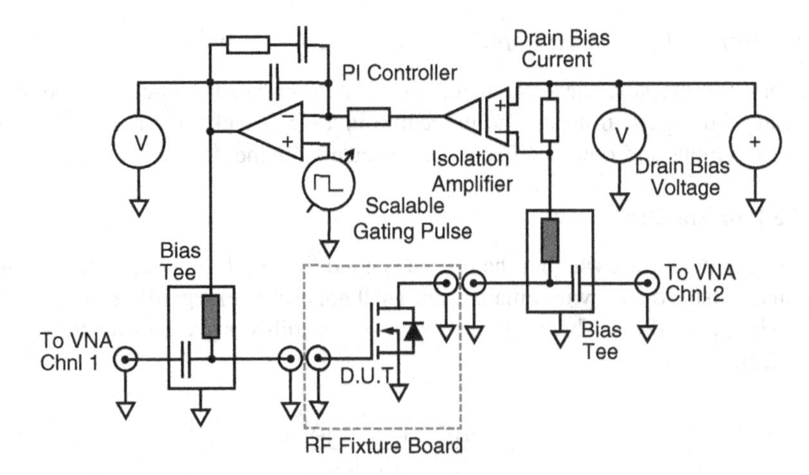

Figure 8.8 Schematic block diagram for transistor bias control under pulsed RF testing

component. In addition, rapid changes in gate voltage can lead to unwanted ringing of the drain and can cause bias instability and oscillations. The choice of bias tees for pulse testing is also critical because the frequency response of the bias tees must satisfy both RF and bias requirements.

A dedicated pulse controller similar to the schematic block diagram shown in Figure 8.8 can be used for RF testing FETs. Using an isolation amplifier, the drain current is measured and used with a proportional-integral (PI) controller to regulate the gate voltage to maintain the drain current. The PI controller is also controlled by the gate pulse to avoid rapid gate transitions.

For higher power RF measurements, the pulse controller can be adapted such that gate voltage can be fed to a second device while the drain bias current for the second device is not measured by the controller [21]. The first device will not be exposed to RF thereby providing a stable reference for the device under RF testing. This setup is shown in Figure 8.9.

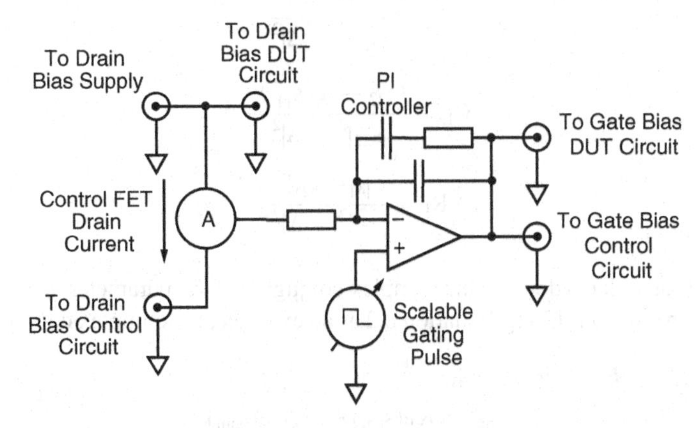

Figure 8.9 Power RF testing using the pulse controller having a reference device and a device under test

8.4.3 Analyzing the S-Parameters

With the ability to measure the s-parameters, the data now needs to be analyzed so that an amplifier can be designed from it. The procedure involves checking for stability issues and determining the input and output reflection coefficients for the device.

8.4.3.1 Test for Stability

It is important to determine whether the device is conditionally or unconditionally stable. An unconditionally stable device will remain stable (will not oscillate) regardless of the impedance presented to its gate or drain. The test for unconditional stability is given by the Rollett stability factor (K) [22]:

$$K = \frac{1 - |s_{11}|^2 - |s_{22}|^2 + |\Delta|^2}{2 \cdot |s_{12} \cdot s_{21}|} \geq 1 \tag{8.5}$$

and

$$|\Delta| \leq 1 \tag{8.6}$$

where

$$\Delta = s_{11} \cdot s_{22} - s_{12} \cdot s_{21} \tag{8.7}$$

If this stability criterion for either K or $|\Delta|$ cannot be satisfied, then the transistor will be defined as conditionally stable. This means that any design using the transistor must avoid the unstable region. Stability circle plots on a Smith chart [8] are used to determine where the unstable regions are located. The stability circles are given by the following equations:

$$C_S = \frac{\left(s_{11} - \Delta \cdot s_{22}^*\right)^*}{|s_{11}|^2 - |\Delta|^2} \tag{8.8}$$

$$R_S = \frac{|s_{12} \cdot s_{21}|}{|s_{11}|^2 - |\Delta|^2} \tag{8.9}$$

$$C_L = \frac{\left(s_{22} - \Delta \cdot s_{11}^*\right)^*}{|s_{22}|^2 - |\Delta|^2} \tag{8.10}$$

$$R_L = \frac{|s_{12} \cdot s_{21}|}{|s_{22}|^2 - |\Delta|^2} \tag{8.11}$$

The superscript asterisk (*) denotes the complex conjugate of the parameter, also known as the reflection of the parameter. Using Equation 8.2 as an example, the complex conjugate for s_{nm} is given as:

$$s_{nm}^* = \mathcal{R}e(s_{nm}) - i \cdot \mathcal{J}m(s_{nm}) \tag{8.12}$$

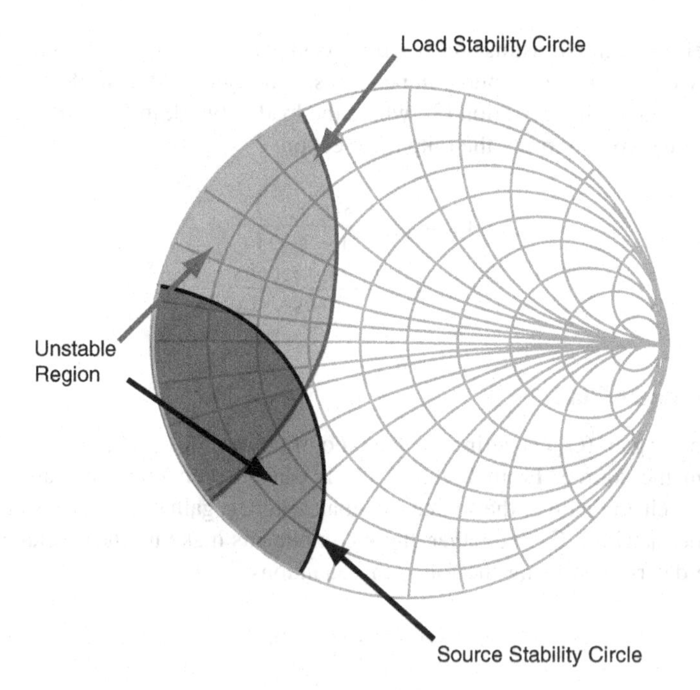

Figure 8.10 Stability circle plot showing unstable regions for both source and load ports

The unstable region falls inside the stability circle, as shown for example in Figure 8.10 (for a tutorial on the use of Smith Charts see reference [8]).

8.4.3.2 Transistor Input and Output Reflection

The RF transistor will ultimately be placed in an amplifier circuit with input and output matching networks that transform the standard source impedance (Z_0) and standard load impedance (Z_0) to desired values (Γ_S and Γ_L), as shown in Figure 8.11. Matching networks will be discussed in detail later in this chapter.

Although they share the same nomenclature, the transistor input and output reflection coefficients are not simply given by s_{11} and s_{22} (reflection coefficients) as one might expect, but are given by Γ_{in} and Γ_{out} for the input (gate circuit) and output (drain circuit) respectively. This

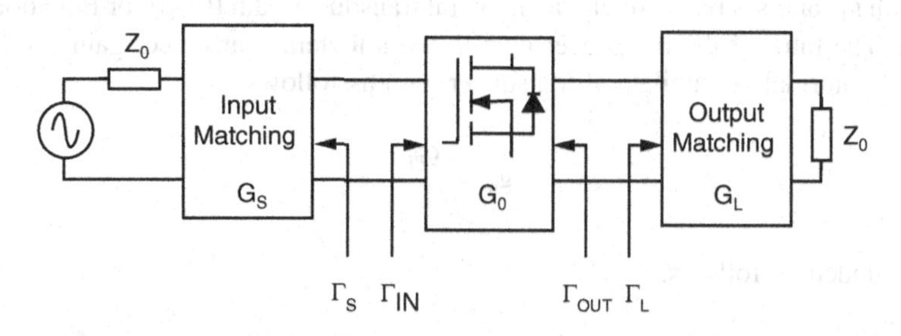

Figure 8.11 Basic amplifier structure with input and output matching with reflection coefficients for the FET and matching networks

is due to the effect of the transmission coefficients s_{12} and s_{21} cross influencing the input and output by way of the load and source impedances. This can be seen in the equations for the transistor input and output reflection [8], where the load network impacts the input reflection, and the source network impacts the output reflection:

$$\Gamma_{in} = s_{11} + \frac{s_{12} \cdot s_{21} \cdot \Gamma_L}{1 - s_{22} \cdot \Gamma_L} \tag{8.13}$$

$$\Gamma_{out} = s_{22} + \frac{s_{12} \cdot s_{21} \cdot \Gamma_S}{1 - s_{11} \cdot \Gamma_S} \tag{8.14}$$

8.4.3.3 Transducer Gain

Transducer power gain (G_T) is defined as the ratio of power delivered to the load to the power available from the source. From Figure 8.11, it can be seen that there are multiple gain components which are: gain of the source side matching G_S, gain of the transistor G_0, and gain of the load side matching G_L. Together, these components make up the transducer gain (G_T), which can be determined using the following equations:

$$G_T = G_S \cdot G_0 \cdot G_L \tag{8.15}$$

and written using s-parameters:

$$G_S = \frac{1 - |\Gamma_S|^2}{|1 - \Gamma_{in} \cdot \Gamma_S|^2} \tag{8.16}$$

$$G_0 = |s_{21}|^2 \tag{8.17}$$

$$G_L = \frac{1 - |\Gamma_L|^2}{|1 - \Gamma_{out} \cdot \Gamma_L|^2} \tag{8.18}$$

8.4.3.4 Unilateral/Bilateral Transistor Test

A unilateral transistor is defined as one where s_{12} is very small relative to s_{21} [8,23]. Since a value of zero for s_{12} is not physically possible, we can perform a test to determine if a transistor can be regarded as unilateral or bilateral. Using Equations 8.16 and 8.18 and substituting Γ_{in} and Γ_{out} with s_{11} and s_{22} respectively, a unilateral transducer gain (G_{TU}) for Equation 8.15 can be derived. The ratio of the transducer gain to the unilateral transducer gain can be used to determine the normalized unilateral transducer gain as follows:

$$g_u = \frac{G_T}{G_{TU}} \tag{8.19}$$

which is bounded as follows:

$$\frac{1}{(1 + U)^2} < g_u < \frac{1}{(1 - U)^2} \tag{8.20}$$

where U is the unilateral figure of merit and given by the following equation:

$$U = \frac{|s_{11}| \cdot |s_{12}| \cdot |s_{21}| \cdot |s_{22}|}{\left(1 - |s_{11}|^2\right) \cdot \left(1 - |s_{22}|^2\right)} \tag{8.21}$$

A transistor can be regarded as unilateral if g_u is within 10% of unity, otherwise it is considered bilateral. Also, a unilateral transistor will always be stable by definition, as the device effectively does not have a feedback mechanism. A unilateral amplifier design is simplified greatly as Γ_{in} and Γ_{out} (Equations 8.13 and 8.14) can be reduced to s_{11} and s_{22}, respectively. The unilateral solution will not be covered in this book, as the procedure to design an amplifier using a bilateral transistor also can be adapted for a unilateral transistor.

8.5 Amplifier Design Using Small-Signal S-Parameters

The essence of an amplifier design is to determine the input (Γ_S) and output (Γ_L) reflection coefficients for the matching networks of the transistor. The basis for the design can be maximum gain, or a specific gain that is within the capability of the transistor.

The example here assumes a bilateral transistor, and the procedure followed will be based on a specific gain requirement. The unique solution for an unconditionally stable transistor that yields the maximum transducer gain (G_{Tmax}) can also be determined using this procedure. The specific maximum transducer gain needs to be determined to know its gain limit. For an unconditionally stable transistor, the maximum transducer gain is given by:

$$G_{Tmax} = \frac{|s_{21}|}{|s_{12}|} \cdot \left(K - \sqrt{K^2 - 1}\right) \tag{8.22}$$

The unconditionally stable transistor maximum transducer gain amplifier solution is defined as a conjugately matched amplifier, where the matching networks are designed to have zero reflection. It will be the complex conjugate of the transistor ports and can be written in the form:

$$\Gamma_{in} = \Gamma_S^* \tag{8.23}$$

$$\Gamma_{out} = \Gamma_L^* \tag{8.24}$$

8.5.1 Conditionally Stable Bilateral Transistor Amplifier Design

The design of an amplifier using a conditionally stable bilateral transistor can involve plotting many gain circles to find an acceptable solution. In this section, we will present a simpler method based on conjugately matching one port of the transistor, and mismatching the other port. The solution can then be adjusted to find a solution where both input (Γ_S) and output (Γ_L) reflection coefficients fall within the stable operating regions. The design procedure will make use of constant gain circles where an arbitrary gain value and reflection coefficient for that port are chosen, and the equations solved to find the other port's reflection coefficient. Amplifier design using feedback networks will not be covered.

8.5.1.1 Available Gain

The amplifier will be designed using the available gain (G_A) approach [11], so that the output network will be conjugately matched with the transistor and the input network mismatched with the transistor. The approach will reduce the amount of reflected power being transmitted back to the input via s_{12} due to any output mismatch. A mismatch on the input results in a significantly lower reflection magnitude than it would if the output were mismatched. Available gain (G_A) is defined as the ratio of the power available from the amplifier to the power available from the source. Substituting Equation 8.24 into Equation 8.18, Equation 8.13 into 8.16, and solving for Equation 8.15 under these conditions yields:

$$G_A = \frac{1 - |\Gamma_S|^2}{|1 - s_{11} \cdot \Gamma_S|^2} \cdot |s_{21}|^2 \cdot \frac{1}{1 - |\Gamma_{out}|^2} \tag{8.25}$$

where the unknown is Γ_S, which will be chosen based on a specific gain requirement for the amplifier. The output will be conjugately matched using the constant available gain circle, G_A. The normalized available power gain (g_A) is given by:

$$g_A = \frac{G_A}{|s_{21}|^2} \tag{8.26}$$

8.5.1.2 Constant Available Gain Circles

Using the available gain, constant available gain circles can be derived [11] and are summarized as follows:

$$C_A = \frac{g_A \cdot \left(s_{11} - \Delta \cdot s_{22}^*\right)^*}{1 + g_A \cdot \left(|s_{11}|^2 - |\Delta|^2\right)} \tag{8.27}$$

$$R_A = \frac{\sqrt{1 - 2 \cdot K \cdot |s_{12} \cdot s_{21}| \cdot g_A + |s_{12} \cdot s_{21}|^2 \cdot g_A^2}}{1 + g_A \cdot \left(|s_{11}|^2 - |\Delta|^2\right)} \tag{8.28}$$

A specific available gain is chosen and the gain circle is plotted. The gain circle gives all possible values for the input reflection coefficient (Γ_S) that will yield this gain. For each value of (Γ_S), a value for (Γ_L) can be determined, and a circle of (Γ_L) can be calculated using Equations 8.14 and 8.24. A range of values where both (Γ_S) and (Γ_L) fall outside their respective stability circles, and that also lie inside the unity circle of the Smith chart, will yield a stable amplifier design. The best choice would be based on values that lie furthest from the stability circles.

8.6 Amplifier Design Example

Next, an amplifier will be designed using an enhancement-mode GaN transistor [19], and the available gain approach. In this example, the s-parameters have been measured using the reference planes shown in Figure 8.6 and will operate at 500 MHz.

At 500 MHz, $V_{DSQ} = 30\,V$ and $I_{DQ} = 500\,mA$, the enhancement-mode GaN transistor (EPC8009 [19]) has the following s-parameter values:

$$\begin{aligned}
s_{11} &= -0.926 \quad -i\cdot 0.157, \quad |s_{11}| = 0.939\\
s_{22} &= -0.658 \quad -i\cdot 0.46, \quad |s_{22}| = 0.803\\
s_{12} &= -0.002 \quad +i\cdot 0.013, \quad |s_{12}| = 0.013\\
s_{21} &= \ \ 5.280 \quad +i\cdot 4.042, \quad |s_{21}| = 6.65
\end{aligned}$$

From the s-parameters, it can be determined whether the transistor is unilateral or bilateral using Equations 8.20 and 8.21.

$$\begin{aligned}
U &= \frac{|s_{11}|\cdot|s_{12}|\cdot|s_{21}|\cdot|s_{22}|}{\left(1-|s_{11}|^2\right)\cdot\left(1-|s_{22}|^2\right)}\\
&= \frac{0.926\cdot 0.013\cdot 6.65\cdot 0.803}{\left(1-0.926^2\right)\cdot\left(1-0.803^2\right)}\\
&= 1.534
\end{aligned} \tag{8.29}$$

and

$$\begin{aligned}
\frac{1}{(1+U)^2} &< g_u < \frac{1}{(1-U)^2}\\
\frac{1}{(1+1.534)^2} &< g_u < \frac{1}{(1-1.534)^2}\\
0.156 &< g_u < 3.5
\end{aligned} \tag{8.30}$$

From the result, it is clear that neither unilateral FOM boundary is within 10% of unity, and therefore, the transistor is considered bilateral.

Next, we need to decide whether the transistor is conditionally or unconditionally stable using Equations 8.5–8.7.

$$\begin{aligned}
\Delta &= s_{11}\cdot s_{22} - s_{12}\cdot s_{21}\\
&= (-0.926 - i\cdot 0.157)\cdot(-0.658 - i\cdot 0.46) - (-0.002 + i\cdot 0.013)\cdot(5.280 + i\cdot 4.042)\\
&= -0.6 + i\cdot 0.472
\end{aligned} \tag{8.31}$$

and

$$|\Delta| = 0.763 \tag{8.32}$$

with the Rollett stability factor

$$\begin{aligned}
K &= \frac{1-|s_{11}|^2-|s_{22}|^2+|\Delta|^2}{2\cdot|s_{12}\cdot s_{21}|}\\
&= \frac{1-0.939^2-0.803^2+0.763^2}{2\cdot|(-0.002+i\cdot 0.013)\cdot(5.280+i\cdot 4.042)|}\\
&= 0.326
\end{aligned} \tag{8.33}$$

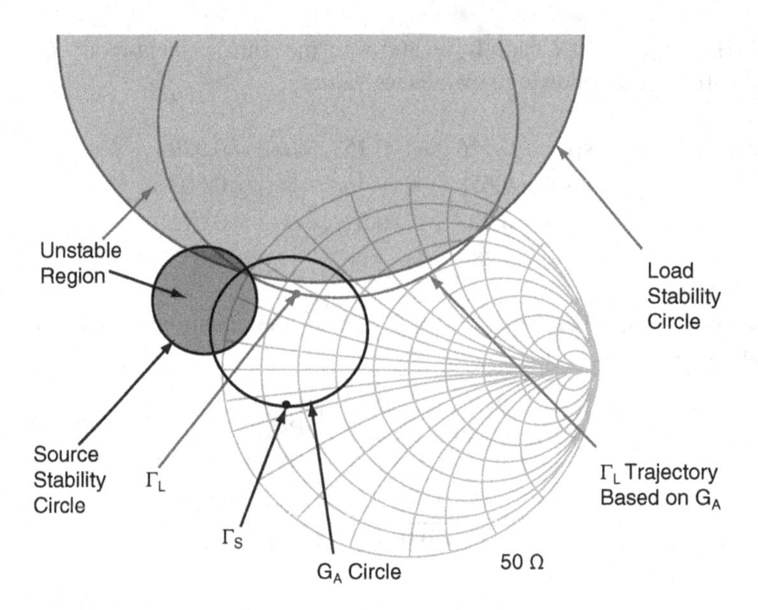

Figure 8.12 Stability circles for EPC8009 [19] at 500 MHz with $V_{DSQ} = 30\,V$ and IDQ = 500 mA together with the available gain circle of 23 dB and load reflection coefficient Γ_L trajectory

The unconditional stability of $K \geq 1$ and $|\Delta| \leq 1$ has not been met for this transistor due to K. Therefore, the device is conditionally stable and the stability circles can be plotted to determine the unstable regions.

The stability circles can be calculated using Equations 8.8–8.11, and are shown in Figure 8.12 with the unstable regions shaded. The amplifier design for (Γ_S) and (Γ_L) need to avoid these regions. As the transistor is conditionally stable, we need to determine a suitable gain that will yield an amplifier that will not oscillate (is always stable). Before we can select a working gain we need to determine the maximum stable gain of the transistor using the following equation:

$$\begin{aligned} G_{MSG} &= \frac{|s_{21}|}{|s_{12}|} \\ &= \frac{6.65}{0.013} \\ &= 520.3 = 27.2\,\text{dB} \end{aligned} \tag{8.34}$$

The maximum stable gain is a simple method to determine the stable gain limit of the transistor; however, it may not yield a workable solution. For this example, a design gain of 200 = 23 dB is selected.

Next, using the gain value selected, the available gain circle is plotted as shown in Figure 8.12, which reveals all values of Γ_S that will yield a gain of 23 dB. Next, a specific value for (Γ_S) can be selected and, using Equations 8.14 and 8.24, a value for Γ_L can be determined. Both Γ_S and Γ_L must lie outside the unstable regions of the stability circles. The trajectory of Γ_L, based on the available gain circle, has been plotted in Figure 8.12 to make it easier to observe if a solution exists. If there are points where both the G_A circle and the Γ_L trajectory lie outside the stability circles and also lie inside the unity Smith chart, then a workable solution exists.

The reflection coefficients at 500 MHz are:

$$\Gamma_S = -0.604 - i \cdot 0.167$$
$$\Gamma_L = -0.557 + i \cdot 0.458$$

Using these reflection coefficients, the amplifier matching networks can be designed. These reflection coefficients are usually provided in the datasheets of RF components over a range of frequencies.

8.6.1 Matching and Bias Tee Network Design

Figure 8.13 shows a block diagram for the RF amplifier design. To accommodate a small heatsink for the transistor, a 50 Ω transmission line 12.25 mm long for the gate circuit and 14 mm long for the drain circuit needs to be added to bring out the terminals before connection to the bias tees and matching networks. The impact of both the transmission lines and bias tees on the reflection coefficients will need to be calculated for adjustment prior to calculating the matching network.

The 50 Ω transmission line rotates the selected impedance (reflection coefficient) around the center of the Smith chart to a new location, since only the phase component changes and not the characteristic impedance. Given that this is the transmission line, the direction of rotation is counter-clockwise. The angle of rotation will depend on the length of the transmission line, transmission line design, and the frequency of operation. The derivation for the electrical angle based on a microstrip is given in [24]. In this example, the electrical phase is 12.46° for the gate transmission line, and 14.24° for the drain transmission line. The rotation on the Smith chart is twice the electrical angle.

The bias tee circuits are used to provide the quiescent supply to the transistor. It must be designed so as to not affect the RF characteristics of the circuit, yet it needs to adequately provide the required bias conditions. The bias tee circuit for the amplifier consists of a second-order passive filter (DC pass – AC block) as shown in Figure 8.14. The impact on the RF circuit

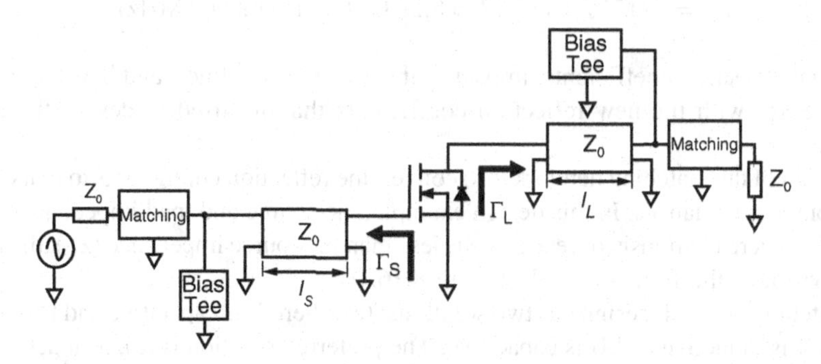

Figure 8.13 Block diagram of the complete amplifier design that includes transmission lines, bias tees, and matching networks

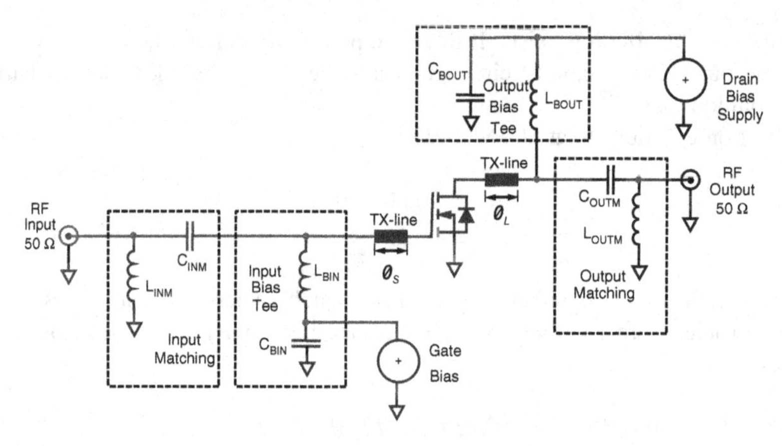

Figure 8.14 Amplifier schematic showing transmission lines, bias tees and matching networks

can be determined as the series combination of the two passive elements, shunting the RF signal at the point of connection. Since the amplifier will be pulsed, additional design considerations need to be made to ensure stable pulse operation of the transistor with some useful design tips given in [25].

The selected bias tee components have the following electrical properties at 500 MHz:

$$
\begin{aligned}
L_{Bin} &= 48.4\,\text{nH}, \text{ESR} = 16\,\Omega \\
C_{Bin} &= 100\,\text{pF} \\
L_{Bout} &= 240.8\,\text{nH}, \text{ESR} = 126\,\Omega \\
C_{Bout} &= 10\,\text{nF}
\end{aligned}
$$

Using the bias tee network values, the calculated input and output reflection coefficients, and the effect of the transmission lines, the new reflection coefficients (Γ_S) and (Γ_L) can be determined. In this example, the results are:

$$
\begin{aligned}
\Gamma_S &= -0.543 - i \cdot 0.414 = 10.45\,\Omega, 19.65\,\text{pF (at 500 MHz)} \\
\Gamma_L &= -0.707 + i \cdot 0.13 = 8.26\,\Omega, 1.41\,\text{nH (at 500 MHz)}
\end{aligned}
$$

The original reflection coefficients, impact of the transmission lines, and bias tees are shown in Figure 8.15, with the new reflection coefficients that are used to design the matching networks.

The design of the matching network is to convert the reflection coefficients to that of the load and the source impedances. In this design example, the source and load impedances are 50 Ω. For the case where the transistor resistance is less than the source impedance (Z_0), the matching network will take the form shown in Figure 8.16.

The matching network design has two solutions: (a) where X is capacitive and B is inductive, (b) where X is inductive and B is capacitive. The preferred solution is (a), as it acts as a high-frequency pass with low-frequency filtering, and is desirable because the transistor has a higher gain in the lower frequencies and any unwanted signal can easily corrupt the amplifier. The

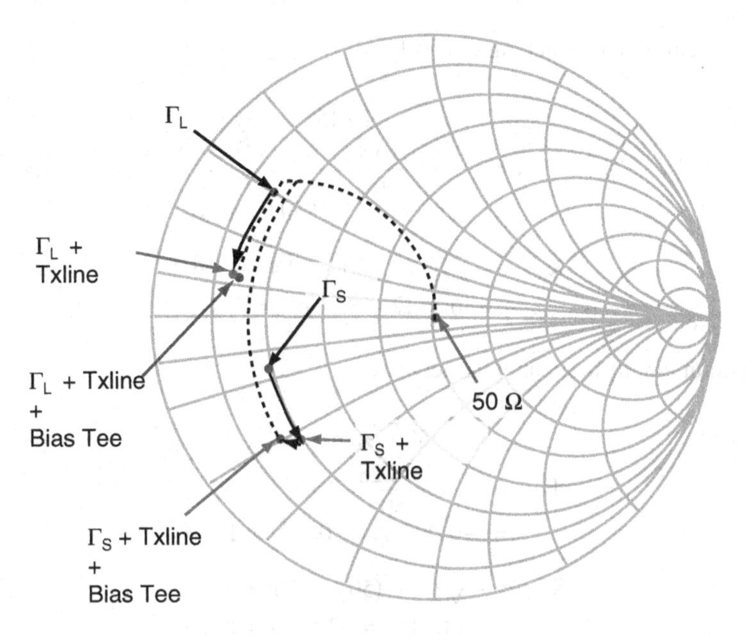

Figure 8.15 Amplifier design of reflection coefficients showing impact of transmission lines, bias tees, and matching network trajectories for each port are also shown

solution for the matching network design for the schematic of Figure 8.16 is given in [8] and repeated in Equations 8.35 and 8.36:

$$B = \pm \frac{\sqrt{\dfrac{Z_0 - R_L}{R_L}}}{Z_0} \tag{8.35}$$

$$X = \pm\sqrt{R_L \cdot (Z_0 - R_L)} - X_L \tag{8.36}$$

Since a DC block is required for both the RF input and output, and this function can be integrated into the matching network, the case with the negative values for X and B in Equations 8.35 and 8.36 can be calculated for the matching networks as follows.

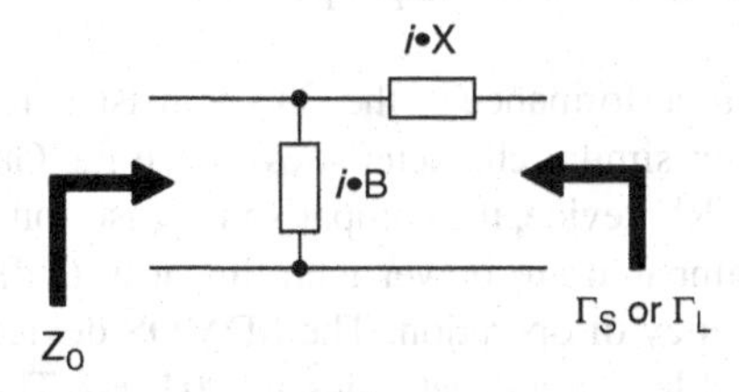

Figure 8.16 Matching network configuration for the amplifier

For the gate circuit the matching network:

$$\begin{aligned} B_{in} &= -\frac{\sqrt{\dfrac{50-10.45}{10.45}}}{50} \\ &= -0.039\ \text{S} = 8.18\ \text{nH} = L_{inM} \end{aligned} \tag{8.37}$$

$$\begin{aligned} X_{in} &= -\sqrt{10.45 \cdot (50-10.45)} - 16.2 \\ &= -36.53\ \Omega = 8.71\ \text{pF} = C_{inM} \end{aligned} \tag{8.38}$$

and for the drain circuit the matching network:

$$\begin{aligned} B_{out} &= -\frac{\sqrt{\dfrac{50-8.26}{8.26}}}{50} \\ &= -0.045\ \text{S} = 7.08\ \text{nH} = L_{outM} \end{aligned} \tag{8.39}$$

$$\begin{aligned} X_{out} &= -\sqrt{8.26 \cdot (50-8.26)} + 4.42 \\ &= -14.14\ \Omega = 22.51\ \text{pF} = C_{outM} \end{aligned} \tag{8.40}$$

The matching network solution trajectories for these solutions have been plotted in Figure 8.15.

8.6.2 Experimental Verification

Based on the design example, a 500 MHz RF amplifier was designed and tested using the EPC8009 [19] eGaN FET. The drain bias current was set to 250 mA and 500 mA, and the drain bias voltage at 30 V. The amplifier was tested in pulse mode with a pulse duration of 240 µs and a repetition rate of 10 Hz. The amplifier was tested using a vector network analyzer with an additional RF amplifier to boost the input RF power to the amplifier, and loaded with a 20 W-capable, 30 dB RF attenuator. An RF power-in to power-out sweep was performed. Figure 8.17 shows the 1 dB compression point for this amplifier at both bias current settings.

Figure 8.18 shows the gain and drain efficiency for the amplifier as a function of output power.

It can be seen from the experimental results that the EPC8009 with 500 mA-drain bias current has a 1 dB compression point at 40.6 dBm (11.6 W) output power, where the power gain is 20.6 dB with drain efficiency of 57.4%. At a drain bias current of 250 mA, the device has a 1 dB compression at 38.4 dBm (6.96 W) output power, where the power gain is 19.3 dB with drain efficiency of 45.9%.

Having determined the RF performance of the GaN transistor, it may be compared to state-of-the-art LDMOS FETs with similar characteristics. Since the GaN FET was designed as a switching device, and not an RF device, the comparison focuses on the differences pertinent to RF designs. The metrics compared are power gain, linearity (1 dB compression,) and drain efficiency at the same frequency of operation. The LDMOS devices selected for comparison are [26,27] based on comparable power capabilities at 500 MHz. The comparison data between the GaN transistor and LDMOS FETs are given in Table 8.3.

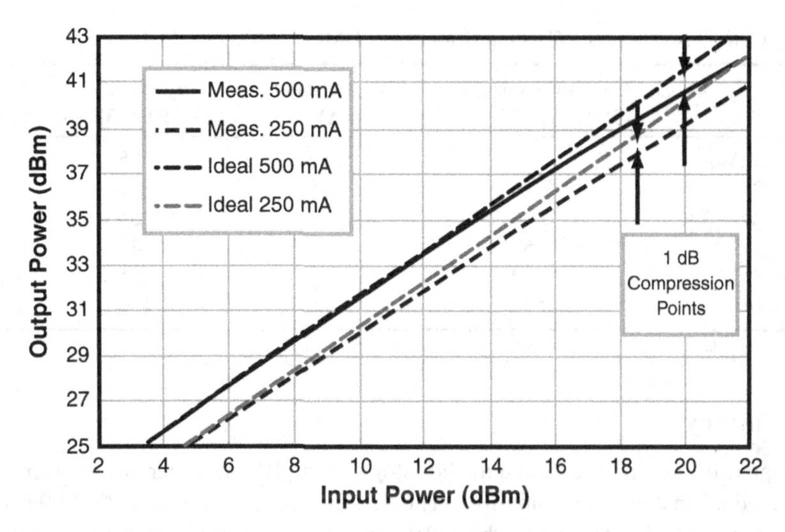

Figure 8.17 Measured 1 dB compression points for the EPC8009-based RF amplifier with 30 V drain bias voltage and 250 mA and 500 mA drain bias currents while operating at 500 MHz

From Table 8.3, it can be seen that the GaN FET has a higher gain than the LDMOS while operating at a higher voltage with comparable drain efficiency despite the higher bias power. The capacitances of the GaN transistor are also significantly lower than the LDMOS FETs, ensuring lower matching impedance conversion.

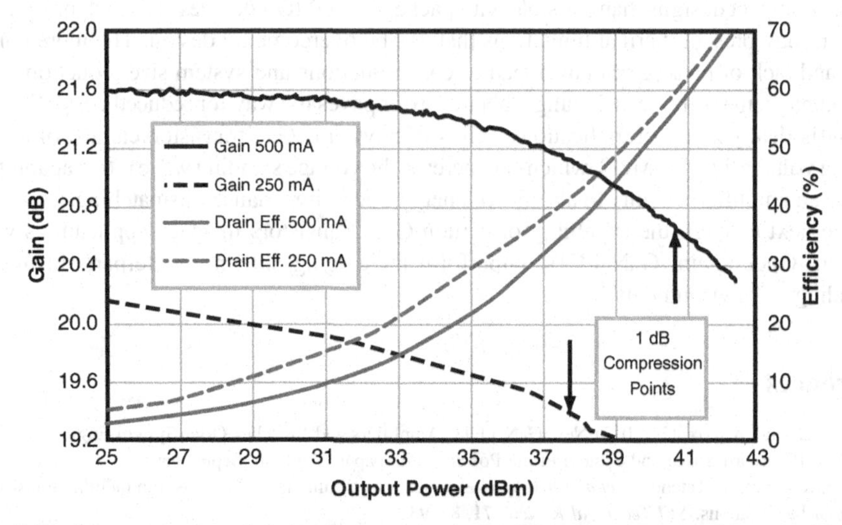

Figure 8.18 Measured gain and drain efficiency for the EPC8009-based RF amplifier with 30 V drain bias voltage and 250 mA and 500 mA drain bias currents while operating at 500 MHz

Table 8.3 Comparison between GaN transistors and LDMOS FET at 500 MHz

Parameter	GaN transistor [19] (500 mA)	GaN transistor [19] (250 mA)	LDMOS FET [26]	LDMOS FET [27]
Output power (W)	11.6	6.96	15	8
1 dB gain (dB)	20.6	19.3	14	13
Drain efficiency (%)	57.4	45.9	55	60
Rated voltage (V)	65	65	40	40
Bias voltage (V)	30	30	12.5	12.5
Bias current (mA)	500	250	150	150

8.7 Summary

In this chapter, the key metrics and a methodology for an RF amplifier design were presented and compared against an actual amplifier. The design was based on the EPC8009 eGaN FET, which was not originally designed to be an RF device. Despite this, the results show excellent RF characteristics with stable gain in excess of 20 dB, and a drain efficiency approaching 60% at the 1 dB compression point. The GaN transistor was compared against two LDMOS FETs with similar RF characteristics, and it was seen that the GaN transistor yields a higher gain than the LDMOS FETs, as well as comparable drain efficiency. Many LDMOS devices are internally matched to enhance their RF performance around a specific frequency. This reduces the usable bandwidth to a few tens of MHz, whereas an LDMOS FET designed for broadband applications can have an operating bandwidth around 100 MHz. GaN transistors, designed as power switching transistors are not internally matched and, as a result, the higher impedances enable the device to span a bandwidth of as much as 3 GHz.

It was also demonstrated that, despite being designed as a switching device, the GaN transistor can easily be connected to a RF circuit using reference planes and microstrip tapers. This allows for more compact designs than possible with packaged LDMOS devices. The lack of a package, however, may present thermal limitations that can be overcome by design. The more compact layout and lack of package can also lead to cost reductions and system size reduction.

The enhancement-mode switching GaN transistor paves the way for reduced cost RF applications and is ideally suited for applications such as MRI systems. GaN transistors can also offer higher blocking voltage than LDMOS, which can increase the voltage standing wave ratio capability and increase an amplifier's ability to absorb RF energy due to impedance mismatching.

In the next chapter, the capability of certain GaN transistors in space applications will be reviewed. Once again, GaN FETs outperform their aging silicon counterparts in the most demanding of environments.

References

1. White, D. and Wilcox, G. (2012) New GaN FETs, Amplifiers and Switches Offer System Engineers a Way to Reduce RF Board Space and System Prime Power, white paper, TriQuint, September.
2. Inoue, K., Sano, S., Tateno, Y. *et al.* (2010) Development of gallium nitride high electron mobility transistors for cellular base stations. *SEI Technical Review*, **71**, 88–93.
3. "Gallium Nitride (GaN) Microwave Transistor Technology For Radar Applications," white paper, Aethercomm, Dec. 2007.

4. Murphy, M. (2011) "NXP goes with GaN," *Compound Semiconductor*, Aug./Sep., pp. 23–26.
5. "GaN devices set benchmarks for power and bandwidth," *Microwave Product Digest*, Feb. 2012, Available from www.mpdigest.com.
6. Ishida, T. (2011) GaN HEMT technologies for space and radio applications. *Microwave Journal*, **54** (8), 57–63.
7. Nitronex (April 2013) Gallium Nitride 28V, 25W RF Power Transistor, NPT1012 datasheet, NDS-025 Rev. 3. Available from www.nitronex.com/NPT1012_Product_Page.html.
8. Pozar, D.M. (2005) *Microwave Engineering*, 3rd edn, J. Wiley.
9. Gonzales, G. (1997) *Microwave Transistor Amplifiers*, 2nd edn, Prentice Hall.
10. Hejhall, R.C. (1993) RF Small Signal Design Using Two-Port Parameters, Motorola, Appl. Note AN215A.
11. Payne, K. (2008) Practical RF Amplifier Design Using the Available Gain Procedure and the Advanced Design System EM/Circuit Co-Simulation Capability, white paper, 5990-3356EN, Agilent Technologies, Available from, www.agilent.com.
12. Lidow, A., Strydom, J., Rooij, M.de., and Ma, Y. (2012) *GaN Transistors for Efficient Power Conversion*, 1st edn, Power Conversion Press, El Segundo.
13. Strydom, J. (Oct. 2012) "eGaN® FET- Silicon Power Shoot-Out Volume 11: Optimizing FET On-Resistance," *Power Electronics Technology*, Available from http://powerelectronics.com/discrete-semis/gan_transistors/egan-fet-silicon-power-shoot-out-volume-11-optimizing-fet-on-resistance-1001/.
14. de Rooij, M. and Strydom, J. (June 2012) "eGaN® FET-Silicon Power Shoot-Out Volume 9: Low Power Wireless Energy Converters," *Power Electronics Technology*. Available from http://powerelectronics.com/discrete-power-semis/egan-fet-silicon-shoot-out-vol-9-wireless-power-converters.
15. Cree, GaN HEMT RF FET, CGH55015, datasheet. Available from http://www.cree.com/RF/Products/SBand-XBand-CBand/Packaged-Discrete-Transistors/CGH55015F2-P2.
16. Engen, G.F. and Hoer, C.A. (December 1979) "Thru-reflect-line: an improved technique for calibrating the dual six-port automatic network analyzer," IEEE Trans. Microwave Theory and Techniques.
17. Agilent, Network Analysis Applying the 8510 TRL Calibration for Non-Coaxial Measurements Product Note 8510-8A.
18. Fleury, J. and Bernard, O. (2001) Designing and Characterizing TRL Fixture Calibration Standards for Device Modeling, Applied Microwave & Wireless Technical Note 13, pp. 26–55.
19. Efficient Power Conversion Corporation (2013) "EPC8009 – Enhancement-mode Power Transistor," EPC8009 datasheet, Sept. Available from http://epc-co.com/epc/documents/datasheets/EPC8009_datasheet.pdf.
20. Rogers Corporation (2014) Rogers 4350 Laminates, datasheet. Available from http://www.rogerscorp.com/acm/products/55/RO4350B-Laminates.aspx.
21. de Rooij, M.A. and Strydom, J.T. (Sep. 2013) Method for Bias Control of a Class A Power RF Amplifier, U.S. patent pending.
22. Rollett, J.M. (1962) Stability and power-gain invariants of linear two ports. *IRE Transactions on Circuit Theory*, **9** (1), 29–32.
23. Orfanidis, S.J. (2014) *Electromagnetic Waves and Antennas*, Available from http://www.ece.rutgers.edu/~orfanidi/ewa/.
24. Bahl, I.J. and Trivedi, D.K. (1977) A designer's guide to microstrip line. *Microwaves*, May 1977, 174–182.
25. Baylis, C., Dunleavy, L., and Clausen, W. (2006) Design of bias tees for a pulsed-bias, pulsed-RF test system using accurate component models. *Microwave Journal*, **49** (10), 68–75.
26. STMicroelectronics, PD55015 – RF Power Transistor, datasheet, Aug. 2011. Available from http://www.st.com/web/en/resource/technical/document/datasheet/CD00128612.pdf.
27. Freescale Semiconductor, RF Power Field Effect Transistor, MRF1518N datasheet, [Rev. 11 June 2009] Available from http://www.freescale.com/files/rf_if/doc/data_sheet/MRF1518N.pdf.

9

GaN Transistors for Space Applications

9.1 Introduction

Radiation in space is generated by many sources within and outside of our solar system. This radiation comes in the form of gamma rays, energetic electrons, protons, and heavier ions, which are all known to cause damage in semiconductors. Over the years, silicon-based devices have been well characterized under various radiation conditions, vulnerabilities have been identified and, to some extent mitigated through design and process improvements. NASA has published guidelines to help designers of satellite systems consistently design for the different environments encountered in various earth orbits [1].

In this chapter, we will look at the research on GaN capabilities under exposure to different types of radiation with a particular focus on GaN transistors used in power conversion. We will then look at actual measurements of commercial grade enhancement mode GaN transistors and compare their capability against radiation-resistant silicon power MOSFETs.

9.2 Failure Mechanisms

An energetic particle can cause damage to a semiconductor in three primary ways: (1) it can cause traps in non-conducting layers, (2) it can cause physical damage to the crystal (displacement damage) or the interface between the crystal and a Schottky barrier, and (3) it can create a cloud of electron–hole pairs that will cause the device to momentarily conduct (and possibly burn out in the process) [2].

Power MOSFETs, in particular, are vulnerable to radiation in two ways. First, gamma (electron) radiation can cause positively charged (hole) traps to develop in the thin gate oxide [2,3]. The addition of positive charges between the gate electrode and the channel reduces the threshold voltage of the n-channel power MOSFET. The second major vulnerability stems from energetic particles that can fully penetrate the semiconductor device. These particles cause the single event effects (SEE), and the failures are known as single event upset (SEU).

GaN Transistors for Efficient Power Conversion, Second Edition.
Alex Lidow, Johan Strydom, Michael de Rooij, and David Reusch.

Companion Website: http://www.wiley.com/go/gan_transistors

In order to understand how these same radiation effects impact the electrical performance of GaN transistors, we need to consider Schottky-based depletion-mode devices (Figure 1.6(a) in Chapter 1), MOS-based depletion-mode devices (Figure 1.6(b) in Chapter 1), and pGaN-based enhancement-mode devices (Figure 1.7 in Chapter 1). All three of these structures use an AlGaN/GaN barrier to generate the 2DEG, but have different gate structures that can lead to different vulnerabilities when exposed to radiation. Cascode-configured GaN transistors will not be considered in this chapter because their behavior would be influenced significantly by the silicon MOSFET's radiation tolerance.

9.3 Standards for Radiation Exposure and Tolerance

The NASA guidelines for transistors [1] set the standard set for a "Rad Hard" device as the ability to withstand a total incident dose (TID) between 200 kRad(Si) and 1 MRad(Si), and an SEU threshold linear energy transfer (LET) of 80–150 $\mathrm{MeV\,mg^{-1}\,cm^{-2}}$. A "Rad" is defined as the mean energy absorbed per unit mass of irradiated material at the point of interest [3].

$$1\,\mathrm{Rad} = 100\,\mathrm{ergs/g}$$

A Rad(Si) would be a measure of absorption in a silicon crystal. In order to maintain comparisons with silicon power MOSFETs, most researchers have elected to use Rad(Si) to report their results for GaN devices.

LET is a measure of the energy deposited by an incident particle (the "stopping power") per unit of track length. A heavier ion, such as Au, would have a higher LET for the same energy than a lighter ion, such as a proton.

9.4 Gamma Radiation Tolerance

The common approach to measuring device tolerance to gamma radiation is to use the decay radiation from a Cobalt-60 (^{60}Co) source. Using this method, testing of depletion-mode Schottky gate transistors designed for RF applications has been reported in [4] and [5], where the results showed no degradation when exposed to up to 600 MRads(Si), and are consistent with results for silicon JFETs [2]. This is not surprising because there are no oxides in the gate structure that can trap charge, and thereby modulate the threshold voltage.

To date, no measurements have been reported for MOS-gated GaN HEMT transistors. There is no reason to believe that their behavior would be significantly different from the behavior of silicon power MOSFETs. Gamma radiation-induced trapping in the insulating gate will cause a negative shift in threshold voltage that could lead to circuit failure.

Enhancement-mode pGaN HEMT transistors have been tested extensively [6,8] with favorable results. In these tests, the devices had either drain-to-source bias during testing, or gate-to-source bias. By exposing devices to radiation while applying bias on each of the transistor terminals, the in-circuit performance in a high radiation environment can be projected. Figure 9.1 is a graph showing the progression of drain-to-source leakage current and threshold voltage of several devices tested up to 1 MRad(Si).

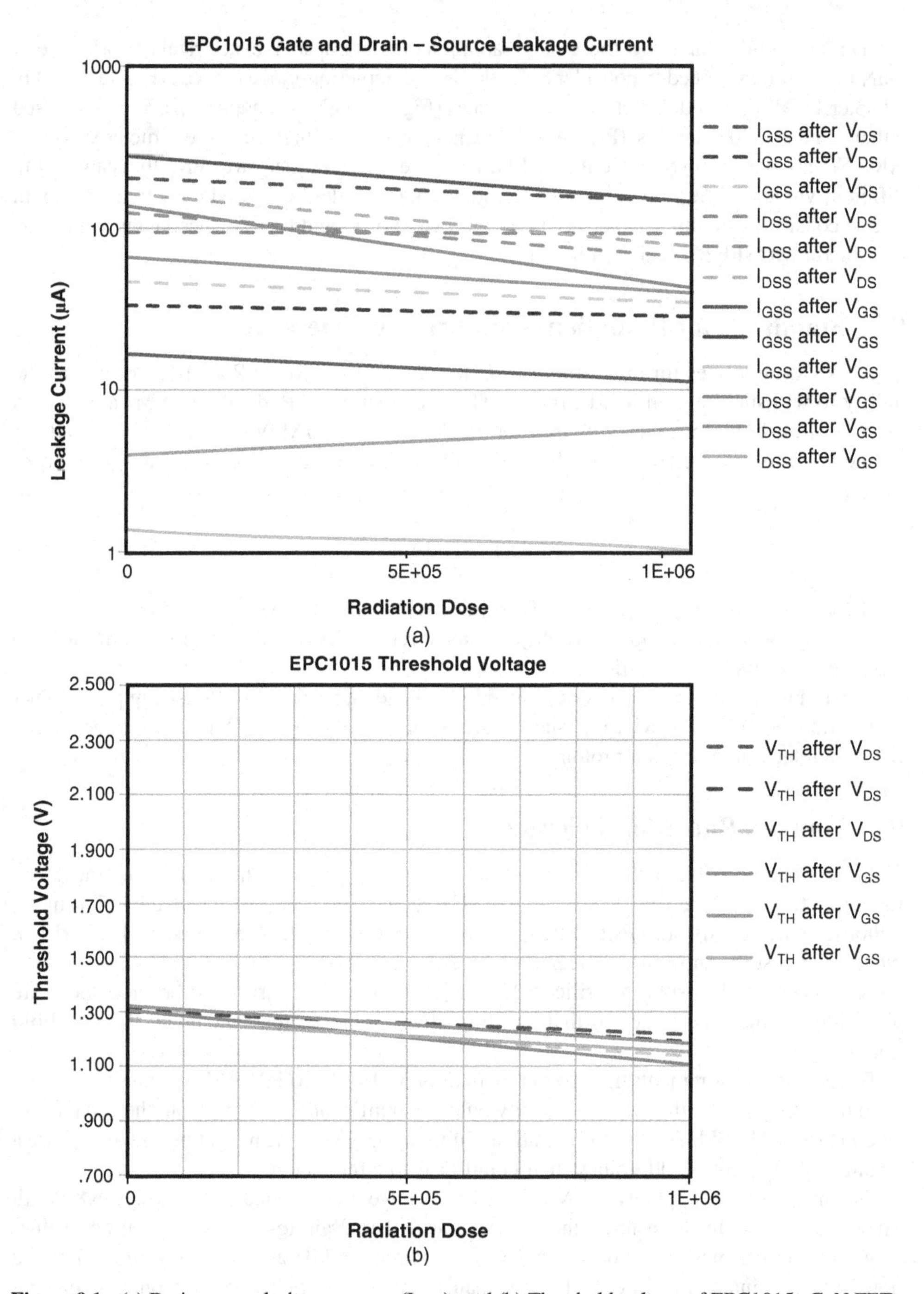

Figure 9.1 (a) Drain-source leakage current (I_{DSS}), and (b) Threshold voltage of EPC1015 eGaN FETs after exposure to 1×10^6 Rad(Si). Devices with 5 V gate-source bias during testing are shown as solid lines. Devices with 80 V drain-source bias during testing are shown as dotted lines [6]

9.5 Single-Event Effects (SEE) Testing

As with gamma radiation testing, there has been significant research on SEE in Schottky gate RF HEMT transistors [7,8], but no MOS gate HEMT device testing has been reported. In the case of a cascode GaN transistor, the reaction to single event radiation would be determined by the combination of the silicon MOSFET and the GaN transistor in their series configuration. Enhancement-mode pGaN gated devices have also had extensive testing [6,9,10].

Failures under SEE testing of power transistors fall into two categories: single event burnout (SEB) and single event gate rupture (SEGR). SEGR is unlikely in a Schottky gate or pGaN gate HEMT device because there are no insulating gate layers that would be the source of failure.

The Schottky gate and pGaN gate devices showed remarkable resistance to SEE radiation. In reference [7], Schottky gate devices showed slow degradation of I_{DSS} when irradiated with bromine ions at 39 LET. Abrupt failures (SEB) were observed in these devices with 70 V drain-to-source bias at a LET of 60 MeV mg^{-1} cm^{-2}. Schottky gate transistors in reference [8] also showed no degradation when exposed to lesser levels of SEE radiation from FE, O, and C atoms. Reference [10] also showed excellent SEE tolerance with the main failure mechanism being the increase of I_{DSS} leakage, the rate of which scaled with the LET and drain-to-source voltage.

The highest level of SEE capability for a GaN HEMT was reported in references [9,11]. Tests showed eGaN FET devices capable of withstanding 1×10^{13} cm^{-2} Au atoms at a LET of 87.2 MeV mg^{-1} cm^{-2} with full-rated voltage applied to 40 V or 100 V rated transistors. The 200 V rated eGaN FETs survived up to 190 V bias and, at that voltage, failures were a result of a slow increase in I_{DSS}. An example of the stability of devices under Au bombardment is shown in Figure 9.2. The basic failure mechanism is the physical displacement of atoms in the crystal due to the large momentum transfer from the incident heavy ions.

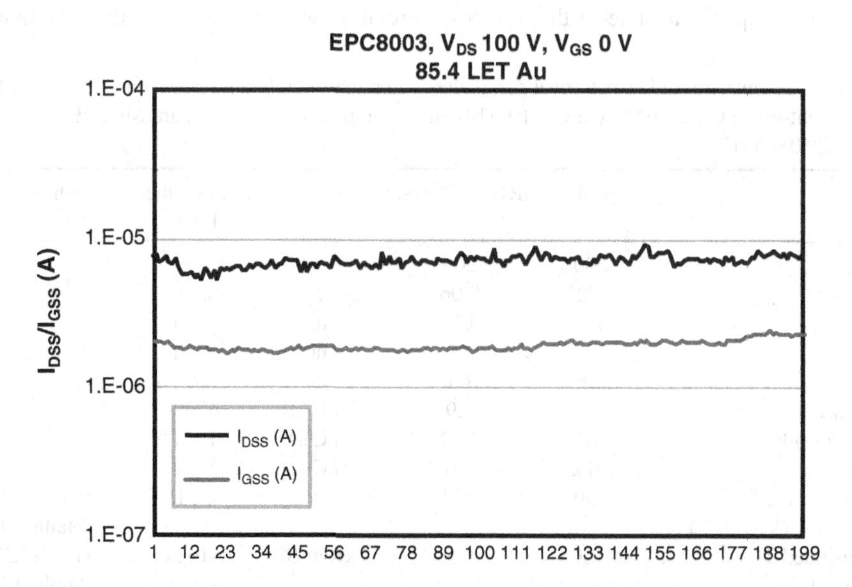

Figure 9.2 I_{DSS} and I_{GSS} progression for MGN8903 GaN transistor during single event testing with $1{\cdot}10^6$ ions/cm^2 Au at 85.4 LET and biased at 100 V_{DSS} [11]

Table 9.1 Comparison of key electrical 200 V rated commercial MOSFET (IPB107N20N3 G) [12] and a comparably rated rad-hard MOSFET (IRHN57250SE) [13]

	IPB107N20N3G	IRHN57250SE	Units	Performance Ratio
Rad tolerant	No	Yes		
BV_{DSS}	200	200	V	1.0
$R_{DS(on)}$	0.011	0.06	Ω	5.5
Q_G	161	132	nC	0.8
Q_{GS}	23	45	nC	2.0
Q_{GD}	8	60	nC	7.5
$Q_G \cdot R_{DS(on)}$	1.8	7.9	nC·Ω	4.5
$(Q_{OSS}+Q_G) \cdot R_{DS(on)}$	2.6	11.2	nC·Ω	4.3
$Q_{GD} \cdot R_{DS(on)}$	0.09	3.6	nC·Ω	41

9.6 Performance Comparison between GaN Transistors and Rad-Hard Si MOSFETs

Schottky gate and pGaN gate HEMT transistors can tolerate high doses of radiation without significant performance degradation, and without the need for any changes to the device design or fabrication process. In the case of the enhancement-mode GaN devices, as shown in Chapters 3–7, they significantly outperform state-of-the-art commercial power MOSFETs. Unlike GaN transistors, power MOSFETs designed to operate in high-radiation environments do not have dynamic switching performance comparable to commercial MOSFETs. Significant compromises in device geometries and fabrication process are needed to harden the final product [3]. Illustrating these significant compromises required of rad-hard MOSFET users, Table 9.1 compares a state-of-the art commercial power MOSFET with a comparable

Table 9.2 Comparison of key electrical parameters and radiation tolerance between a 200 V rated enhancement-mode GaN HEMT (EPC2010) [14] and a comparably rated rad-hard MOSFET (IRHN57250SE) [13]

	EPC2010	IRHN57250SE	Units	Performance Ratio	Method
BV_{DSS}	200	200	V		
$R_{DS(on)}$	0.025	0.06	Ω	2 : 1	
Q_G	7.5	132	nC	18 : 1	
Q_{GS}	2	45	nC	23 : 1	
Q_{GD}	2.6	60	nC	23 : 1	
$Q_G \cdot R_{DS(on)}$	0.19	7.9	nC·Ω	42 : 1	
$(Q_{OSS}+Q_G) \cdot R_{DS(on)}$	1.12	11.2	nC·Ω	10 : 1	
$Q_{GD} \cdot R_{DS(on)}$	0.065	3.6	nC·Ω	55 : 1	
Demonstrated SEE SOA at 84 LET ($V_G = 0$ V)	190	200	V	1 : 1	MIL-STD750E Method 1080
Demonstrated TID capability	>1000	100	kRad(Si)	>10 : 1	MIL-STD750E Method 1019

radiation-tolerant version [12,13]. The last two rows in the table compare the soft-switching figures of merit (FOM) and the hard-switching FOM first discussed in Chapter 6 and 7, respectively. The latest commercial MOSFET are 4–5 times superior to the rad-hard MOSFET in these two key measures of device performance in switching converters.

Moving on to radiation-tolerant transistors, Table 9.2 compares the performance of a 200 V enhancement-mode GaN HEMT [14] and a 200 V rad-hard MOSFET [13,15]. The GaN HEMT has comparable SEE capability, 10 times the gamma radiation tolerance (TID), a hard-switching FOM ($R_{DS(on)} \cdot Q_{GD}$) 50 times superior, and a soft-switching FOM ($R_{DS(on)} \cdot (Q_G + Q_{OSS})$) 10 times superior to the comparable rad-hard MOSFET. Any power conversion system would operate with significantly lower losses and higher gamma radiation tolerance using these enhancement-mode GaN transistors.

9.7 Summary

GaN transistors have been tested under heavy ion bombardment and gamma irradiation. These devices demonstrate readiness for use in the most stringent of radiation environments and far exceed the capabilities of silicon power MOSFETs. The problem that designers encounter with silicon MOSFETs is that they must choose between radiation tolerance and electrical performance. Commercial MOSFETs have thick gate oxides and so trap a lot of charge, resulting in large shifts in the threshold voltage and eventual failure at relatively low total-dose exposure. The radiation-hardened MOSFETs available have FOMs several times worse than their commercial counterparts, leading to either low efficiency or large size (due to the low switching frequency). Enhancement-mode GaN transistors give designers a new capability with electrical performance superior to the cutting-edge Si MOSFETs, and radiation tolerance at least as high as the best rad-hard power MOSFETs available. These GaN transistors bring a combination of electrical and radiation performance that establishes a new state of the art.

The tools and techniques discussed in Chapters 3–7 lay the foundation for the applications that are discussed in Chapter 10. The applications selected are among the first to widely use GaN transistors for power conversion.

References

1. Space Radiation Effects on Electronic Components in Low Earth Orbit, NASA Practice No. PD-ED-1258.
2. Messenger, G.C. and Ash, M.S. (1986) *The Effects of Radiation on Electronic Systems*, Van Nostrand Reinhold Company, New York, NY.
3. Ma, T.P. and Dressendorfer, P.V. (1989) *Ionizing Radiation Effects in MOS Devices and Circuits*, John Wiley and Sons, Inc., New York.
4. Aktas, O., Kuliev, A., Kumar, V. *et al.* (2004) ^{60}Co gamma radiation effects on DC, RF, and pulsed I–V characteristics of AlGaN/GaN HEMTs. *Solid-State Electronics*, **48**, 471–475.
5. McClory, J.W. (2008) The Effect Of Radiation on the Electrical Properties of Aluminum Gallium Nitride/Gallium Nitride Heterostructures, Ph.D. dissertation, The Air Force Institute of Technology, Wright-Patterson Air Force Base, Ohio, June.
6. Lidow, A., Witcher, J.B., and Smalley, K. (March 2011) Enhancement mode gallium nitride (eGaN®) FET characteristics under long-term stress, *GOMAC Tech Conference*, Orlando Florida.
7. Bazzoli, S., Girard, S., Ferlet-Cavrois, V. *et al.* (2007) SEE sensitivity of a COTS GaN transistor and silicon MOSFETs, 9th European Conference on Radiation and Its Effects on Components and Systems, RADECS 2007.
8. Sonia, G., Brunner, F., Denker, A. *et al.* (2006) Proton and heavy ion irradiation effects on AlGaN/GaN HFET devices. *IEEE Transactions on Nuclear Science*, **53** (6)
9. Lidow, A. and Smalley, K. (2012) Radiation tolerant enhancement mode gallium nitride (eGaN®) FET characteristics, GOMAC Tech Conference, Las Vegas, Nevada, March 2012.

10. Lidow, A., Strydom, J., and Rearwin, M. (2014) Radiation tolerant enhancement mode gallium nitride (eGaN®) FETs for high-frequency DC-DC conversion, *GOMAC Tech Conference*, Charleston, South Carolina, April 2014.
11. Kuboyama, S., Maru, A., Shindou, H. *et al.* (2011) Single-event damages caused by heavy ions observed in AlGaN/GaN HEMTs. *IEEE Transactions on Nuclear Science*, **58** (6)
12. Infineon (July 2011) "OptiMOS™3 Power Transistor," IPB107N20N3 G datasheet.
13. International Rectifier (Dec. 2011) "Radiation hardened power MOSFET surface mount (SMD-1)," IRHN57250SE.
14. Efficient Power Conversion Corporation "EPC2010 – Enhancement-mode Power Transistor," EPC2010 datasheet, Jul. 2011 [Revised Feb. 2013]. Available from http://epc-co.com/epc/documents/datasheets/EPC2010_datasheet.pdf.
15. Strydom, J., Lidow, A., and Goti, T. (2013) Radiation Tolerant enhancement mode gallium nitride (eGaN®) FETs in DC-DC converters, GOMAC Tech Conference, Las Vegas, Nevada, March 2013.

10

Application Examples

10.1 Introduction

This chapter presents some application examples where GaN transistors are already making inroads. As presented in the previous chapters, these inroads stem directly from the GaN transistor's relative improvement in figures of merit (FOM) over the silicon MOSFET, be it in hard-switch or soft-switching applications. GaN transistors offer the potential to improve performance over the aging population of Si MOSFETs, enabling a new generation of power converters that offer higher frequencies, efficiencies, and densities than ever before.

10.2 Non-Isolated DC-DC Converters

Non-isolated, point-of-load (POL) converters are found in computers, telecommunication systems, handheld electronics, and many other applications. They are perhaps the most common way to convert from one DC voltage to a different DC voltage. With the ever-increasing power demands of modern technologies, combined with the desire for smaller size and lower power consumption, POL converter designs must drive toward higher power density, efficiency, and transient response capability in order to meet these evolving system demands.

The majority of POLs are non-isolated step-down buck converters, often with a large step-down ratio from as much as 60 V input to below 1 V output. The most straightforward way to improve power density and transient response capability is to increase switching frequency. This enables a volume reduction in the passive components, as well as a faster reaction time to transient spikes of voltage and current from the source or the load. The practical issue with today's silicon-based solutions is that increasing switching frequency *decreases* efficiency as a result of higher switching losses in the aging power MOSFETs, thus limiting these solutions to the range of a couple hundred kilohertz to a megahertz. GaN transistors greatly relieve this constraint by easily operating well above these frequencies without loss of system efficiency.

Shown in Figure 10.1 are the efficiencies and power losses of GaN transistors and Si MOSFETs when applied in 1.2 V_{OUT} buck converters for various input voltages at switching frequencies of 500 kHz and 1 MHz. As frequency and voltage increase, so do the gains offered by GaN transistors. For a 12 V_{IN} design operating at 1 MHz, GaN transistors can improve

GaN Transistors for Efficient Power Conversion, Second Edition.
Alex Lidow, Johan Strydom, Michael de Rooij, and David Reusch.

Companion Website: http://www.wiley.com/go/gan_transistors

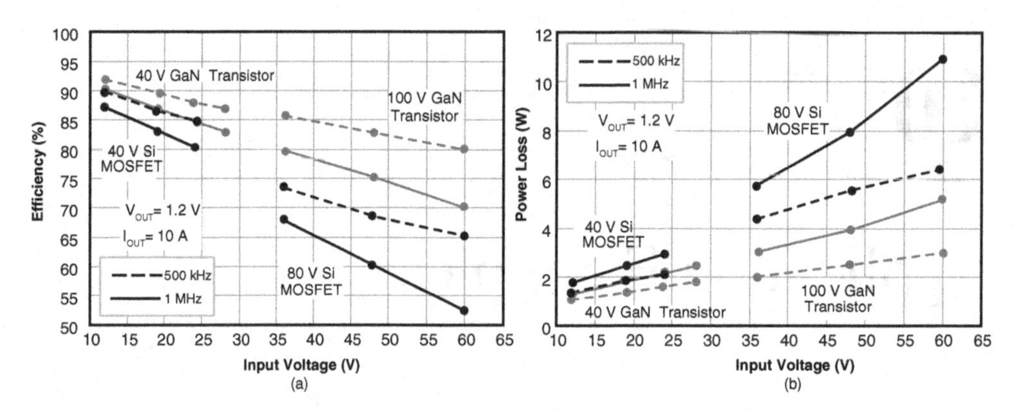

Figure 10.1 (a) Efficiency and (b) power loss comparisons for GaN transistors and Si MOSFETs in a synchronous buck converter ($V_{OUT} = 1.2$ V, $I_{OUT} = 10$ A, 40 V GaN transistors: control device: EPC2015 synchronous rectifier: EPC2015, 40 V MOSFETS: control device: BSZ097N04LSG synchronous rectifier: BSZ040N04LSG,100 V GaN transistors: control device: EPC2001, synchronous rectifier: EPC2001, 80 V MOSFETS: control device: BSZ123N08NS3G, synchronous rectifier: BSZ123N08NS3G)

efficiency by around 3%. For a 60 V_{IN} design at 1 MHz, GaN transistors can improve efficiency by around 18%. This section will demonstrate the advantages offered by GaN transistors in common POL applications ranging from 60 V_{IN} to 12 V_{IN}.

10.2.1 12 V_{IN} – 1.2 V_{OUT} Buck Converter

In this section, a POL converter designed for one of the most common requirements will be discussed, 12 V_{IN} and 1.2 V_{OUT}. This is both a cost-sensitive and an efficiency-sensitive high-volume application. In Chapters 4 and 6, the importance of minimizing parasitics and providing an optimal layout for GaN transistors was discussed, In this section, these techniques will be applied to a converter operating at 1 MHz to yield a benchmark efficiency.

To compare the performance of the optimal power loop discussed in Chapter 4 (Section 4.4) with conventional lateral or vertical layouts, four separate designs were created. The overall thickness of the board and the distance between the top layer and the first inner layer in the board (inner layer distance) varies as illustrated in Figure 10.2. The component layouts remained unchanged from those used earlier in Section 4.4. All designs have four layers with two-ounce copper thickness, and their variables are given in Table 10.1.

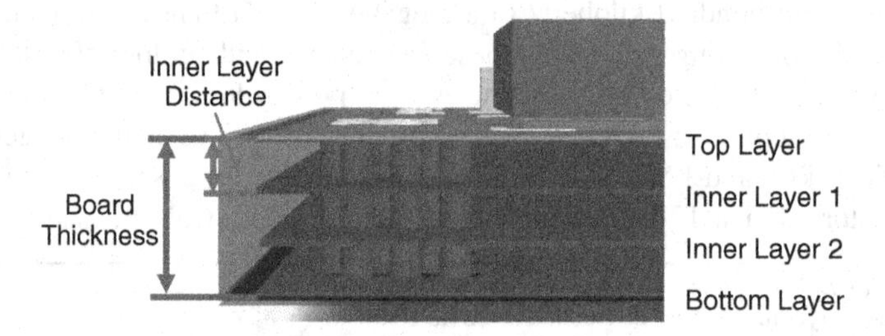

Figure 10.2 PCB cross-section drawing of board thickness and inner layer distance

Table 10.1 Board variables for layout comparison

	Board thickness (mils)	First inner layer distance (mils)
Design 1	31	4
Design 2	31	12
Design 3	62	4
Design 4	62	26

Simulated values of the high-frequency loop inductance for varying board thicknesses and inner layer distance are presented in Figure 10.3. From the data, it can be seen that for the lateral power loop the board thickness has little impact on the high-frequency loop inductance, while the inner layer distance (the distance from the power loop to the shield layer) significantly impacts the inductance. For the vertical power loop, the inner layer distance has very little impact on the inductance of the design, while the board thickness significantly increases the inductance by as much as 80% when the board thickness is doubled from 31 to 62 mils (0.8 mm to 1.5 mm).

For the optimal layout, the design shares the traits of the lateral power loop by showing little dependence on board thickness and a strong dependence on inner layer distance. This design provides a significant reduction in loop inductance by the removal of the shield layer and a reduced physical size of the power loop, traits similar to the vertical power loop design. Combining the strengths of both conventional designs, and limiting the weaknesses, this design provides a significant reduction in inductance compared to the best lateral and vertical power loops.

The power loss for the four board thicknesses and the three different loop layouts is shown in Figure 10.4. From this data, it can be seen that, for similar parasitic inductances, the power loss of the lateral loop is higher than the vertical loop. This increased loss in the lateral power loop

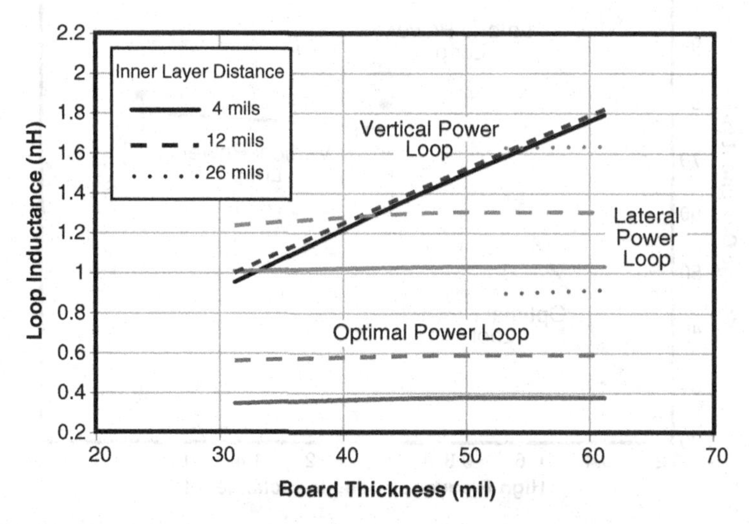

Figure 10.3 Simulated high-frequency loop inductance values for lateral, vertical, and optimal power loops with different board thickness and inner layer distance

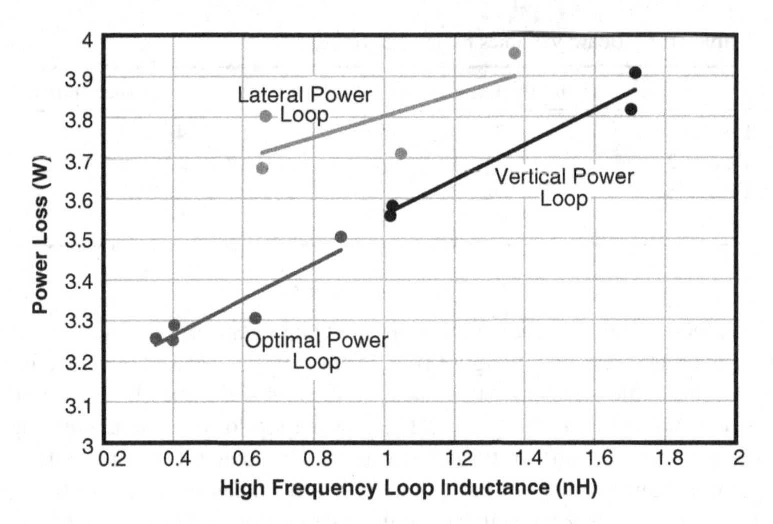

Figure 10.4 Power loss plot for lateral, vertical, and optimal power loop designs. (V_{IN} = 12 V, V_{OUT} = 1.2 V, I_{OUT} = 20 A, f_{sw} = 1 MHz, L = 300 nH, control device: EPC2015, synchronous rectifier: EPC2015)

can be attributed to the additional loss in the shield layer due to eddy currents, which is not required in the vertical or optimal power loop.

The voltage overshoot for the different designs is shown in Figure 10.5. As loop inductance increases up to 1.4 nH, so does the voltage overshoot. Beyond 1.4 nH, the voltage overshoot does not significantly increase. This can be explained by Figure 10.6 that shows the measured switching speed of the different designs. As the loop inductance increases, the dv/dt of the

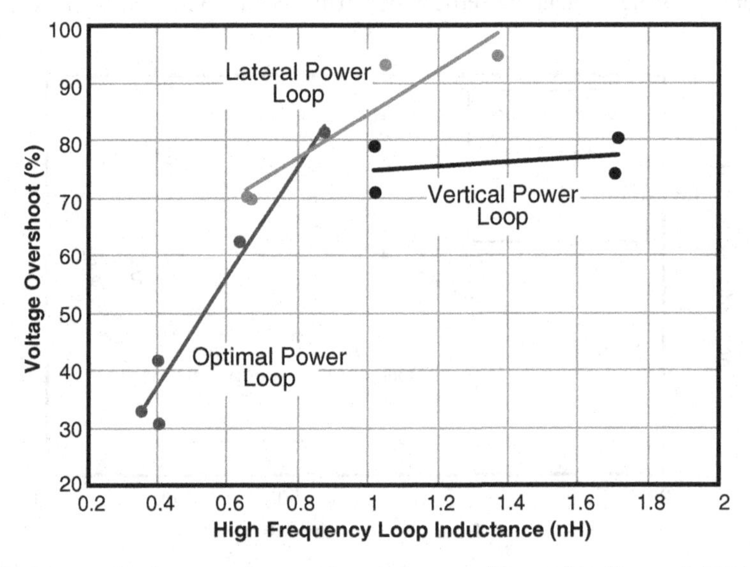

Figure 10.5 Measured voltage overshoot vs. loop inductance (V_{IN} = 12 V, V_{OUT} = 1.2 V, I_{OUT} = 20 A, f_{sw} = 1 MHz, L = 300 nH, control device: EPC2015, synchronous rectifier: EPC2015)

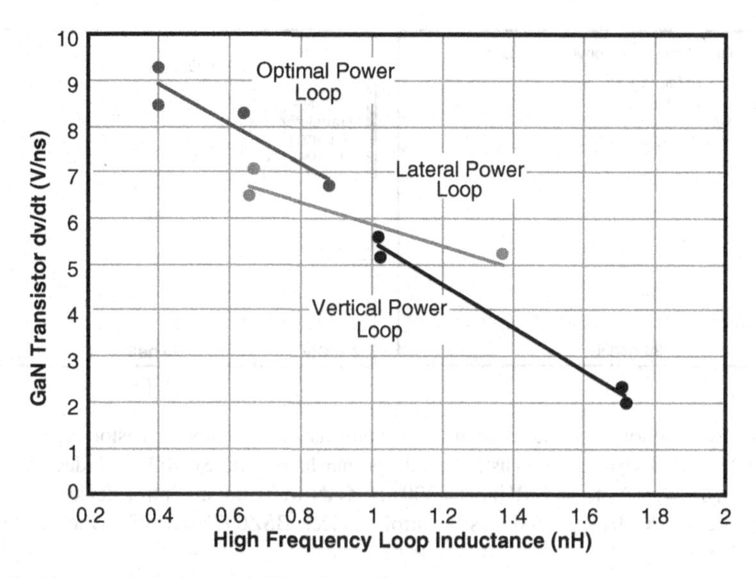

Figure 10.6 Measured device switching speed vs. loop inductance ($V_{IN} = 12\,V$, $V_{OUT} = 1.2\,V$, $I_{OUT} = 20\,A$, $f_{sw} = 1\,MHz$, $L = 300\,nH$, control device: EPC2015, synchronous rectifier: EPC2015)

device decreases significantly. This results in higher power loss, but helps limit voltage overshoot. For the two vertical loop designs with the highest loop inductance, the switching speed is reduced by over 60% when compared to all the other designs.

Shown in Figure 10.7 are the efficiency results of design 1 (Table 10.1) for the three GaN transistor based designs compared to a silicon implementation utilizing a vertical power loop with the smallest commercial package, a 3 × 3 mm TSDSON-8. For the Si MOSFET design,

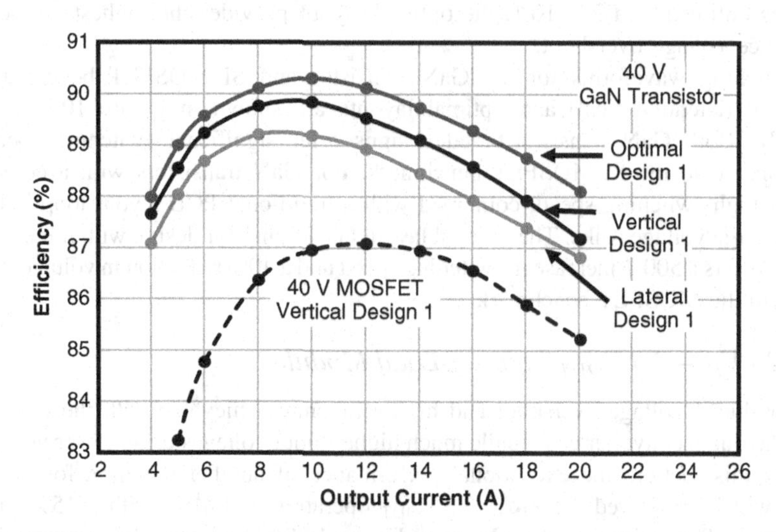

Figure 10.7 Efficiency comparisons for design 1 in Table 10.1 ($V_{IN} = 12\,V$, $V_{OUT} = 1.2\,V$, $f_{sw} = 1\,MHz$, $L = 300\,nH$, GaN transistors: control device: EPC2015, synchronous rectifier: EPC2015, MOSFETs: control device: BSZ097N04LSG, synchronous rectifier: BSZ040N04LSG)

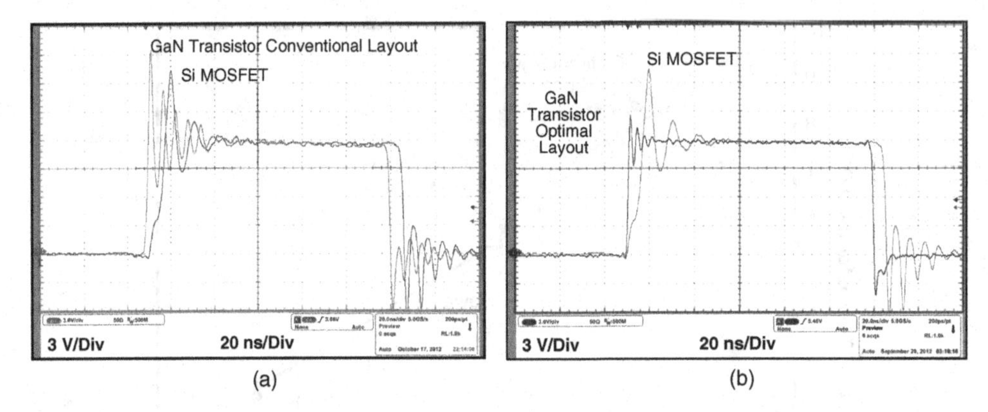

Figure 10.8 Synchronous rectifier switching waveforms of (a) GaN transistor-based conventional layout vs. Si MOSFET (b) GaN transistor-based optimal layout vs. Si MOSFET design ($V_{IN} = 12$ V, $V_{OUT} = 1.2$ V, $I_{OUT} = 20$ A, $f_{sw} = 1$ MHz, L = 300 nH, GaN transistors: control device: EPC2015, synchronous rectifier: EPC2015, MOSFETs: control device: BSZ097N04LSG, synchronous rectifier: BSZ040N04LS G)

the high-frequency loop inductance was measured to be around 2 nH, compared to 1 nH for a similar power loop using GaN transistors. This is due to the large packaging inductance of the Si MOSFET dominating the loop design. As a result of the superior FOM and packaging of the GaN transistors, all of the power loop structures outperform the Si MOSFET benchmark design. With the optimal power loop, the GaN transistor designs can be improved even further, achieving almost 3% full load and 3.5% peak efficiency improvements.

For the different GaN transistor designs, the optimal power loop provides a 0.8% and 1% full load efficiency improvement over the vertical and lateral power loops, respectively. For all of the designs outlined in Table 10.1, the optimal layout provides the highest efficiency and lowest device voltage overshoot.

The switching waveforms for the GaN transistor and Si MOSFET benchmark-based conventional (lateral/vertical) and optimal layouts are shown in Figure 10.8(a) and (b), respectively. Both GaN transistor-based designs offer significant switching speed gains when compared to the Si MOSFET benchmark. For GaN transistors with a conventional layout, the high switching speed, combined with a traditional PCB layout loop inductance, results in a large voltage spike. The optimal layout GaN transistor design with minimized loop inductance offers a 500% increase in switching speed and a 40% reduction in voltage overshoot compared to the MOSFET benchmark.

10.2.2 28 V_{IN} – 3.3 V_{OUT} Point-of-Load Module

With the reduced voltage overshoot and high efficiency achievable with the optimal GaN transistor layout, a converter can handle much higher input voltages using lower voltage-rated devices. In this section, the exceptional performance of another common format of POL converter will be reviewed: 28 V_{IN} – 3.3 V_{OUT} operating at 1 MHz with a 15 A maximum output current. The entire circuit is shown in Figure 10.9(a) and occupies an area of 0.25 in^2 (0.4 cm^2). The peak overshoot at full load and at 28 V input is around 3 V, as illustrated in the oscillogram of Figure 10.9(b), easily allowing the use of 40 V GaN transistors. As shown in

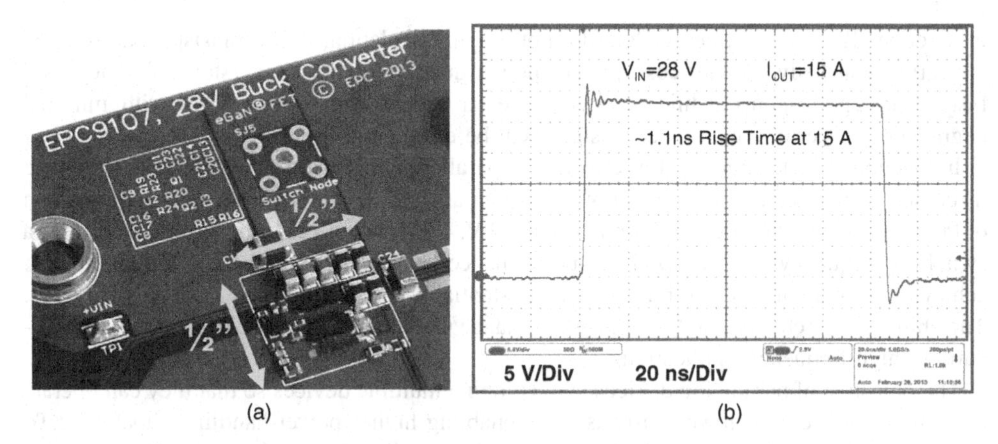

Figure 10.9 (a) Photo of a GaN transistor-based POL module (b) Synchronous rectifier switching waveform (V_{IN} = 28 V, V_{OUT} = 3.3 V, I_{OUT} = 15 A, f_{sw} = 1 MHz, GaN transistors: control device: EPC2015, synchronous rectifier: EPC2015)

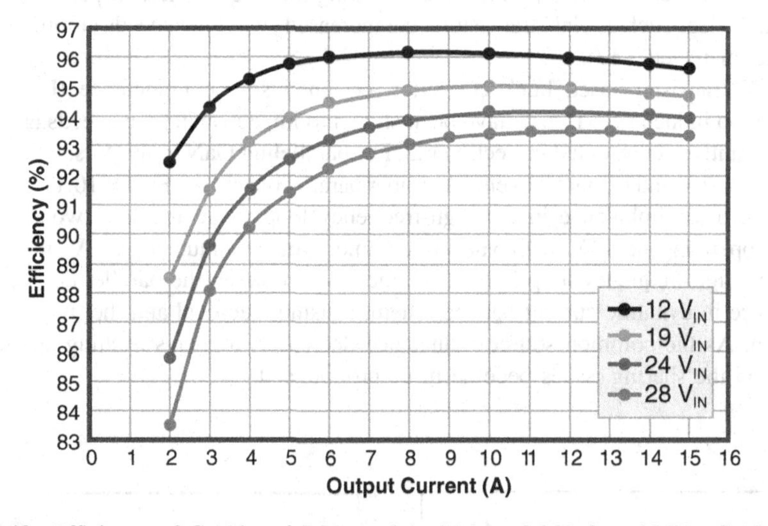

Figure 10.10 Efficiency of GaN-based POL module (V_{OUT} = 3.3 V, f_{sw} = 1 MHz, GaN transistors: control device: EPC2015, synchronous rectifier: EPC2015)

Figure 10.10, this design demonstrates efficiency above 96% for a 12 V-input and over 93% for a 28 V input.

10.2.3 48 V_{IN} – 12 V_{OUT} Buck Converter with Parallel GaN Transistors for High-Current Applications

Common to many computer servers is a 48 V power distribution bus. In many cases there is a secondary distribution bus at 12 V (more on this in Section 10.3). As a result, there is a large need for DC-DC converters that step down from 48 V_{IN} to 12 V_{OUT}. Many of these systems require isolation between the input and output, and those isolated converters will be discussed in Section 10.3. For systems that do not require isolation, GaN-transistor-based buck converters can deliver lower cost, greater power density, higher conversion efficiency, and faster transient response while achieving greater output current capability with minimal degradation in performance. Such designs will be considered in this section.

In Section 4.5, techniques for effectively paralleling high-speed GaN transistors were discussed. In this section, the impact of in-circuit parasitics on performance and a comparison of PCB layouts will be examined for a 48 V to 12 V, 480 W, 40 A buck converter operating at a switching frequency of 300 kHz. High-performance GaN transistors operated in parallel can achieve higher frequency and power with substantially higher efficiency than Si MOSFETs. The ability to effectively parallel high-performance GaN transistors enables a variety of high-current, high-frequency applications.

The objective of paralleling devices is to combine multiple devices so that they can operate as a single device with lower on-resistance, enabling higher power-handling capability. To effectively parallel devices, each device in a switch cluster should equally share current dynamically, and equally divide switching-related losses in steady state. The introduction of imbalanced in-circuit parasitics between parallel devices leads to uneven power sharing and degraded electrical and thermal performance, limiting the effectiveness of paralleling [1]. For high-speed devices such as GaN transistors, the increased switching speeds amplify the impact of parasitic mismatches [2].

In a GaN transistor-based buck converter, common source inductance (L_S) and high-frequency loop inductance (L_{Loop}) have been shown to impact switching speeds and performance significantly, as presented in Section 6.3. For paralleling GaN transistors, these parasitics must be minimized and balanced to ensure proper parallel operation. Figure 10.11(a) shows the impact of parasitic imbalance in the high-frequency loop inductance for two parallel GaN transistors operating at 48 V with various common source inductances. As the difference between the high-frequency loop inductance increases between the parallel devices, so does the difference in dynamic current between them, causing electrical and thermal performance degradation. As the common source inductance decreases, higher switching speeds can be achieved and the sharing issues become more pronounced.

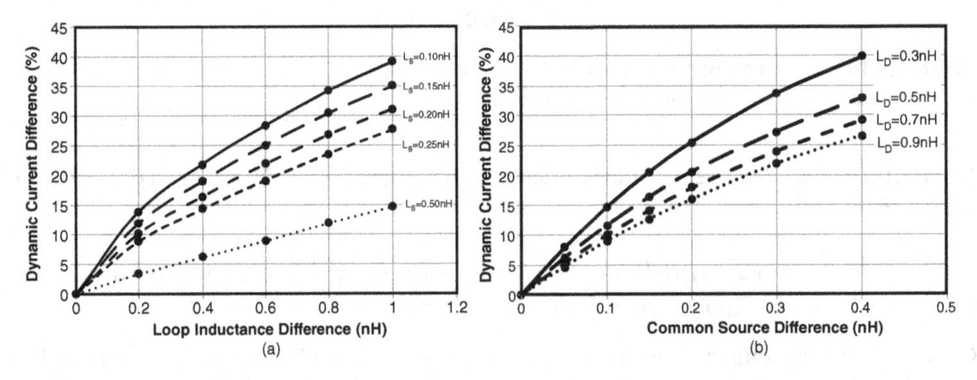

Figure 10.11 Impact of (a) high-frequency loop inductance (b) common source inductance parasitic imbalance on transistor dynamic current sharing for a $V_{IN} = 48$ V, $I_{OUT} = 25$ A, GaN transistor-based buck converter with two devices operating in parallel (GaN transistors: EPC 2001)

Figure 10.11(b) shows the dynamic current difference resulting from parasitic imbalance in the common source inductance for two parallel GaN transistors operating at 48 V with various high-frequency loop inductances. Similar to loop inductance imbalance, as common source inductance varies, current sharing worsens. This trend is magnified as loop inductance decreases and capable switching speeds increase.

To improve the parallel performance of high-speed GaN devices, the parasitic imbalance contributed by the PCB layout must be minimized. Two different parallel layouts will be reviewed, each based on an optimal layout, and an assessment given of their ability to provide parallel performance similar to the optimal single transistor design. Each half-bridge design contains four devices in parallel for the control switch (T_{1-4}) and the synchronous rectifier (SR_{1-4}). The design was tested in a buck converter configuration operating with input of 48 V and output of 12 V with a switching frequency of 300 kHz. In total, eight 100-V EPC2001 GaN transistors were used to achieve an output power up to 480 W and an output current of up to 40 A.

The first parallel design is shown in Figure 10.12(a) using the paralleling layout techniques shown in Figure 4.8 in Section 4.5. In this layout the four GaN transistors are located in close proximity to operate as a "single" power device, with a single high-frequency power loop as shown in Figure 4.10. As mentioned in Section 4.5, this layout may not be the best design for paralleling a large number of GaN transistors. The drawbacks of this layout are that the devices will have imbalanced parasitics, leading to current sharing and thermal issues.

The second parallel design is shown in Figure 10.12(b) and utilizes four distributed high-frequency power loops as was shown in Figure 4.12. This design is expected to provide the lowest overall parasitics for each device pair and, most importantly, provide the best balancing of the parasitic elements, ensuring proper parallel operation.

The switch-node voltage waveforms of the synchronous rectifiers for the two designs are shown in Figure 10.13. The switch-node waveforms for the single high-frequency power loop design are shown in Figure 10.13(a), the voltage transitions for the innermost and outermost devices show an almost 2 ns switching time difference, which equates to about 25% of the total switching time. The voltages were measured using the techniques described in Section 5.4 (specifically Figure 5.20) with individual connection pads for each of the devices. This voltage difference demonstrates the parasitic imbalance in this PCB layout, which leads to inefficient paralleling, resulting in current sharing and thermal issues.

The switch-node waveforms for the symmetrical four high-frequency power loop design are shown in Figure 10.13(b). The voltage transitions for the devices are almost identical,

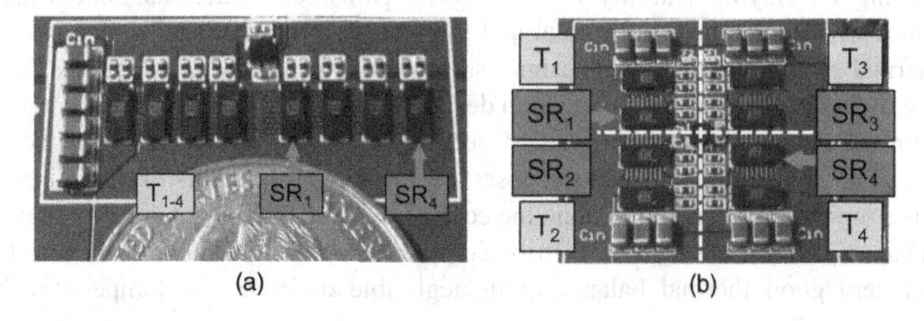

Figure 10.12 Four parallel transistor GaN layouts with (a) single high-frequency power loop (b) four distributed high-frequency power loops

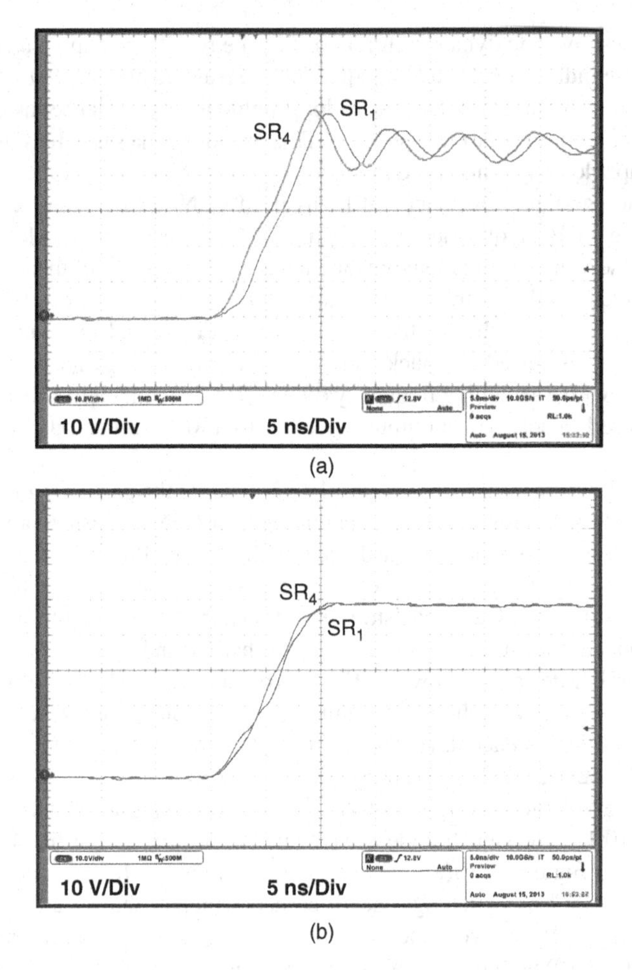

Figure 10.13 Switching waveforms of parallel GaN transistors for (a) single high-frequency power loop (b) four distributed high-frequency power loop designs ($V_{IN} = 48$ V, $V_{OUT} = 12$ V, $I_{OUT} = 30$ A, $f_{sw} = 300$ kHz, $L = 3.3$ μH, GaN transistors control device/synchronous rectifier: 100 V EPC2001)

demonstrating this layout's ability to balance the parasitics well, thus improving overall performance by offering better electrical and thermal performance.

The thermal evaluation of the two designs, shown in Figure 10.14, demonstrates the thermal imbalance of the single high-frequency loop design. Figure 10.14(a) shows that a hot spot will develop on the devices handling a greatest portion of the power as a result of the parasitic imbalance. The control switch located closest to the input capacitors, T_1, has a maximum temperature more than 10 °C higher than the control switch device located furthest away from the input capacitors, T_4. For the four distributed power loop design, shown in Figure 10.14(b), there is a very good thermal balance, with negligible difference in temperature between devices.

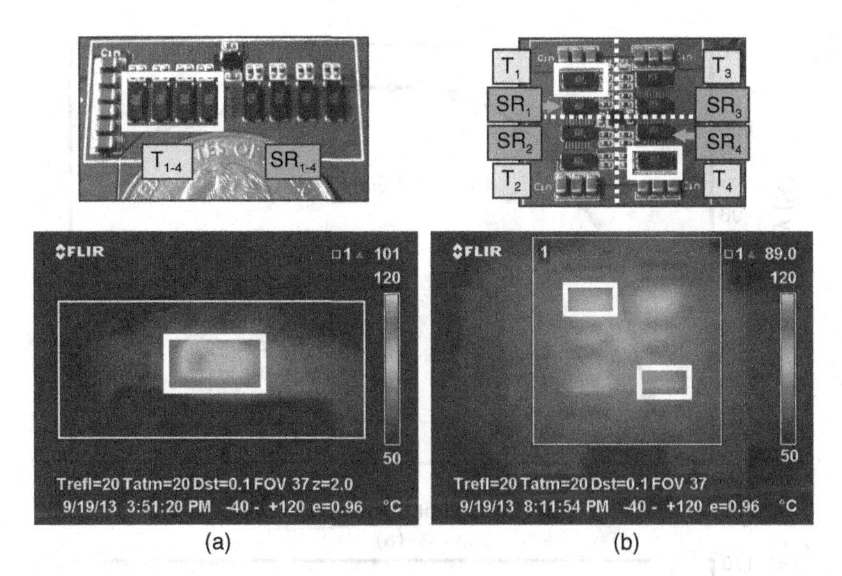

Figure 10.14 Thermal measurements of parallel GaN transistors layouts with a (a) single high-frequency power loop (b) four distributed high-frequency power loops ($V_{IN} = 48$ V, $V_{OUT} = 12$ V, $I_{OUT} = 30$ A, $f_{sw} = 300$ kHz, L = 3.3 μH, GaN transistors: control device/synchronous rectifier: 100 V EPC2001)

By offering lower individual parasitics and better parasitic balance, the four high-frequency loop design is more effective at paralleling. This results in better electrical and thermal performance, as shown in Figure 10.15. The distributed high-frequency loop design offers a 0.2% gain in efficiency at 40 A, and has an almost constant 10 °C improvement in the maximum transistor temperature.

The switching waveforms for an optimal PCB design using a single GaN transistor, two parallel GaN transistors, and four parallel GaN transistors are shown in Figure 10.16. The parallel designs operate like a single device with lower resistance and slower switching speed in proportion to the number of devices connected in parallel.

The ability of GaN transistors to increase switching frequencies and achieve higher power handling through the effective paralleling methods discussed in this section allows for the exploration of new opportunities in popular applications such as the intermediate bus architecture (IBA), whose power architecture will be discussed in the next section. Traditionally, IBAs have required isolation to not only provide user safety, but also to reduce the power handled by the primary devices as the primary current is reduced by the effective turns ratio of the transformer. A growing number of applications are removing the isolation requirement, enabling the bulky transformer and complex control circuits to be removed. High-performance GaN devices, operated in parallel, can handle the higher frequency and power with substantially higher efficiency than Si MOSFETs. Shown in Figure 10.17 is the efficiency comparison between the 300 kHz parallel buck converter and published performance of state-of-the-art eighth-brick Si MOSFET-based bus converters. The GaN transistor-based solution offers a

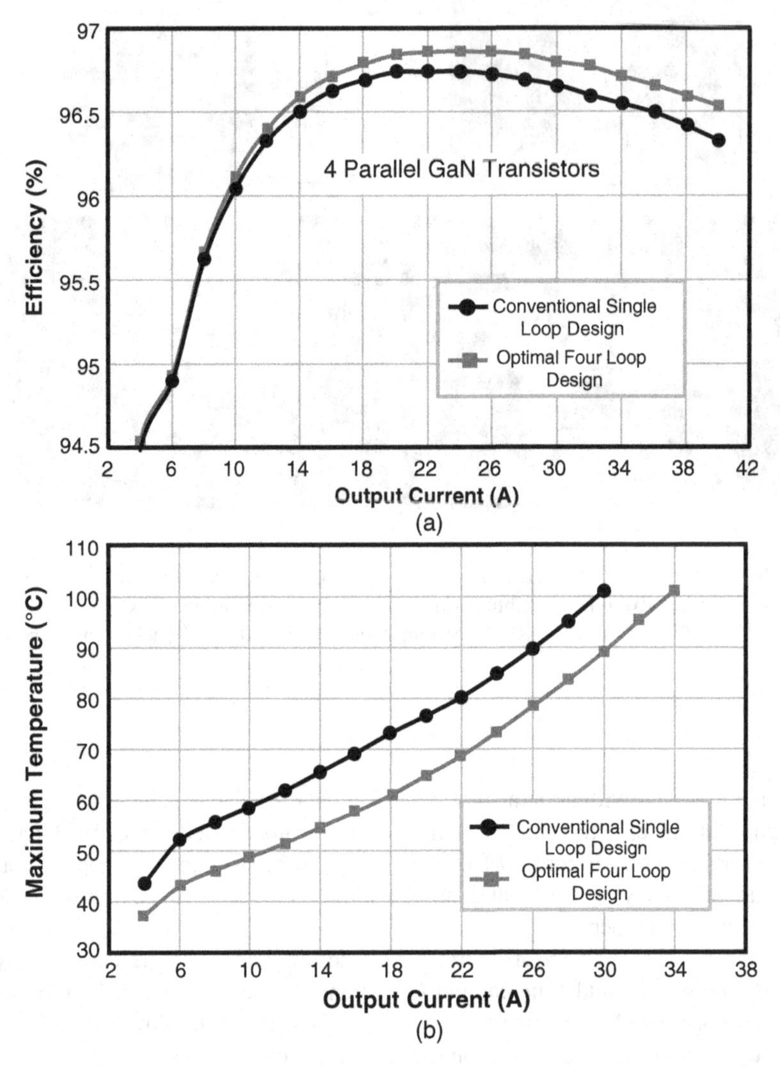

Figure 10.15 (a) Efficiency and (b) thermal comparison for conventional and optimal parallel GaN transistor designs ($V_{IN} = 48\,V$, $V_{OUT} = 12\,V$, $f_{sw} = 300\,kHz$, $L = 3.3\,\mu H$, control switch: 4× EPC2001, synchronous rectifier: 4× EPC2001)

gain of over 2% in peak efficiency and a 50% improvement in power density over the traditional Si brick converter, assuming a constant power loss limitation.

This section demonstrated the advantages offered by GaN transistors in non-isolated, point-of-load applications ranging from 60 V_{IN} to 12 V_{IN}. These converters are commonly found in computers, telecommunication systems, handheld electronics, and many other applications. As shown, GaN transistors enable smaller size and lower power consumption. The increased switching frequency also enables faster transient response. These new capabilities are necessary for meeting the demands for next-generation digital loads.

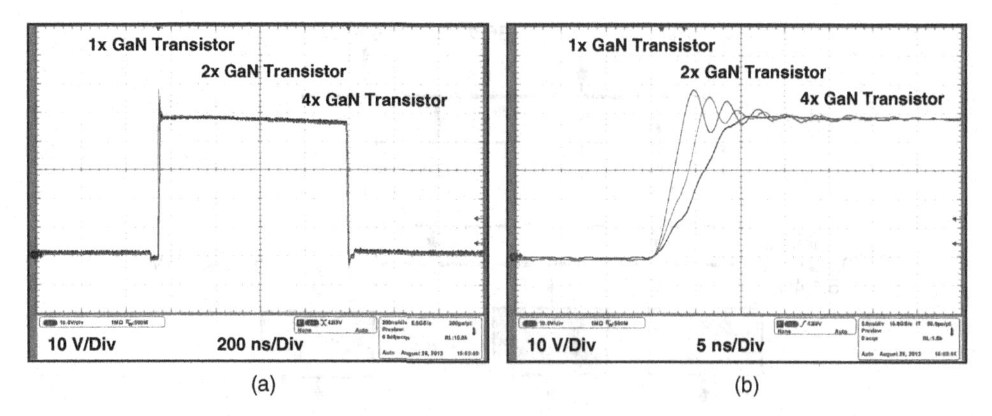

Figure 10.16 Switching waveforms of optimal layout with 1, 2, and 4 parallel GaN transistors ($V_{IN} = 48$ V, $V_{OUT} = 12$ V, $I_{OUT} = 30$ A/(number of GaN transistors) $f_{sw} = 300$ kHz, L = 3.3 μH, GaN transistors: control device/synchronous rectifier: 100 V EPC2001)

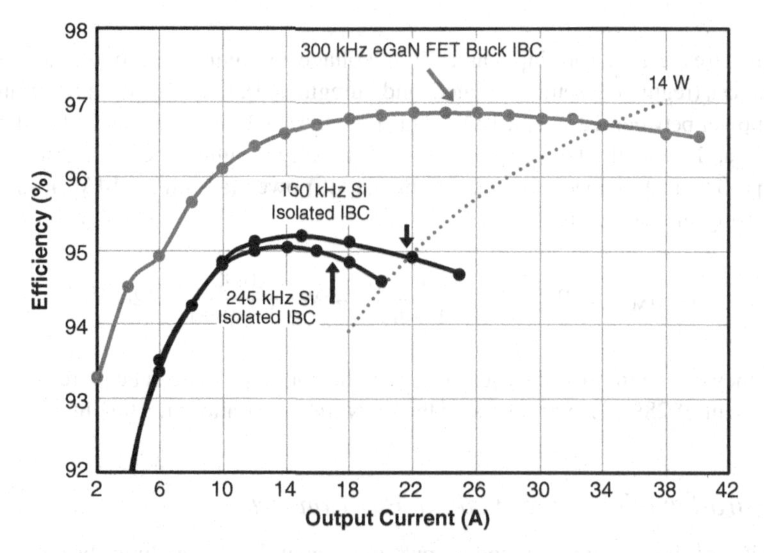

Figure 10.17 Efficiency comparison for conventional Si-based isolated brick DC-DC converter and a GaN transistor non-isolated design for a fully regulated IBC ($V_{IN} = 48$ V, $V_{OUT} = 12$ V)

10.3 Isolated DC-DC Converters

Isolated DC-DC converters are widely used in computing and telecommunication systems as part of the IBA approach. They are available in a variety of standard sizes, input and output voltage ranges, and topologies as shown in Figure 10.18. Their modularity, power density, reliability, and versatility have simplified the isolated power supply market. Since the size of these "brick" converters is strictly defined, designers are continually innovating ways to increase their output power density. Although these ideas are numerous and varied, they are all related to system efficiency. Consider an eighth-brick converter as an example: an eighth-brick converter must have a dimension of 0.9 inches (2.3 cm) by 2.3 inches (5.8 cm).

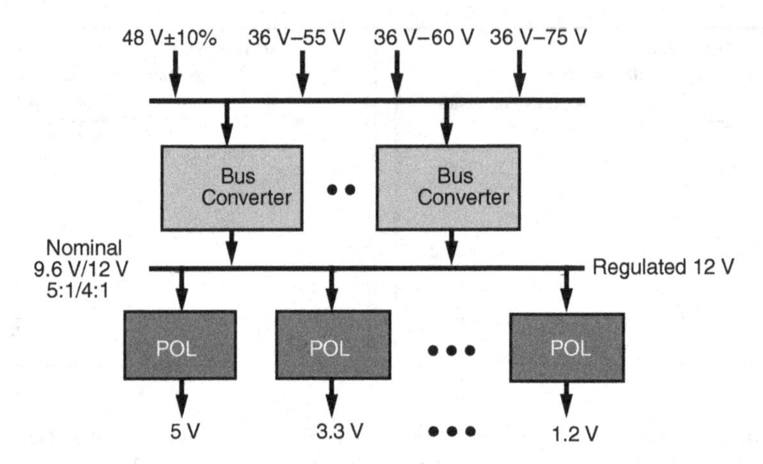

Figure 10.18 IBA showing voltage ranges for bus converters

Although there are numerous input and output voltage configurations, topologies, and output range tolerances (regulated, semi-regulated, and unregulated), they all have a maximum power loss at full power between 12 W and 14 W. This is a physical limit based on the fixed volume of the converter and the method of heat extraction. Thus, for an eighth-brick converter that is 90% efficient ($\eta = 0.9$) at full load, the maximum output power (assuming 14 W losses) can be calculated using the following equation:

$$P_{OUTMAX} = P_{MAXLOSS} \cdot \frac{\eta}{1-\eta} = 14\ \text{W} \cdot \frac{0.90}{1-0.90} = 126\ \text{W} \tag{10.1}$$

If the efficiency can be improved by just 2%, then the output power can be increased to 160 W, an improvement of 28%. Lower losses of the GaN transistor allow for this increase in output power.

10.3.1 Hard-Switching Intermediate Bus Converters

The majority of bus converters today use traditional hard-switching bridge topologies operating in the relatively low frequency range of 150–250 kHz in order to maximize efficiency. At these lower switching frequencies, the isolation transformer and output inductor are very bulky and occupy a large portion of the board area. To improve the power density of the converter, the operating frequency must be increased to process more power through the inductor and transformer. However, as the switching frequency increases for silicon-based converters, losses from the MOSFET body diode conduction, reverse recovery, and transistor switching losses increase, thus limiting the output power capability of the converter. These losses have forced power density improvements to come from changes in design optimization and topology, rather than increases in transistor switching frequencies. Fortunately, with the advent of GaN transistors, designers have the opportunity to overcome the limitations of the silicon switching device. The lower switching losses and the absence of any reverse recovery charge of GaN transistors enable higher operating frequencies for higher output power density.

10.3.1.1 Eighth-Brick Converter Using GaN Transistors

To illustrate the improvements that GaN transistors offer in this application, a high-switching-frequency GaN transistor-based eighth-brick bus converter was constructed. For the eighth-brick converter, a full-bridge primary-side converter with a full-bridge synchronous rectifier was chosen as shown in Figure 10.19. The actual converter is shown in Figure 10.20 and compared side by side to a silicon-based converter. To the skilled designer, the significant amount of "green" space (unfilled PCB area) in the GaN-based converter could be exploited further to improve efficiency.

Efficiency results for the GaN transistor-based converter compared to the same MOSFET-based brick converter are shown in Figure 10.21. Despite the GaN transistor converter operating at 33% higher frequency, it is able to produce 15% more output power for the same power loss. As the GaN transistor-based converter has not been optimized in either topology, thermal design or switching frequency, further improvement is possible.

Since this initial GaN transistor-based brick comparison, topological developments for even higher power MOSFET-based eighth-brick converters have continued to squeeze more performance out of MOSFET-based systems. In Figure 10.22, the output power of three different generations of regulator eighth-brick converters are plotted versus their switching frequency. What is clear from this figure is the continuing downward trend in switching frequency as well as an increase in overall converter loss (given in brackets below each point). This overall increased converter loss places additional strain on the thermal design of these converters, which require complex thermal derating curves. Based on FOM advantages and comparisons of many different topologies, it is assured the GaN transistors will outperform MOSFETs in each and every hard-switching intermediate bus converter.

10.3.1.2 Isolated PoE-PSE Converters

The power over Ethernet (PoE) standard has been evolving over the last few years. The main focus is the systematic increase of power with each new class and type. According to the IEEE 802.3 standard on PoE [3], the power source equipment (PSE) requires an output voltage of 44–57 V for PoE type 1, and 50–57 V for PoE type 2 (PoE+). It also has to be capable of delivering 15.4 W (type 1) or 25.5 W (type 2) per port on the PSE Ethernet switch. For the PSE, the output requires some form of regulation, but tight regulation is not required. There is also an increase in minimum voltage to account for the increase in maximum line droop with the increased power level. With 24, 36, or even 48 ports per Ethernet switch being typical, the total PSE supply power requirement can be as high as 1.2 kW. This drives the need for the higher efficiency and higher power density converters achievable with GaN.

It is difficult to remove more than 35 W of losses from a typical PoE-PSE half-brick converter, even with significant airflow or base plate. In Figure 10.23, the resultant output power that is achievable in a half-brick is plotted versus the required minimum full load efficiency. Since most commercial half-brick PSE converters already have about 95% efficiency, even one-half of one percent efficiency improvement is important and can increase output power by as much as 100 W. As with other brick converters, the most important aspect is cost per watt ($/W), thus, by increasing brick efficiency and therefore output power with GaN transistors, the price of the module per watt is reduced.

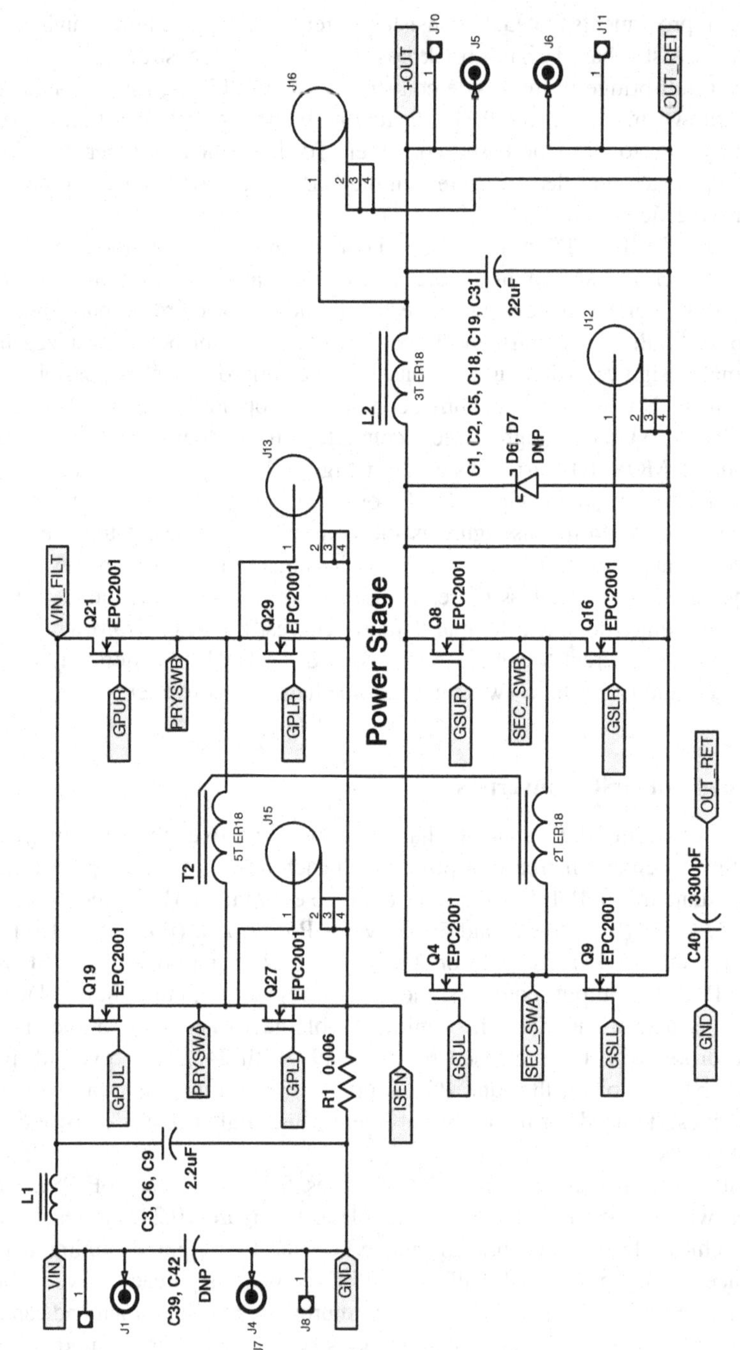

Figure 10.19 200 W, eighth-brick, fully regulated, full-bridge primary with full-bridge synchronous rectification using GaN transistors

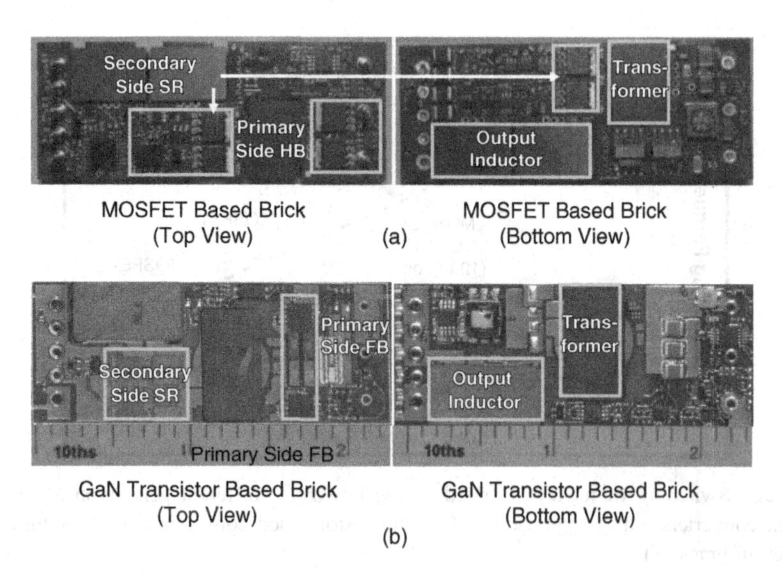

Figure 10.20 Comparison between the 48 V to 12 V GaN transistor-based eighth-brick converter and comparable silicon-based converter (a) MOSFET top view (b) MOSFET bottom view (c) GaN transistor top view (d) GaN transistor bottom view (scale in inches)

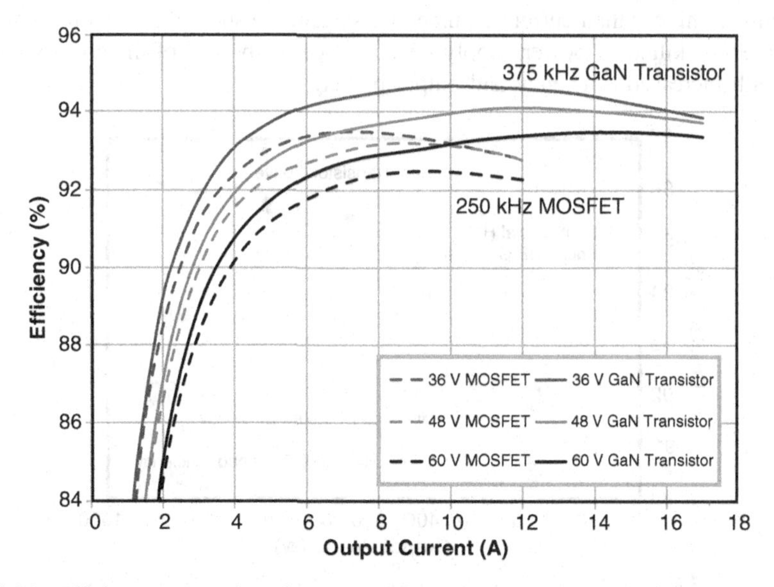

Figure 10.21 Efficiency comparison between a GaN transistor and MOSFET-based eighth-brick converter

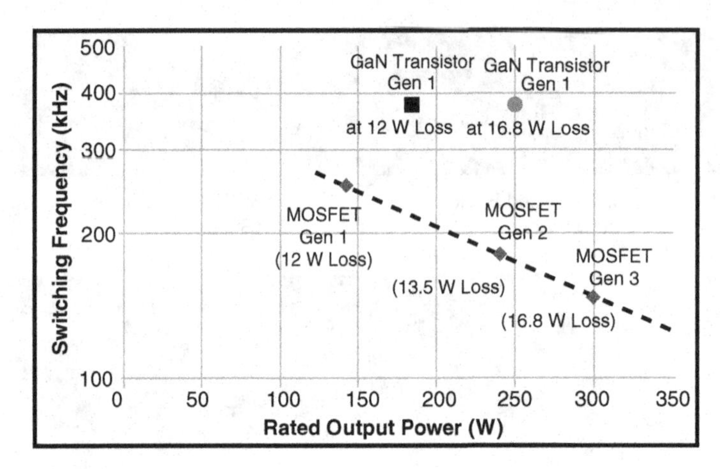

Figure 10.22 Switching frequency vs. rated output power for different generation MOSFET-based eighth-brick converters with comparison to GaN transistor-based converter (converter losses at rated power given in brackets)

It is difficult to compare half-brick PoE-PSE converters because there are a significant number of variations between current commercial designs. Each manufacturer's advances in topology, materials, construction, layout, and other innovations have allowed ever-greater output power and enabled them to differentiate themselves in the market. With each generation of power supply, an increase in output power level is achieved as each manufacturer improves their design in terms of structure, layout, and topology.

A two-phase, interleaved converter design with single-stage conversion was selected for this GaN transistor demonstration. The design goal was to deliberately push the operating frequency much higher than current commercial systems to show that GaN transistors could enable someone skilled in power supply design to develop state-of-the-art next-generation products with increased efficiency and output power.

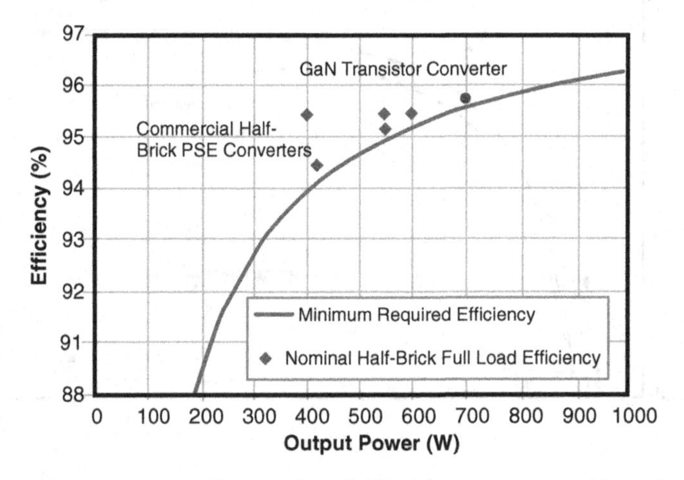

Figure 10.23 Minimum required efficiency for a half-brick converter to achieve the specific output power (assuming a maximum power loss of 35 W)

For larger brick sizes, such as the half-brick, the resultant output power levels and overall converter losses are high enough that multiple power devices are usually required for each switch, both from a thermal requirement and from a minimum available $R_{DS(on)}$ (largest die size) perspective. Thus, if the converter is split into two converters (each for half the power), the overall power device count is not affected. The added cost and size of using an increased number of inductors and transformers is also small, as these components are smaller, and an interleaving of the power converters generally allows for a reduction in the required output capacitance. Furthermore, the size and height restrictions of the bricks mean that a single high-power transformer is height limited with a less optimal magnetic path-length than two smaller transformer cores. The remaining differences, gate drive and control, are likely the determining factors.

For 48–53 V GaN transistor-based half-brick PSE converter, a phase-shifted full-bridge (PSFB) converter with a full-bridge synchronous rectifier (FBSR) topology was chosen, as shown in Figure 10.24 (a more complete schematic is shown in Figure 10.25). Not only does the interleaving of two phases avoid the complexity associated with paralleling devices, but the use of two separate converters conceptually allows for phase shedding to improve light-load efficiency. Efficiency results, showing that light-load efficiency can be improved by at least 2% for one- and two-phase operation, are plotted in Figure 10.26. Each converter operates at a 250 kHz device switching frequency, resulting in an output ripple frequency of 1 MHz. Figure 10.27 shows that, with the increase in switching frequency and the relatively small GaN transistor device size, two of these converters can be constructed within the available volume constraints. The choice of transformer turns ratio (4:7) means that, at 60 V_{IN}, the secondary-side winding voltage (not including switching spike) would be about 105 V, and therefore, 200 V-devices were used on the secondary side with 100 V-devices on the primary side. Unlike conventional brick designs, the magnetic components are not integrated within the main PCB, but are separate PCBs. Not only does this reduce the number of layers required for the main PCB, but it also allows the use of conventional surface-mount inductors for the output

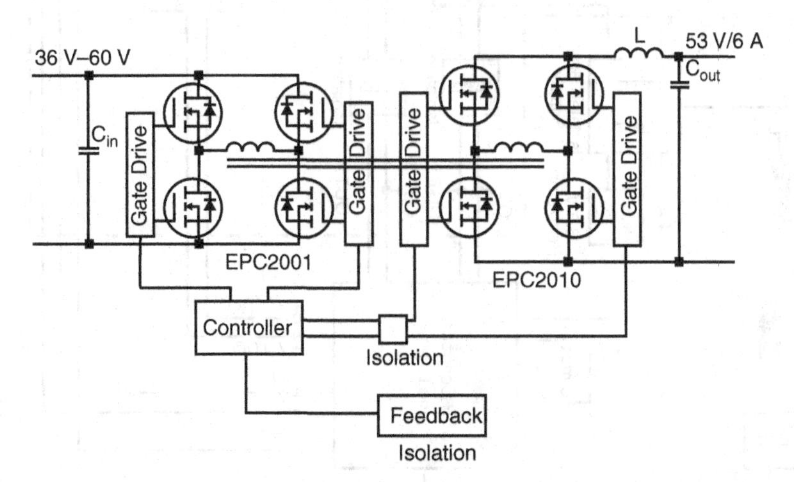

Figure 10.24 A 350 W fully regulated, phase-shifted, full-bridge (PSFB) topology, with full-bridge synchronous rectification (FBSR) using GaN transistors (two 250 kHz converters interleaved for the half-brick design)

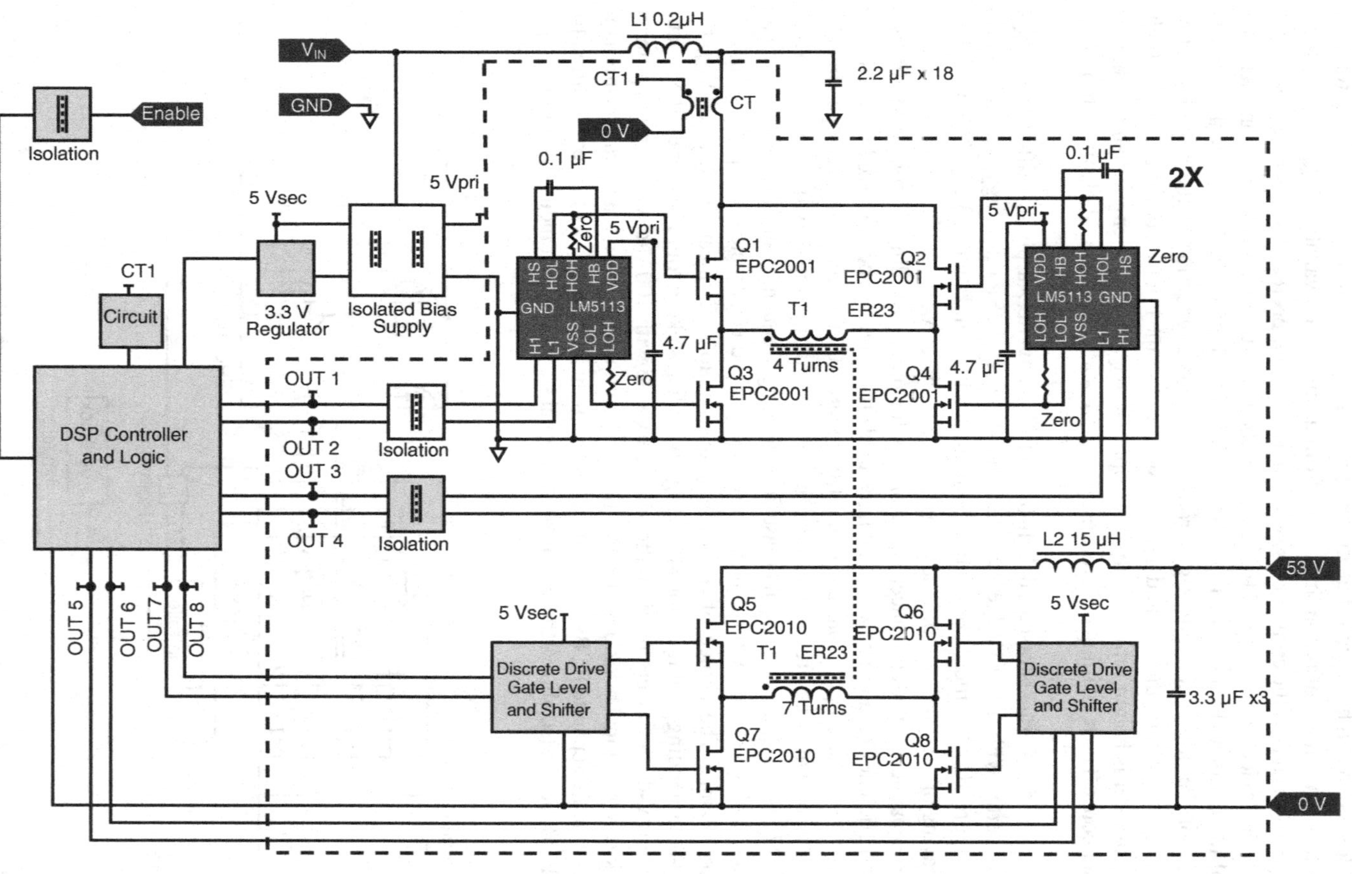

Figure 10.25 Simplified schematic of the GaN transistor-based, eighth-brick 38 V–60 V_{IN} to 53 V_{OUT}, 700 W converter operating at 250 kHz

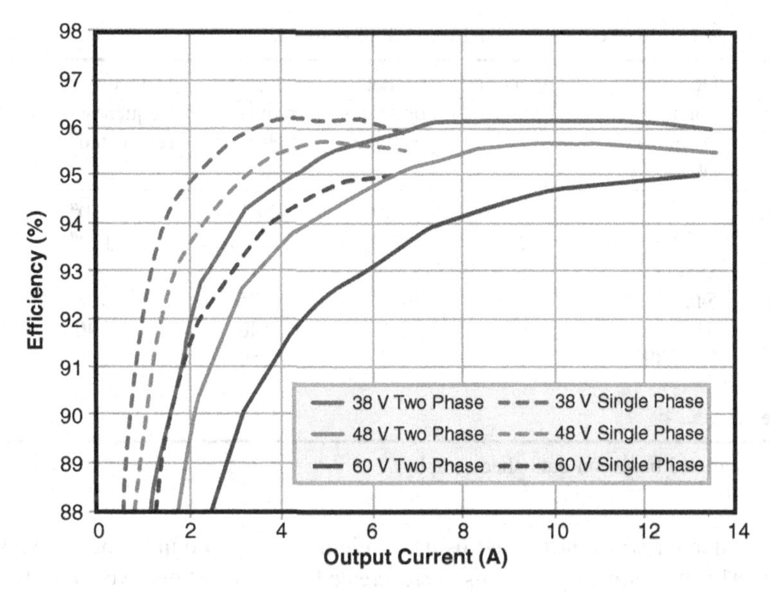

Figure 10.26 GaN transistor-based, half-brick PSE converter efficiency results showing both single-phase (half the converter powered down) and normal two-phase operation

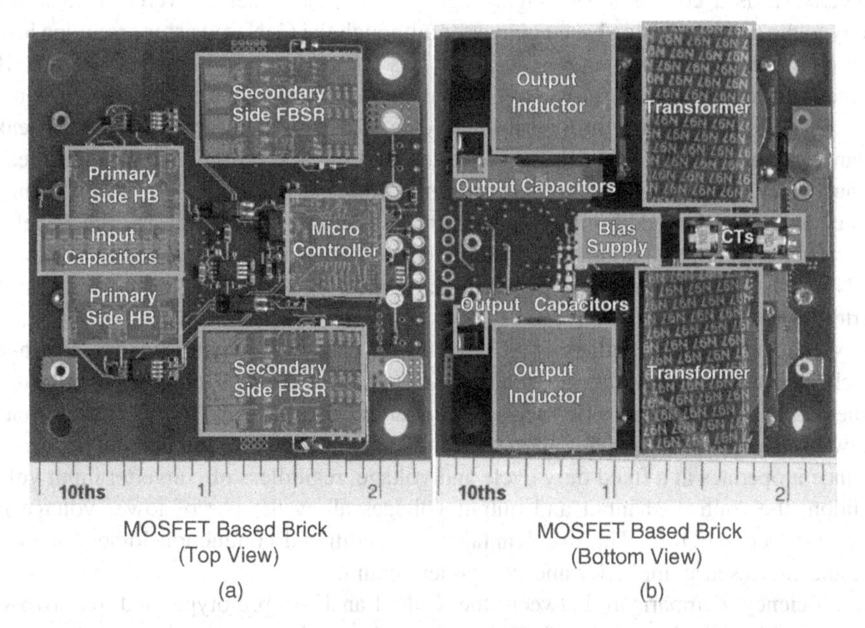

Figure 10.27 GaN transistor-based 48 V to 53 V, half-brick PSE converter (a) top view (b) bottom view (scale in inches)

Table 10.2 Comparison of commercial half-brick PSE converters

	Output voltage (V)/power (W)	Conversion stages	Parallel converters per stage	Datasheet frequency (kHz)	Device switching frequency (kHz) (estimated[a])	Output ripple (kHz)
Converter A	53/400	2	1/2	300/300	150[a]	300
Converter B	50/600	2	2/2	120/240	120[a]	n/a
Converter C	53/424	1	1	270	135	270
Converter D	54/550	1	1	275	138	275
Converter E	54/550	1	1	140	140	280
GaN transistor prototype	53.5/**700**	1	2	250	250	**1000**

[a] Frequency estimated based on data sheet graphs.

filters. The converter was constructed using an eight-layer, two-ounce copper thickness per layer PCB. The transformer windings were created by laminating two eight-layer PCBs together (in parallel) within the winding window.

The GaN transistor-based half-brick PSE converter can be compared with similar 48–53 V fully regulated commercial half-brick converters. These commercial converters span a range of topologies and configurations as listed in Table 10.2. To emphasize how the GaN transistor-based prototype compares to these converters, two products (B and D in Table 10.2) have been selected to highlight the overall results.

Converter D is a conventional single-stage, single-transformer converter with a similar topology to the GaN transistor-based converter (although the GaN transistor prototype has two parallel converters). The efficiency and power loss comparisons in Figures 10.28 and 10.29 show the light-load efficiency advantage that is possible with a lower switching frequency. Losses can be improved at light load through careful design of the core losses and leakage inductance. In the GaN transistor-based converter, the core was designed to minimize leakage inductance and switch at 75% higher switching frequency, resulting in lower light-load efficiency produced by additional magnetic losses. As the load increases, the benefits of GaN transistors become apparent, with the efficiencies becoming similar at about 50% load and the GaN transistor-based prototype producing 25% more power at full load for a similar total converter loss.

The second commercial half-brick, Converter B, used for comparison has a two-stage approach. Although the two-stage approach is different from the GaN transistor approach, both have the output power split into two separate converters, operating in parallel. The advantages of the two-stage approach is that it allows efficiency optimization of the unregulated isolation stage since it operates at a fixed duty cycle and voltage, regardless of converter input voltage. In addition, the controlled input and output voltages allow the use of lower voltage rated devices with better FOMs. The disadvantages are additional conduction losses for the two stages, and increased complexity and component count.

The efficiency comparison between the GaN transistor prototype and the two-stage converter is shown in Figure 10.30. This clearly shows the optimization process that has gone into Converter B as the peak efficiency is achieved at the nominal input of 48 V. The

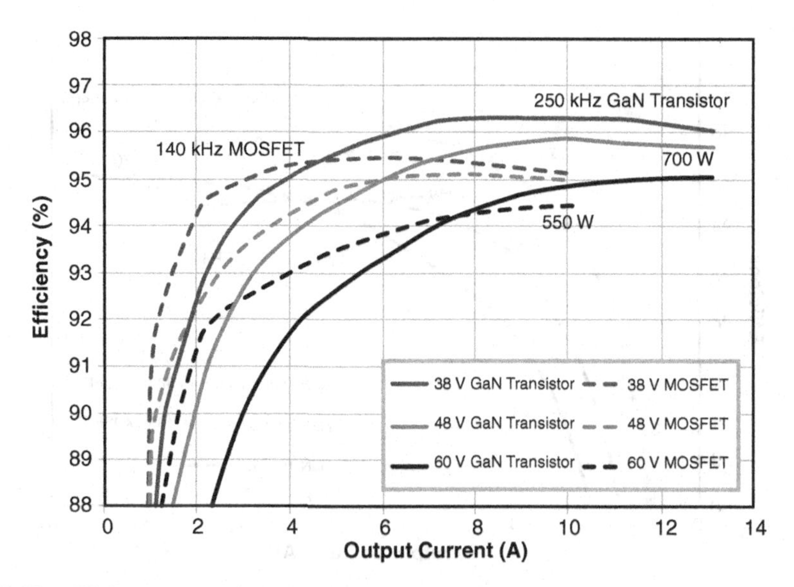

Figure 10.28 Efficiency comparison between half-brick PSE converters showing a GaN transistor-based prototype vs. Converter D – a commercial MOSFET-based solution

distinctions in topology are highlighted best by comparing the 38 V (low-line) input voltage results: Since the two-stage circuit employs a boost regulation stage, this is actually a worst-case condition (conduction loss increased, with no appreciable reduction in switching loss), while for the traditional single-stage approach, low-line is the best case, as switching loss is minimized. The two-stage converter's power losses at low-line reach almost 50 W, about

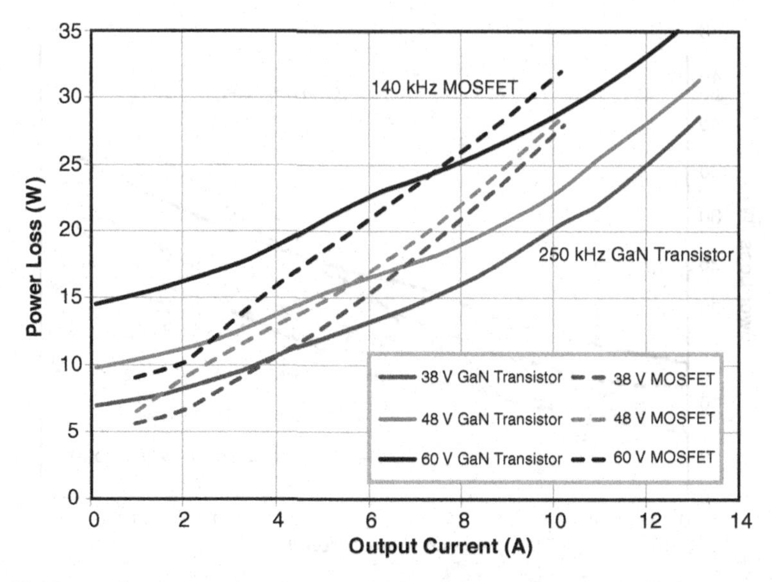

Figure 10.29 Power loss comparison between half-brick PSE converters showing a GaN transistor-based prototype vs. Converter D

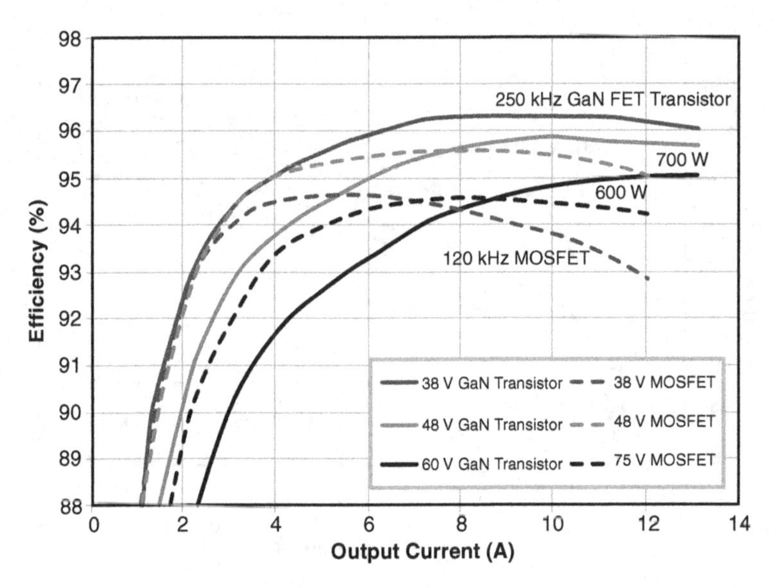

Figure 10.30 Efficiency comparison between half-brick PSE converters showing a GaN transistor-based prototype vs. Converter B – a two-stage MOSFET-based solution

double that of the GaN transistor converter under the same conditions, with the 75 V (high-line) input losses 15% higher than the GaN transistor converter, as shown in Figure 10.31.

With the emergence of GaN transistors, designers have the opportunity to overcome the limitations of the aging silicon MOSFET and reverse the backward trend of reducing switching frequency in hard-switching applications to increase the output power density of fixed

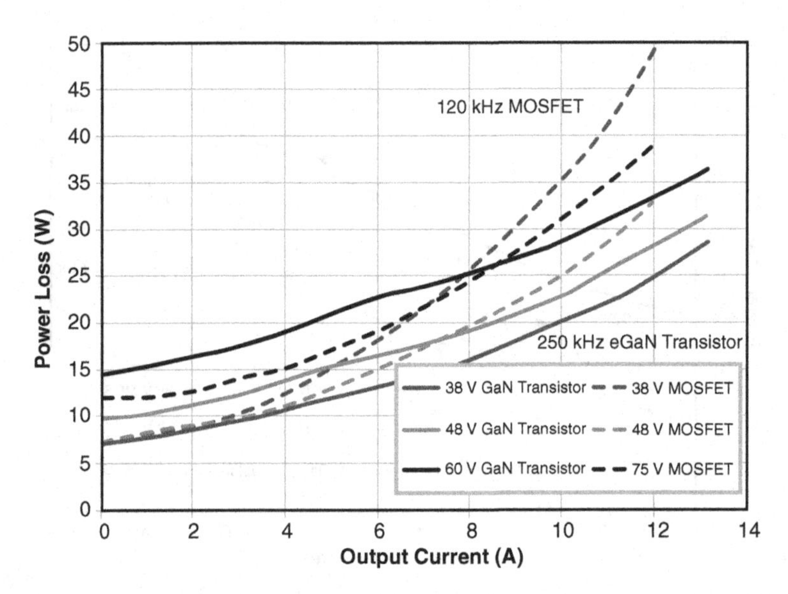

Figure 10.31 Power loss comparison between half-brick PSE converters showing a GaN transistor-based prototype vs. Converter B

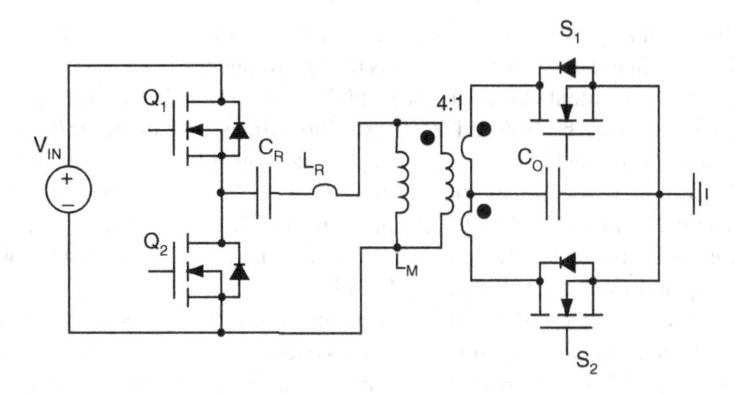

Figure 10.32 LLC resonant converter

form-factor bus converters. The designer can rely upon the lower losses and higher switching frequency capability of the GaN transistor to accomplish much higher power density bus converters.

10.3.2 A 400 V LLC Resonant Converter

The LLC resonant converter, shown in Figure 10.32, is a popular topology in 400 V DC-DC front-end converters. The LLC converter offers reduced switching loss by providing ZVS, and reduced current at turn-off in the primary-side devices, and ZCS on the secondary devices [4,5]. These benefits are similar to the resonant 48 V topology discussed in Chapter 7. The reduction in switching losses makes the LLC resonant converter suitable for high-switching frequencies.

The main advantages of GaN transistors in high-frequency LLC resonant converters are (a) the reduced gate charge, providing lower gate drive losses, and (b) the reduced output charge that enables a reduction in ZVS dead-time or required ZVS current. To evaluate the performance improvements possible with GaN transistors in higher voltage resonant applications, the characteristics of 600 V GaN cascode transistors and Si MOSFETs are compared in Table 10.3.

From Table 10.3, it can be seen that high-voltage GaN transistors offer a significant reduction in the key FOMs for resonant converters as discussed in Chapter 7. The GaN

Table 10.3 Device comparison between GaN transistor (TPH3006PS) and Si MOSFET (FCP190N60E) primary-side devices for 400 V LLC resonant converter

Parameter	GaN transistor[a]	Si MOSFET[b]	FOM ratio
Voltage rating (V_{DSS})	600 V	600 V	
$R_{DS(on)}$	150 mΩ at 8 V	160 mΩ at 10 V	
Q_G	9.3 nC at 8 V[a]	63 nC at 10 V	
Q_{OSS} at 480 V	52.8 nC[a]	85.4 nC[b]	
$Q_G \cdot R_{DS(on)}$	1395 pC·Ω	10,080 pC·Ω	7.23 × reduction
$Q_{OSS} \cdot R_{DS(on)}$	7920 pC·Ω	13,664 pC·Ω	1.73 × reduction
FOM_{SS} $(Q_{OSS} + Q_G) \cdot R_{DS(on)}$	9315 pC·Ω	23,744 pC·Ω	2.55 × reduction

[a] All values calculated from TPH3006PS datasheet and [6].
[b] All values calculated from FCP190N60E datasheet.

transistor delivers an improvement of more than seven times in gate charge FOM over the Si MOSFET, resulting in lower gate driving losses in high-frequency designs. The GaN transistor has a 1.73 times improvement in output charge FOM over the Si MOSFET, resulting in lower conduction losses offered by reduced ZVS transition times and lower ZVS currents. The resonant- and soft-switching FOM_{SS} is around 2.55 times lower for a GaN transistor than for an Si MOSFET, and this will correlate directly into lower losses and improved high-frequency LLC performance [7]. For a 400 V_{IN} to 12 V_{OUT}, 300 W LLC converter, operating at 1 MHz, GaN transistors demonstrated a 1% gain in full-load efficiency and an almost 5% gain at 15% load when compared to state-of-the-art Si MOSFETs [8].

This section has shown the value of GaN transistors in isolated DC-DC converters which are widely used in computing and telecommunication systems as part of the IBA approach. Since the size of these "brick" converters is strictly defined, an increase in efficiency brought about with the use of GaN transistors will significantly increase efficiency, output power, and system power density.

10.4 Class-D Audio

Class-D audio amplifiers using GaN transistors, which have lower conduction losses, faster switching speed and zero reverse recovery losses, provide an efficiency improvement over MOSFETs. In addition, the higher efficiency can, in many cases, eliminate bulky and expensive heatsinks that add to system costs and size. Beyond improved efficiency, however, GaN transistors provide improved sonic quality, critical in audio applications. Since the actual sound interpretation can be subjective, the use of measurable quantities such as total harmonic distortion (THD) and damping factor (DF) are used to validate the improvements provided by GaN transistors.

10.4.1 Total Harmonic Distortion (THD)

Class-D audio amplifiers are buck converters that operate with both positive and negative currents and a wide-ranging duty cycle. The switching-node voltage will either commutate at the turn-off of one device, or at the turn-on of the other, as shown in Figure 10.33 for a sinusoidal output. The variability in device commutation, which is due to the polarity of the

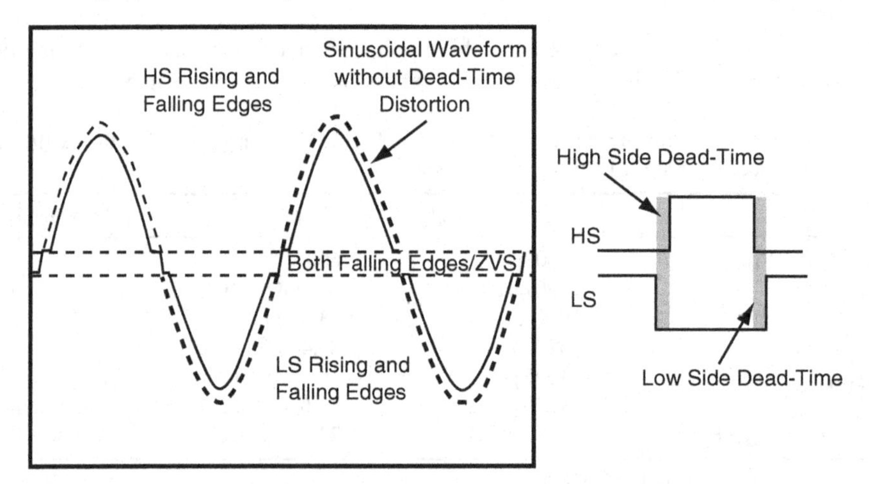

Figure 10.33 Distortion of a sinusoidal output due to changes in output pulse width with duty cycle and current

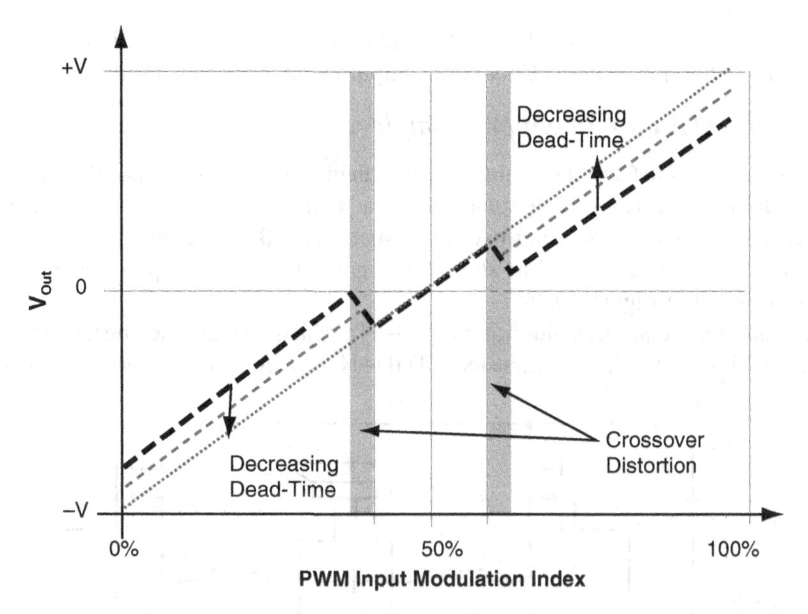

Figure 10.34 Theoretical Class-D converter input-to-output relationship showing crossover distortion regions

device current, results in crossover distortion, as shown in Figure 10.34. To put the impact of distortions on audio quality into perspective, a 0.01% THD corresponds to a 10 mV error on a 100 V_{DC} bus, or 0.25 ns change in pulse width at a switching frequency of 400 kHz [9].

To minimize this distortion, the amount of dead-time must be minimized. With the slower Si MOSFET, this value is typically no less than 25 ns [10]. Using GaN transistors, these dead-times can be readily reduced to around 5 ns, thus offering a significant reduction in open loop THD. Other important factors that affect THD are body diode reverse recovery, switch-node ringing and overshoot, device turn-on and turn-off delay, and finite rise and fall times [9]. For each and every one of these factors, the GaN transistor offers a reduction that directly relates to a corresponding decrease in THD.

10.4.2 Damping Factor (DF)

Another important performance parameter in Class-D audio is damping factor. This is the ratio of the speaker, or load impedance (Z_{Load}), to the Class-D amplifier output, or source impedance (Z_{Source}), and mathematically can be written as:

$$DF = \frac{Z_{Load}}{Z_{Source}} \tag{10.2}$$

A higher damping factor is better. As the speaker and amplifier form a voltage divider network, any significant variation in the output voltage due to changes in output current will cause distortion, with a higher damping factor minimizing distortion. Although the damping factor is determined mainly by the passive filter components and control loop gain, using a switching device with lower on-resistance will improve the damping factor directly. Since Class-D is a hard-switching converter, the optimum die size is a tradeoff between conduction and switching losses. Selecting the optimum die size GaN transistors will result in a lower on-resistance device with both lower switching loss and lower conduction loss than a similarly optimized MOSFET [11]. Thus, the resulting GaN-based amplifier will have with a better damping factor.

10.4.3 Class-D Audio Amplifier Example

Using GaN transistors in Class-D amplifiers offers an improvement in THD, DF, and efficiency over MOSFET designs. To illustrate some of these advantages, a 150 W into 8 Ω, or 250 W into 4 Ω, stereo Class-D amplifier, using two bridge-tied load (BTL) outputs from a ±27 V split supply, was built for one channel as shown in Figure 10.35 [12]. A photo of the complete converter is shown in Figure 10.36.

The total harmonic distortion plus noise (THD+N), which includes the impact of noise floor, is shown in Figure 10.37. The light-load THD+N is dominated by noise related to the

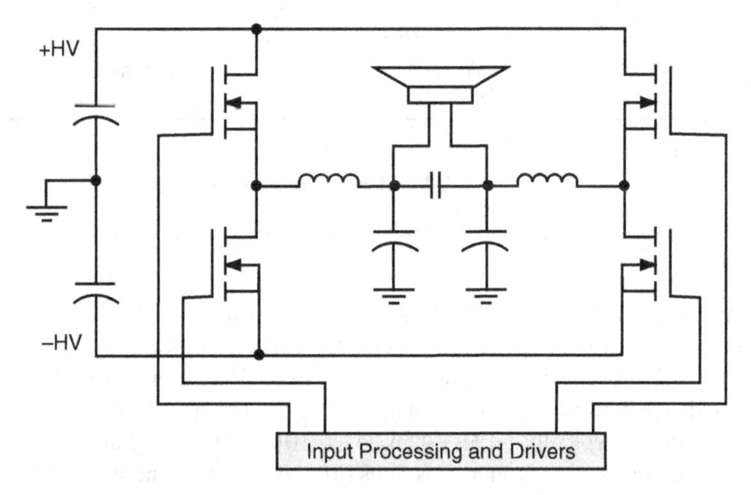

Figure 10.35 Diagram for a single channel of a BTL Class-D amplifier with split supply (±HV)

Figure 10.36 Complete Class-D converter showing output power stage and half-bridge circuits including GaN transistors and gate drivers. No heatsink is required up to 250 W into 4 Ω per channel

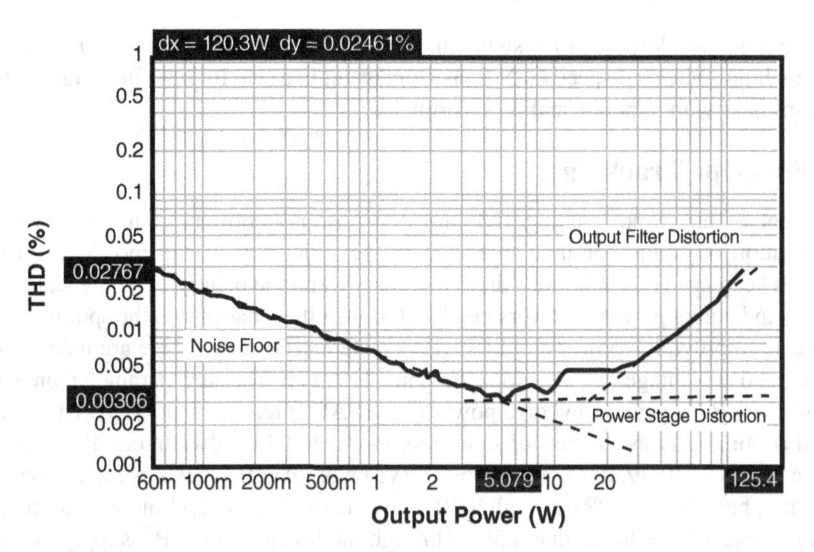

Figure 10.37 THD+N vs. output power at 1 kHz, ±27 V supply, into 8 Ω load. The minimum THD+N is 0.003% at 5 W, and 0.028% at 125 W

pre-amplifier and control loop rather than the GaN transistor-based power stage. Similarly, heavy-load THD+N is dominated by the distortion in the passive output filters, as they are not compensated for within the feedback control loop.

To determine the dynamic noise ratio (DNR) of the GaN transistor amplifier stage, the pre-amplifier volume is turned down during measurement to minimize its noise impact. The resultant DNR is 110 dB when referenced against a full-load power of 168 W [12].

Since the GaN transistor switch-node commutation occurs in around 1–2 ns, and the nominal dead-time is set to around 5 ns, it is possible to achieve these THD results without cross conduction (shoot-through). The resultant power stage efficiency, including output filter, is around 96% @ 150 W into 8 Ω and 91.5% @ 250 W into 4 Ω, as shown in Figure 10.38, and enables the converter to operate without heatsink over the entire load range.

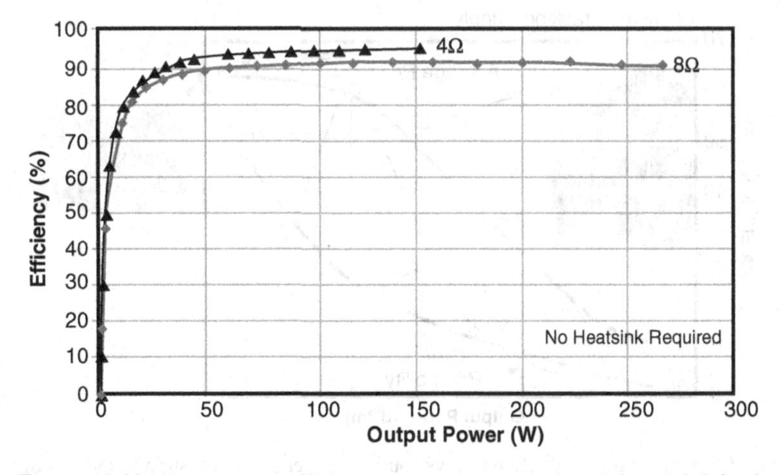

Figure 10.38 Power stage efficiency vs. output power into 4 Ω and 8 Ω loads

Lower conduction losses, faster switching speed, and zero reverse recovery losses made possible by high switching-speed GaN transistors result in a step forward in designing Class-D audio amplifiers with superior audio performance.

10.5 Envelope Tracking

The concept of envelope tracking (ET) for RF power amplifiers is not new. As global communication demands continue to increase, the efficiency of the mobile infrastructure struggles to keep up. The continuous drive to boost cell phone battery life, base station energy efficiency, and output power from very costly RF transmitters has placed the spotlight on ET as the means to improve RF power amplifier (PA) system efficiency. There are many papers on the basics and advantages of envelope tracking [13–17]. The key parameter on which to concentrate is a PA's peak-to-average power ratio (PAPR) requirements [13]. This PAPR has increased continuously as a means of squeezing more digital bandwidth out of RF amplifiers. As shown in Figure 10.39, it is possible to achieve peak PA efficiencies as high as 65% with a fixed supply, but given PAPRs as high as 10, such as used in 4G LTE networks, the average efficiency is likely to be lower than 25%. Through modulation of the PA supply voltage, this can be improved to over 50%, essentially doubling the efficiency and reducing by two-thirds the power losses. This not only reduces power consumption, but also lowers the cost of operation, cooling requirements, and system size [18].

In practice, however, the wide bandwidth of the envelope signal (a 4G LTE signal requires a modulation bandwidth of 20 MHz [19]) makes the tracking implementation difficult due to a degradation of efficiency of the modulating power supply with increasing bandwidths and switching frequency [20]. Thus, developing a high-frequency, high-efficiency modulator is required for realizing a practical ET solution. With the commercial availability of high-performance GaN transistors, a high-frequency high-efficiency envelope tracking power delivery system is now viable. Previous high-frequency buck converters using GaN transistors [21] realized impressive high-frequency buck converter efficiency results, as shown in Figure 10.40. Similar work [22] shows improvement of 20–30% in buck converter efficiency using GaN transistors versus silicon MOSFETs. Despite the excellent high-frequency

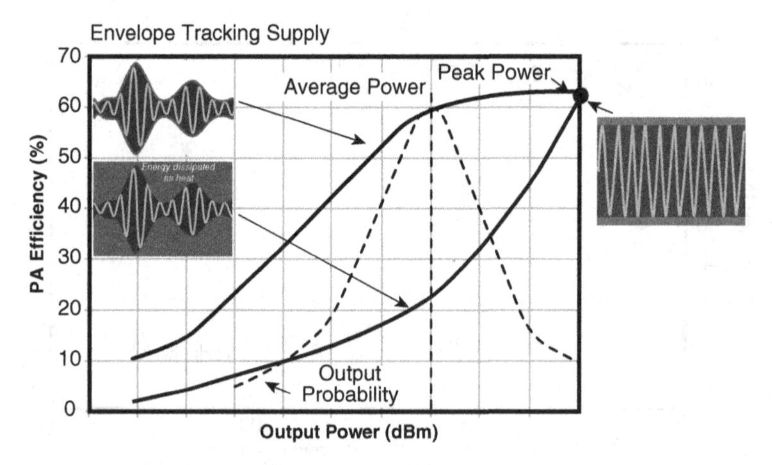

Figure 10.39 Conceptual PA efficiency vs. output power for fixed supply and ET operation

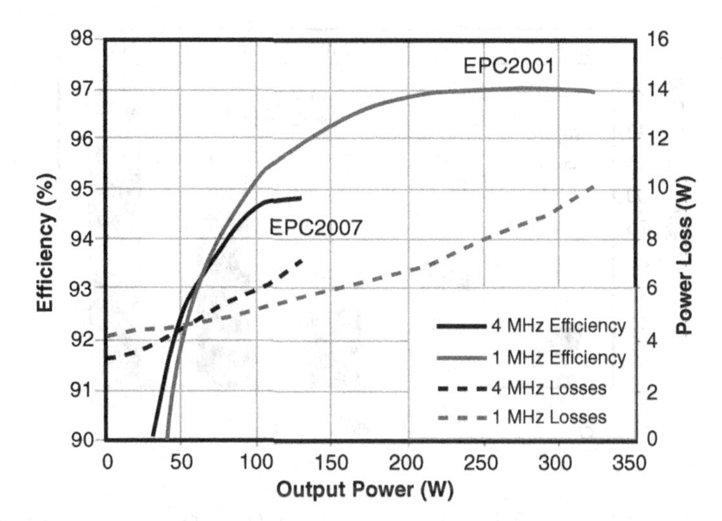

Figure 10.40 Efficiency and loss results for EPC2001 and EPC2007 GaN transistor buck demonstrators operating both at 45 V_{IN}, 22 V_{OUT} [21]

performance, the device active area in these early implementations was relatively large and not optimal for base station applications. A more practical ET solution requires lower power levels per phase at much higher frequencies. This need for smaller active area GaN transistors was addressed through the development devices that are one order of magnitude smaller (higher $R_{DS(on)}$) than the devices used in [21,22].

10.5.1 High-Frequency GaN Transistors

To support ET supplies requires devices with lower capacitance and higher on-resistance. Not only does this require an excellent hard-switching FOM, but also layout and package characteristics that maximize in-circuit performance. Such devices are available as wafer level chip scale packaged (WLCSP) GaN transistors [23] with the device pin-out shown in Figure 10.41. The FOM_{HS} in Figure 10.42 shows that the 65 V rated EPC8005 [24] compares favorably with similar on-resistance state-of-the-art 30 V MOSFETs. From Figure 10.42, it can be seen that the EPC8005 has about half the FOM_{HS} of MOSFETs, while its voltage rating is more than double. For reference, the 100 V GaN transistors used in [21] are comparable to the best 30 V MOSFETs.

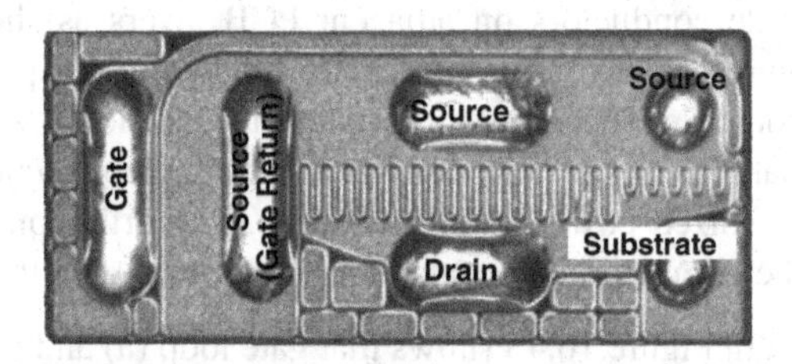

Figure 10.41 The EPC8000 GaN transistor-series WLCSP die showing the bump side with pin-out locations

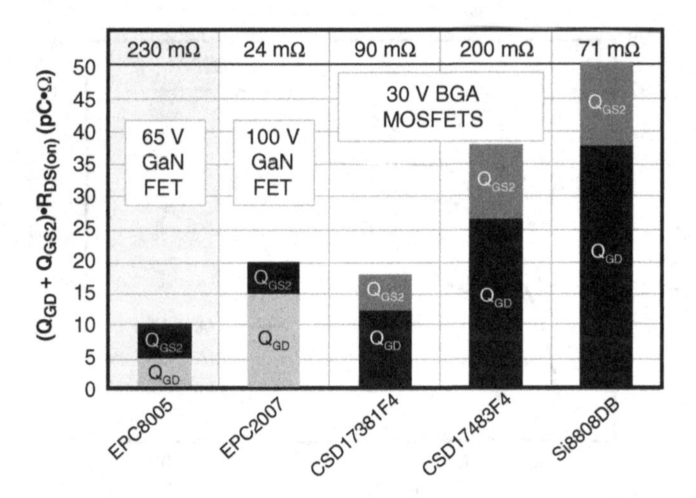

Figure 10.42 Hard-switching FOM comparison between GaN transistors and similar BGA Si MOSFETs

Although a significantly better FOM_{HS} is important, these devices also have a number of features that further improve in-circuit performance. These can be summarized as follows:

1. Complete dv/dt immunity: As discussed in Chapter 3, an important metric for dv/dt immunity is the Miller ratio, which is an indicator of how susceptible gates are to turning back on at high dv/dt [25]. The Miller ratio (Q_{GD}/Q_{GS1}) is reduced to below 0.4, well below the requirement of 1.0.
2. Reduced Q_{GD}: As noted in Section 6.2.1.1, Q_{GD} is the main device parameter that determines switching losses. It has been reduced to almost half that of a similarly scaled GaN transistor used for higher power DC-DC converters, and, as we learned in Section 6.2.1.1, this is the main device parameter that determines voltage-related switching losses.
3. Separate gate return (source) connection: The separate source connection for the gate drive circuit limits the common source inductance to inside the device itself. As discussed in Chapters 4 and 6, the reduction in common source inductance is critical to high-frequency switching performance.
4. Wider parallel gate drive connections: The wider connections for the gate circuit connection significantly reduce the inductance of the connection to the gate circuit. Furthermore, placing the gate and separate gate return terminals parallel to each other allows for low-inductance PCB interconnection to the driver. This is accomplished by routing both return terminals through wide conductors on adjacent PCB layers as shown in Figure 10.42, following the optimum loop layout design first presented in Chapter 4 [26].
5. Parallel drain and source connections, orthogonal to gate loop: As with the gate drive connections above, parallel connection pads allow wide interconnection traces for improved PCB layout with minimized power loop inductance. The orthogonal layout of these two loops also reduces the interaction of the gate circuit current with the drain circuit current.

The half-bridge layout in Figure 10.43 shows that gate loop (a) and power loop (b) currents (arrows) flow in opposite directions on adjacent layers, to help reduce the overall loop inductances through magnetic flux cancellation, as discussed in Chapter 4. Furthermore, these

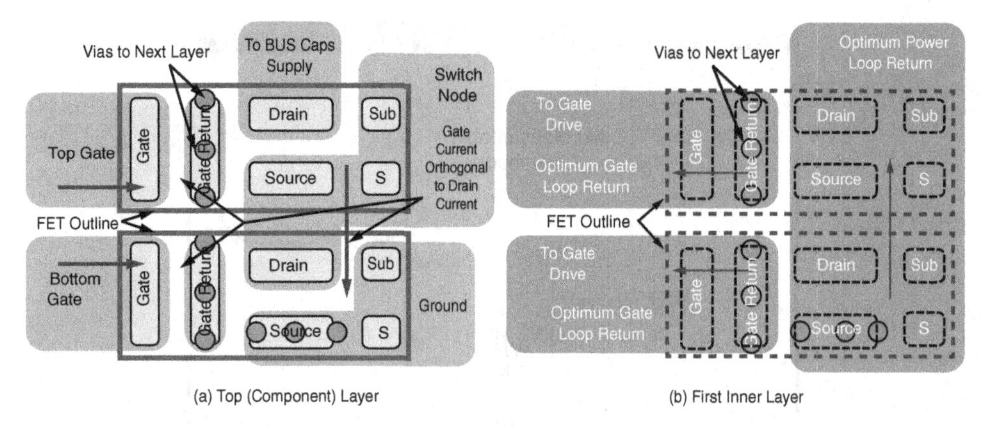

Figure 10.43 Optimal layout design for a half-bridge topology using an EPC8000-series GaN transistor. (a) top (component) layer and (b) first inner layer

traces are kept as wide as possible, while the interlayer distance between these layers is minimized, both of which will further reduce loop inductance.

10.5.2 Envelope Tracking Experimental Results

To demonstrate the capability of GaN transistors in ET applications, a 10 MHz-buck converter operating at a fixed input of 42 V and an output voltage of 20 V was constructed, as shown in Figure 10.44. The turn-on switching waveform for the 10 MHz-buck converter is shown in Figure 10.45. A significant bump in the rise time can be seen during the current commutation interval, while the dv/dt interval shows a slew rate as high as 75 V/ns, resulting in an overall rise time of around 1 ns. The efficiency curve, shown in Figure 10.46, indicates a peak efficiency of 89% at an output power of 35 W.

10.5.3 Gate Driver Limitations

Analysis of the power loss breakdown of this converter [27] shows that a significant portion of the overall converter losses can be attributed to the gate driver and related components, as shown in Figure 10.47. The most significant of these are the internal gate driver IC (presented in Chapter 3) capacitance between the high-side floating driver and ground (which is connected

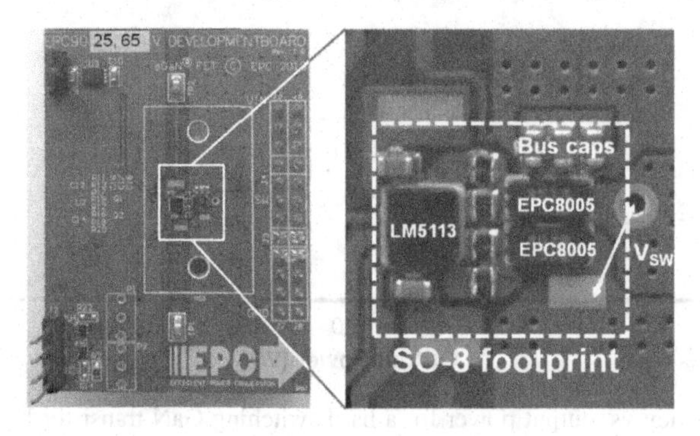

Figure 10.44 Photo of the evaluation board showing the EPC8005 [24] devices and LM5113 gate driver [47]

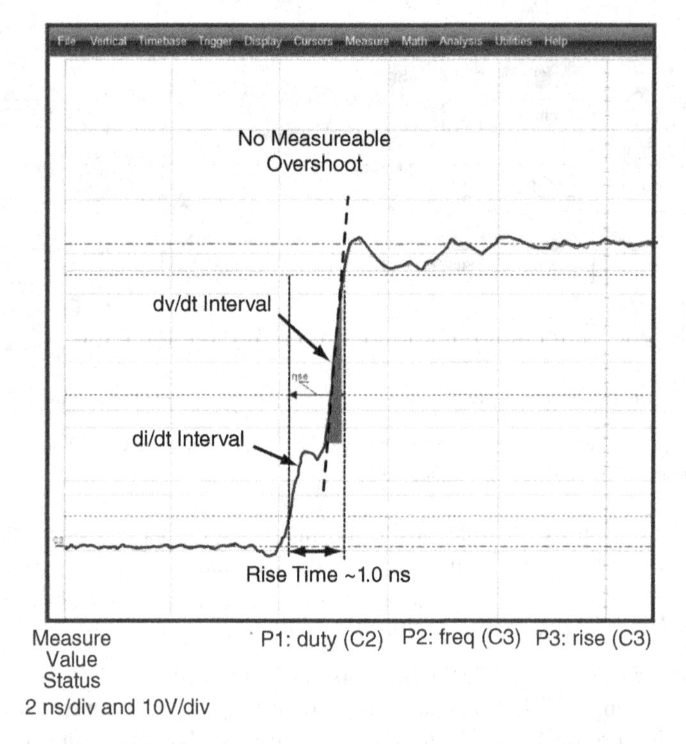

Figure 10.45 Hard-switching switch node waveform of the EPC8005 [24] showing rising edge rise-time of around 1 ns ($f_{sw} = 10$ MHz, $V_{IN} = 42$ V, $V_{OUT} = 20$ V, $I_{out} = 1$ A)

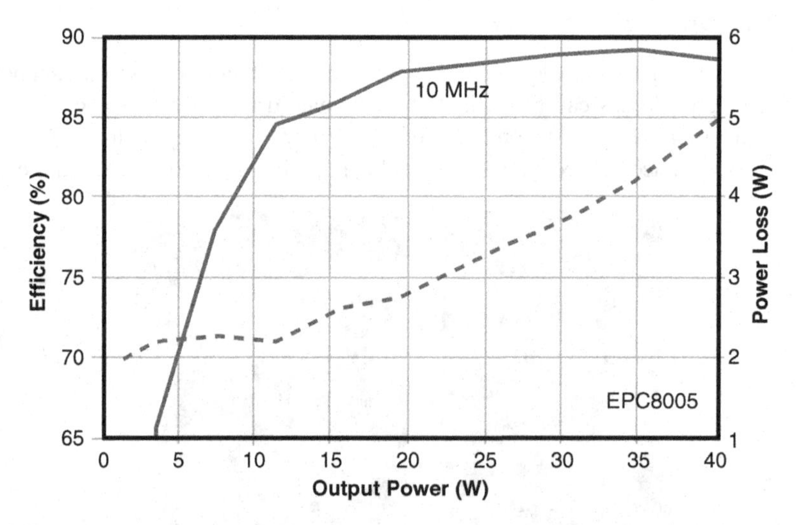

Figure 10.46 Efficiency vs. output power for a hard-switching GaN transistor-based (EPC8005 [24]) buck converter ($f_{sw} = 10$ MHz, $V_{IN} = 42$ V, $V_{OUT} = 20$ V)

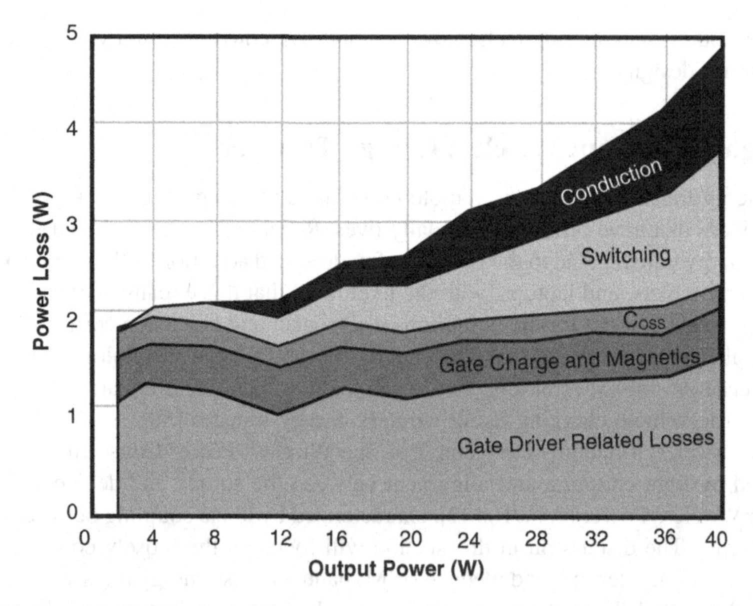

Figure 10.47 Breakdown of estimated loss components for a hard-switching GaN transistor-based buck converter showing impact of the gate driver on overall losses ($f_{sw} = 10\,\text{MHz}$, $V_{IN} = 42\,\text{V}$, $V_{OUT} = 20\,\text{V}$)

in parallel with the lower device C_{OSS}, and the associated losses were discussed in Chapter 6), and the reverse recovery losses of the bootstrap diode used to generate the floating high-side power supply. With improvements in the gate driver design and bootstrap diode, these additional driver-related losses can be mitigated, with the potential converter efficiency approaching that shown in Figure 10.48.

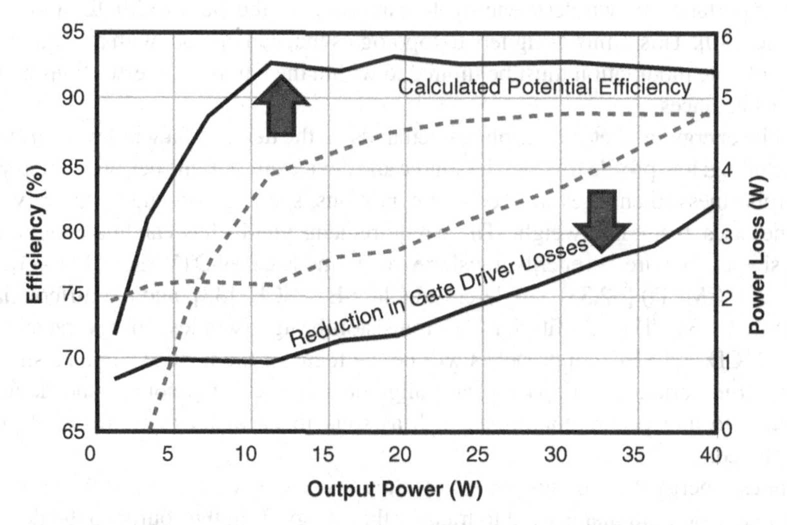

Figure 10.48 GaN transistor-based buck converter efficiency and power loss vs. output power showing actual results (dashed lines) and calculated values based on potential improvements in driver capacitance, bootstrap diode recovery, and internal driver switching losses ($f_{sw} = 10\,\text{MHz}$, $V_{IN} = 42\,\text{V}$, $V_{OUT} = 20\,\text{V}$)

Gallium nitride is an enabling technology for both ET converters and wide bandwidth RF power amplifier designs.

10.6 Highly Resonant Wireless Energy Transfer

Wireless energy transfer enables the remote powering and charging of the myriad of battery powered devices that have infiltrated our daily lives. Recent advances in the basic technology of wireless energy transfer lead to the promise of widespread adoption as the means of charging our cell phones, tablets, and laptops, with the likelihood that this expansion will continue into our homes as a replacement for the ubiquitous wall socket and extension cord. Although most charging applications today are low power, less than 50 W, the technology does allow for higher power levels into several kW. For example, auto manufacturers are already commercializing electric vehicle charging using wireless energy transfer [60].

Two main wireless standards have emerged, the Wireless Power Consortium (WPC) [28], characterized by tight coupling and alignment between the source and device units, and the Alliance for Wireless Power (A4WP) [29], characterized by loose coupling between the source and device units. The discussion in this section will focus on the loosely coupled 6.78 MHz unlicensed industrial, scientific and medical (ISM) band wireless energy transfer solution based on the A4WP standard. This loosely coupled approach eliminates the need for close alignment between the sending and receiving units. Cell phones can be charged while remaining in the user's pocket. Tablets can be charged by being placed anywhere on a desktop equipped with a thin sending coil. Conceivably, transmitting coils could be located in floor tile in order to power common household appliances. Electric vehicles can be charged by simply driving over a charging mat located on the floor of the garage.

Non-communication operation in the 6.78 MHz ISM band is governed by regulatory standards for intentional radiators, which places restrictions on the radiated energy. Of particular importance to wireless energy transmission is the bandwidth limit of $\pm$15 kHz for the carrier [30]. This limits designers to topologies that can operate with a fixed frequency, and any amplitude modulation must be limited to within the bandwidth restrictions and system component tolerances.

The mobile energy market further places demands on the design of a wireless energy transfer system, including: low profile for both the source and device units, high energy efficiency, ease of use, high robustness to changes in operating conditions, such as load and coupling variations, and in some cases, being lightweight. These requirements yield a few candidate topologies that can be considered for wireless energy transfer: voltage-mode Class-D (VMCD) [31–33], current-mode Class-D (CMCD) [32,33], single-ended Class-E (SECE) [33], and the differential-mode Class-E (DMCE) [34]. The benefits that GaN transistors bring to wireless energy transfer systems using the VMCD and SECE approaches will be the focus of the next section. To simplify the discussion, large variations in spacing and alignment between the source and device units, and immunity to unexpected conditions such as solid metal or magnetic material proximity will be excluded.

The wireless energy transfer system architecture is centered on the use of flat spiral coils to generate a magnetic field that is used to transfer the energy from the source to the device. This structure forms a transformer with low coupling coefficient, as shown in Figure 10.49. The equivalent circuit can be used to design the system based on physical design parameters and can be used to predict the performance of the wireless energy transfer system.

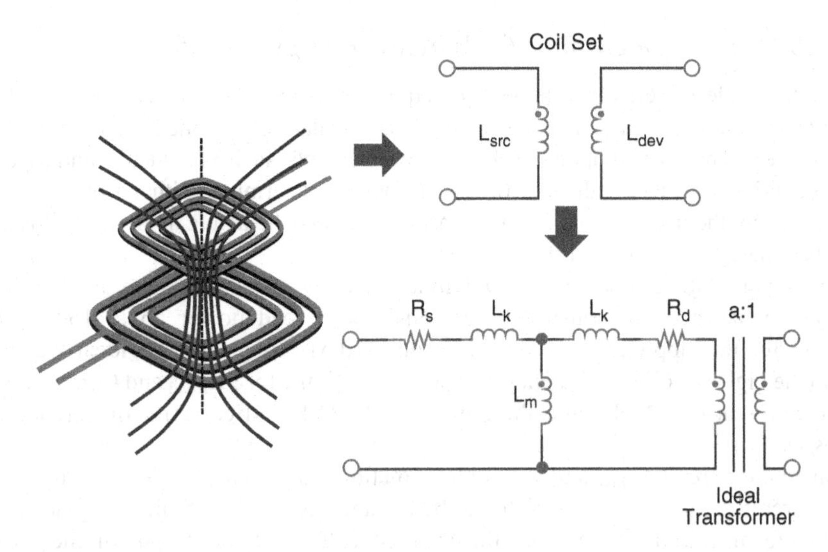

Figure 10.49 Wireless energy transfer coils (left) with equivalent transformer circuit (right)

This transformer has a leakage inductance (L_k) that can be significantly larger than the magnetizing inductance (L_m). An analysis of the transformer model under these conditions reveals that the primary-side leakage inductance almost solely determines the efficiency of energy transfer to the secondary side [35]. To overcome the leakage inductance, use is made of resonance to increase the voltage across the leakage inductance, and hence the magnetizing inductance [29,36,37], with a resulting increase in power delivery.

To simplify the discussion, a schematic reduction to a single element, Z_{Load}, will be considered as shown in Figure 10.50, where Z_{Load} represents the DC-load resistance (R_{Load}), DC-smoothing capacitor (C_{out}), rectifier, device matching (C_{devs}, C_{devp}, L_{devs}), device coil (L_{dev}), and source coil (L_{src}).

The subsequent design and discussions of the various topologies will assume this equivalent circuit represented by Z_{Load}. Only losses associated with this portion of the circuit will be considered for overall system efficiency, but not in the evaluation of the performance comparison between the various topologies.

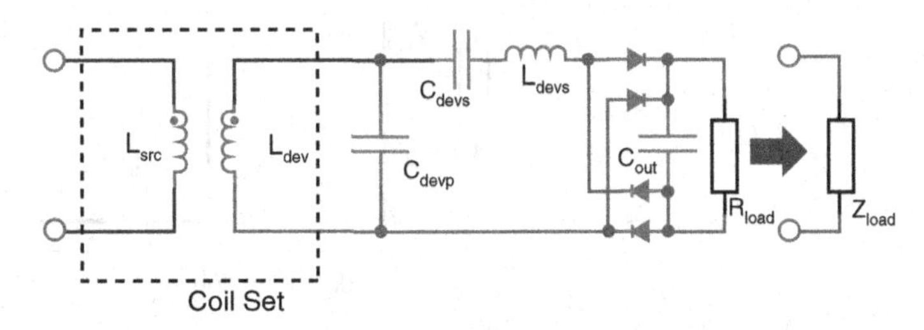

Figure 10.50 Single element reduction of the wireless energy transfer system that includes the source coil, device coil, device matching, rectifier, DC capacitor, and load resistor

10.6.1 Design Considerations for Wireless Energy Transfer

Based on the wireless energy transfer system requirements, the main power circuit needs to be efficient enough to not require a heatsink. The loss calculations and measurements will focus on both active and passive components. For passive elements, such as inductors and capacitors, equivalent series-resistance values at 6.78 MHz will be determined either from s-parameter data provided by the manufacturer or by measurement. For the active devices, the losses will include the energy required to drive the gate.

A traditional voltage-mode Class-D wireless energy transfer system is the simplest topology and has been demonstrated previously with load power up to 15 W [38,39]. The circuit for this approach is shown in Figure 10.51 along with the ideal waveforms. Owing to the high switching frequency and device C_{OSS}, the load Z_{Load} and C_s must be tuned to be inductive at 6.78 MHz, and therefore, enable ZVS and corresponding reduction in C_{OSS} losses.

The matching circuit (L_{mat} and C_{mat}) also functions to increase the voltage to the load-resonant circuit (C_s and Z_{Load}), which can be advantageous when limits are placed on the input voltage magnitude, given that the average voltage at the output of the amplifier (switch-node) is half the supply voltage (V_{DD}). However, the matching inductor will carry the full current of the load, and therefore will have high associated losses. Furthermore, the circuit is sensitive to load resistance variation, which can be due to load or coupling changes, as the matching network becomes an integral part of the tuned-resonant circuit. This can shift the ideal operating inductance point needed to maintain proper ZVS. Another disadvantage of the voltage-mode Class-D topology is the need for a high-side gate driver. Using the bootstrap supply approach has limits for the maximum input capacitance of the devices, as the low-side device on-time is very short for charging the bootstrap capacitor, and the capacitance of the bootstrap circuit itself can even contribute to discharging the bootstrap capacitor, as discussed in Section 10.5.3.

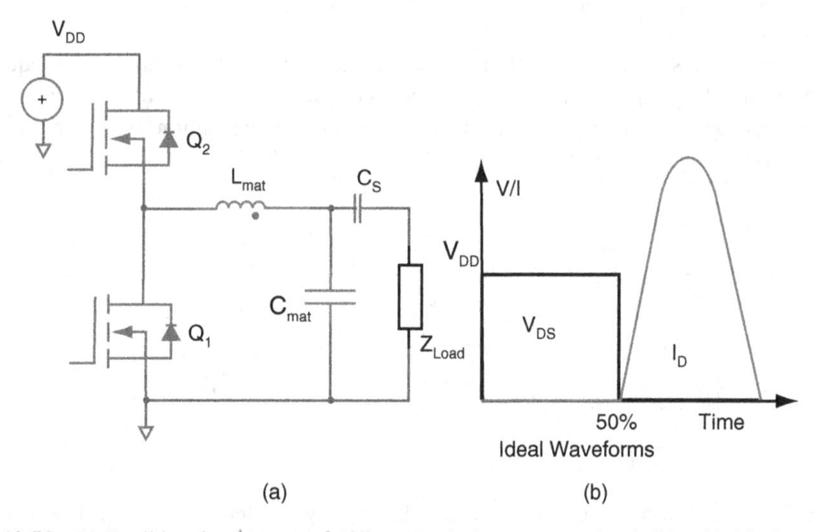

Figure 10.51 (a) Traditional voltage-mode Class-D wireless energy circuit (b) with ideal waveforms

10.6.2 Wireless Energy Transfer Examples

In this section, two alternative topologies suitable for wireless energy transfer using GaN transistors will be analyzed: (1) a ZVS voltage-mode Class-D and, (2) a single-ended Class-E. The target is 30 W load power without the need for forced air-cooling or a heatsink. Both amplifier topologies fall under a sub-class of resonant converters and, as such, comparison between devices will follow the similar metrics discussed in Chapter 7.

10.6.2.1 ZVS Voltage Mode Class-D Amplifier Example

In this section, a variation to the traditional voltage-mode Class-D amplifier suitable for wireless energy transfer is introduced, the ZVS VMCD [40]. The schematic and ideal waveforms for this topology are shown in Figure 10.52.

The ZVS voltage-mode Class-D converter works by adding a non-resonant L-C tank circuit (L_{ZVS} and C_{ZVS}) to the output of the amplifier that operates as a no-load buck converter. The function of this tank circuit is to absorb the C_{OSS} of the devices by providing a current that will allow the circuit to self-commutate the switch-node with the necessary dead-time between the gate signals.

The load can then be tuned to appear resistive at 6.78 MHz, which may even be slightly capacitive as long as the ZVS tank can yield sufficient current to maintain ZVS for the devices. The circuit allows for greater manufacturing variation of the load coils and their tuning, yet can maintain a high-energy transfer characteristic. The theoretical waveforms for the switching devices of the ZVS VMCD amplifier are shown in Figure 10.53(b) and the two components are split out for the ZVS tank circuit and the resonant load circuit in Figure 10.53(a). Load variations have only a minimal impact on the ZVS tank circuit as long as the deviations remain well below the peak current in L_{ZVS}. The ZVS VMCD, therefore, always ensures proper switching for the devices, and hence, maintains low losses.

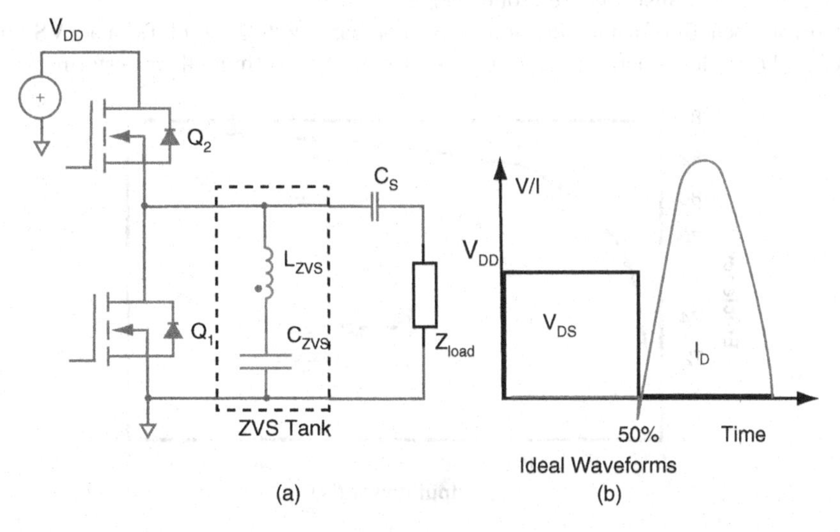

Figure 10.52 (a) ZVS voltage-mode Class-D wireless energy circuit (b) with ideal waveforms

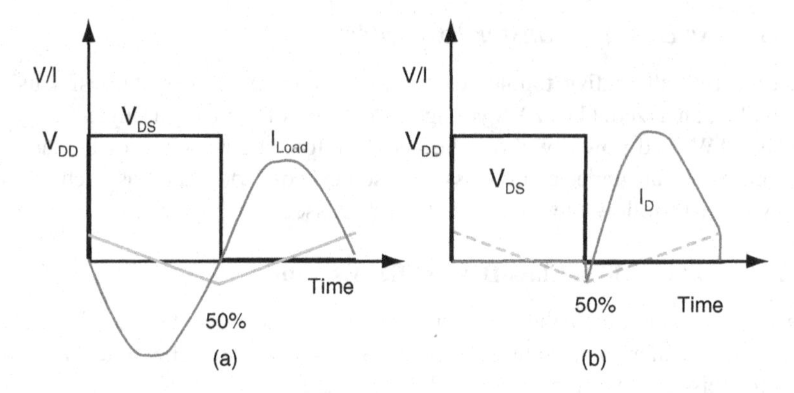

Figure 10.53 (a) ZVS voltage-mode Class-D circuit ideal operating waveforms (b) and component waveforms

A ZVS Class-D wireless transfer system was built using enhancement-mode GaN transistors [41], and tested using the same coil set [42] that was used in [38,39], making it possible to compare performance with other versions of the wireless energy transfer systems. Values of 300 nH for L_{ZVS} and 1 μF for C_{ZVS} were chosen with corresponding dead-time of 3.2 ns at 36 V input. The coil set was tuned to resonance with C_s at 6.78 MHz. The measured system efficiency (input supply to output load), including gate power is shown in Figure 10.54 for a 35.4 Ω load and 23.6 Ω load. The system efficiency peaks at 83.7%, at just above 35 W load power.

Because it is very difficult to accurately determine the breakdown of losses in the system, the best method to show the performance of GaN transistors is to use thermal imaging. Figure 10.55 shows a thermal image of the devices operating with 35 W load, alongside a photograph of the setup. No heatsink or forced air-cooling was used.

From the thermal image, it can be seen that the gate driver is the hottest component, at around 59 °C, and the GaN transistors are around 50 °C.

To show the benefit of using GaN transistors compared with MOSFETs in a ZVS voltage-mode Class-D wireless energy transfer system, we need to examine their performance in the

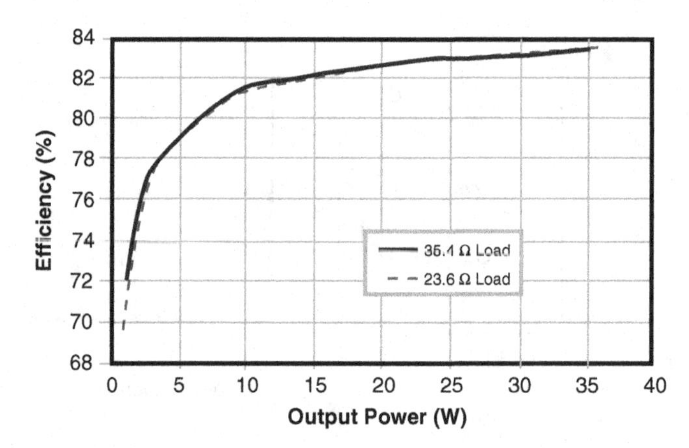

Figure 10.54 Measured efficiency for the ZVS voltage-mode Class-D wireless energy transfer system

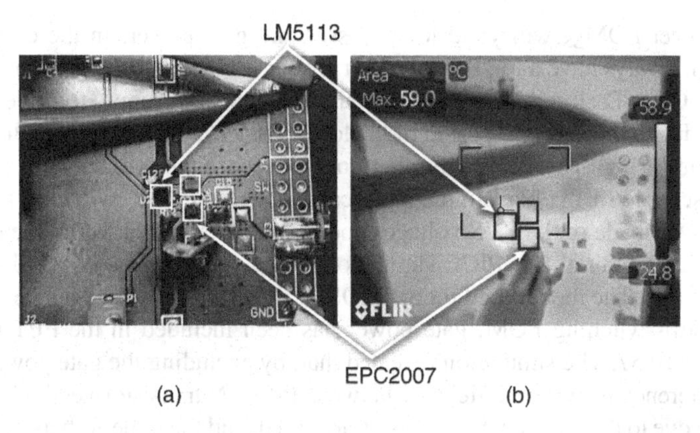

Figure 10.55 Thermal image (b) of the ZVS voltage-mode Class-D wireless energy transfer system operating at 35 W load power alongside a photograph (a) of the setup. Parts shown are: GaN Transistor (EPC2007 [41]) and GaN transistor driver (LM5113TM [47])

circuit, rather than comparing total system performance. There are two main reasons: (1) the device C_{OSS} is absorbed into the circuit, and its impact will become diminished overall, and (2) the losses in the balance of the system, represented by Z_{Load}, dominate the system performance. The GaN transistor accounts for less than 2% of the system efficiency at 30 W.

For the ZVS voltage-mode Class-D topology, the MOSFET chosen for comparison is the FDMC8622 [51], which has a similar Q_{OSS} value, and the same voltage rating as the EPC2007 tested [41]. Regardless of whether a MOSFET is selected based on similar $R_{DS(on)}$ or Q_{OSS}, the results will show similar differences.

The ZVS voltage-mode Class-D topology falls under the soft-switching converter category, and so a soft-switching FOM comparison will be made between the devices as shown in Figure 10.56. The converter is constrained by C_{OSS} due to circulating energy, and therefore any

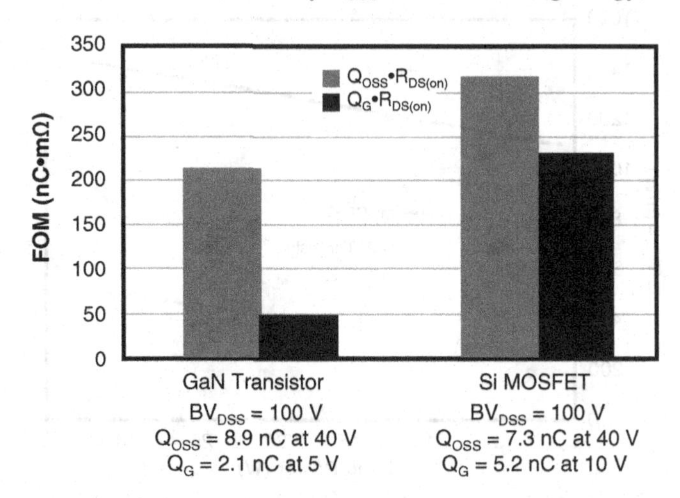

Figure 10.56 FOM comparison between the GaN transistor and MOSFET suitable for the ZVS voltage-mode Class-D wireless energy transfer circuit. Parts shown are: GaN transistor (EPC2007 [41]) and MOSFET (FDMC8622 [51])

devices with lower FOM_{SS} will yield lower losses and gate power. In the case of the GaN transistor, it will always have a lower $R_{DS(on)}$ for the same C_{OSS}, and a significantly lower Q_G. In addition, the Class-D circuit is limited when using a bootstrap supply to drive the high-side device. This is due to the short on-period of the lower device at 6.78 MHz, and the gate charge requirements for the high-side device. Furthermore, the capacitance of the bootstrap diode can cause some discharge of the bootstrap supply capacitor during the switching transition.

An analysis was made of the ZVS voltage-mode Class-D circuit based on the circuit shown in Figure 10.52. The initial LTSPICE simulation of the full circuit was verified against the experimental results. The model used for the MOSFET [51] was provided by the manufacturer. Based on the soft-switching FOM, gate power has been included in the FET power and is shown in Figure 10.57. The simulation revealed that, by excluding the gate power, there is no appreciable difference in system efficiency between the GaN transistor version and MOSFET version. This is due to the way in which C_{OSS} is absorbed, and the tradeoff between $R_{DS(on)}$ and the timing and magnitude of the ZVS tank circuit impact.

The difference between the GaN transistor and the MOSFET is based on gate power consumption and reveals that GaN transistors will have a larger impact on converter efficiency at lower output power levels. The total device power difference is near constant at around 900 mW over the entire load power range.

In the case of the voltage mode Class-D, with a half-bridge topology, providing the gate drive signal to the upper device requires some form of level shifting circuit. Wireless energy transfer systems typically have power levels below 30 W, so isolation circuits are not an option as they increase system cost and complexity. A bootstrap supply works well, but it must be designed to operate at the high frequency. The gate charge requirements, gate driver parasitic elements, and gate driver energy requirements for the upper device circuit mean that all the energy requirements for the upper transistor must be transferred in the short charging time available. Given that MOSFETs have as high as four times the gate charge requirements of

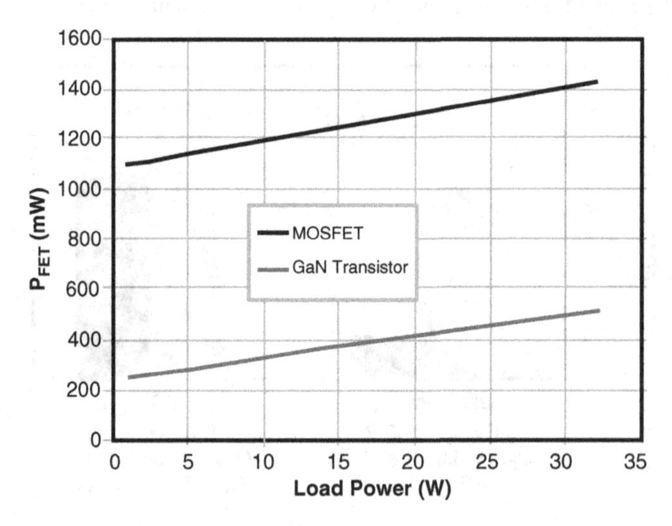

Figure 10.57 Total FET power (including gate power) for the ZVS voltage-mode Class-D circuit comparison between the GaN transistor and MOSFET. Parts shown are: GaN transistor (EPC2007 [41]) and MOSFET (FDMC8622 [51])

GaN transistors, MOSFETs will reach an upper frequency operating limit at a significantly lower frequency than GaN devices.

10.6.2.2 Single-Ended Class-E Example

The next amplifier that will be investigated is the single-ended Class-E. The schematic and ideal waveforms for this topology are shown in Figure 10.58. The design of a Class-E amplifier is well documented [32–34,43–46] and, for this example, requires the load inductance of Z_{Load} to be resonated by C_s to yield only the resistive portion of Z_{Load}. This, in turn, defines the supply voltage (V_{DD}) needed for a specific load power specification.

In the case of wireless energy transfer using a Class-E amplifier, the design must be based initially on the highest coupling coefficient and lowest load resistance for the load. This will ensure that the amplifier will always operate with manageable power dissipation in the device with any load and coupling variation. Operating with a load resistance below the design point will significantly increase the power losses generated in the device. The Class-E amplifier experiences losses in the RF choke (L_{RFck}), the device (Q_1), and the extra inductor (L_e). The device experiences very low losses at the turn-on event with both zero current and zero voltage present under ideal operating conditions. The turn-off event occurs as a ZVS event. Deviations from the ideal operating point, due to load resistance changes, can quickly induce high switching losses resulting from incomplete voltage- or current-resonant transitions in the matching network (L_e and C_{sh}). The choice of device must have a lower or equal C_{OSS} than the required shunt capacitance (C_{sh}) from the design equations. If the design requires a shunt capacitance that is smaller than C_{OSS}, the design cannot be realized.

A single-ended Class-E wireless transfer system was built using GaN transistors [50], and tested using the same coil set [42] as was used for the ZVS Class-D example. The design used the classic Sokal equations [48] with the quality factor set to infinity [49] for simplification purposes. Operation at 6.78 MHz is low enough to make this a valid assumption. A value of 360 nH for L_e and 150 pF for C_{sh} was chosen for operation up to 30 V input. The coil set was tuned to resonance with C_s at 6.78 MHz. The measured system efficiency (input supply to

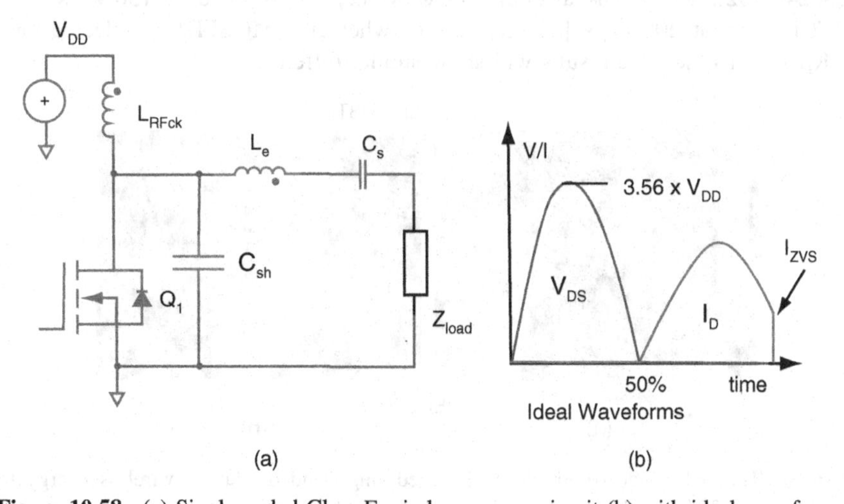

Figure 10.58 (a) Single-ended Class-E wireless energy circuit (b) with ideal waveforms

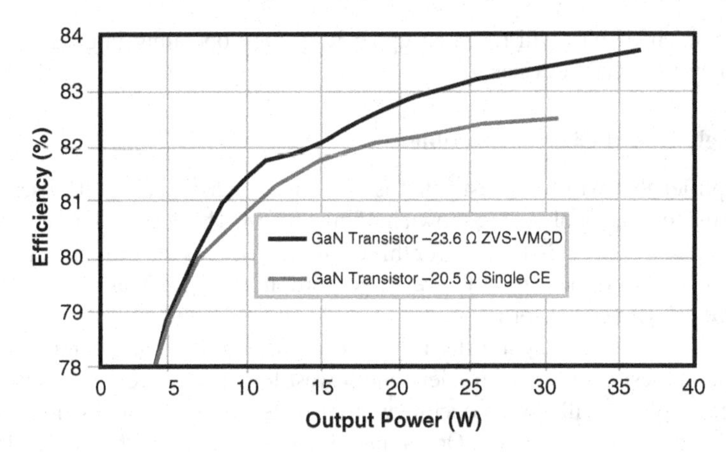

Figure 10.59 Measured efficiency for the single-ended Class-E wireless energy transfer system with the performance of the ZVS-VMCD for comparison

output load), including gate power, is shown in Figure 10.59 for a 20.5 Ω load. The system efficiency peaks at 82.5%, with a load power of just above 30 W. The performance of the ZVS voltage-mode Class-D system is also shown on Figure 10.59 for comparison.

Figure 10.60 shows the thermal image of the devices operating with 30 W load alongside a photograph of the setup. No heatsink or forced air-cooling was used.

From the thermal image, it can be seen that the device is the hottest component, at around 44 °C, and that the gate driver is around 35 °C. Compared to the ZVS voltage-mode case, the gate driver is driving only a single device, which accounts for the large difference in temperature compared with the ZVS Class-D design.

Again, we can show the benefit of using GaN transistors over MOSFETs in a single-ended Class-E wireless energy transfer system in the same way as for the ZVS voltage-mode Class-D. For the single-ended Class-E topology, the MOSFET chosen for comparison is the FDMC86248 [52], which has a similar Q_{OSS} value, but is rated at 150 V, whereas the EPC2012 is rated at 200 V [50]. Regardless of whether a MOSFET is selected based on similar $R_{DS(on)}$ or Q_{OSS}, the results will show similar differences.

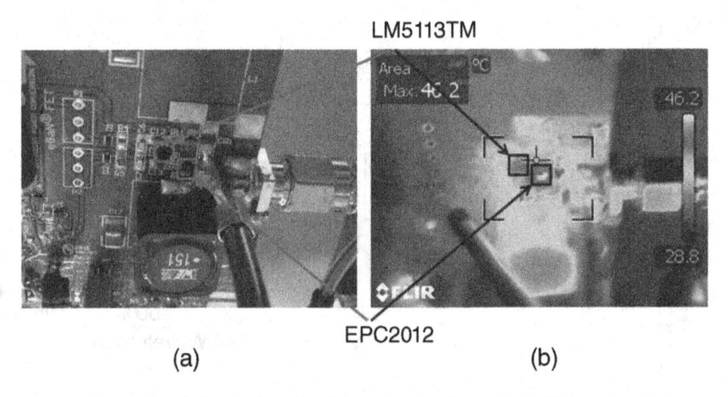

Figure 10.60 Thermal image (b) of the GaN-based single-ended Class-E wireless energy transfer system operating at 30 W load power with (a) photograph of the setup

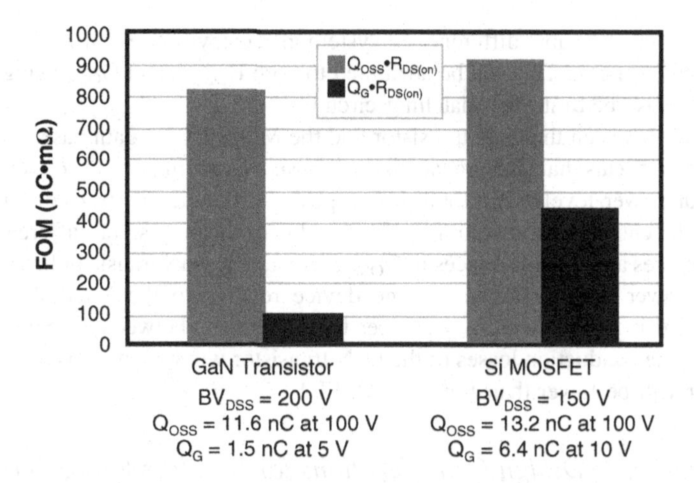

Figure 10.61 FOM comparison between the GaN transistor and MOSFET suitable for the single-ended Class-E wireless energy transfer circuit. Parts shown are: GaN transistor (EPC2012 [50]) and MOSFET (FDMC86248 [52])

The single-ended Class-E topology falls under the soft-switching converter category and so a soft-switching figure of merit comparison will be made between the devices as shown in Figure 10.61.

An analysis was made of the single-ended Class-E circuit, based on the circuit shown in Figure 10.60. The initial LTSPICE simulation of the full circuit was verified against the experimental results. The model used for the MOSFET [52] in the equivalent simulation was provided by the manufacturer. Based on the soft-switching FOM, gate power has been included in the FET power as shown in Figure 10.62. The simulation revealed that, by excluding the gate

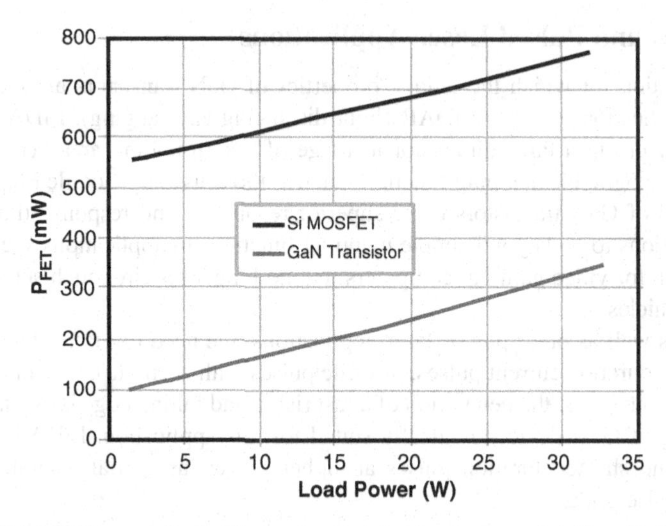

Figure 10.62 Total FET power (including gate power) for the single-ended Class-E circuit comparison between the GaN transistor and MOSFET. Parts shown are: GaN transistor (EPC2012 [50]) and MOSFET (FDMC86248 [52])

power, there is no appreciable difference in system efficiency between the GaN FET version and the MOSFET version. This can be attributed to both $R_{DS(on)}$ and Q_{OSS} being similar, and the C_{OSS} being absorbed into the matching circuit.

The difference between the GaN transistor and the MOSFET is again based on gate power consumption and reveals that GaN transistors will have a larger impact on converter efficiency at lower output power levels. The total device power difference is near constant at around 430 mW over the entire load power range. This is due to the very small difference in $R_{DS(on)}$ between the devices and the differences in C_{OSS} current. The GaN transistor has a lower Q_{OSS}, and therefore lower current flowing in the device relative to the external capacitor C_{sh}. Alternatively, if using equivalent C_{OSS} devices for comparison between the MOSFET and the GaN transistor, the conduction losses of the GaN transistor will be lower, as the $R_{DS(on)}$ of the GaN transistor will be lower than for the MOSFET.

10.6.3 Summary of Design Considerations for Wireless Energy Transfer

In this section, highly resonant, loosely coupled wireless energy transfer systems were demonstrated using GaN transistors in ZVS voltage-mode Class-D and single-ended Class-E topologies. It was shown that at low power output, up to 30 W, GaN transistors show a significant improvement in conversion efficiency, largely due to lower gate drive power consumption. At higher output power the advantages of GaN would be even more pronounced.

In the case of the voltage mode Cass-D amplifier, with a half-bridge topology, the lower gate charge allows a bootstrap supply to be used at a much higher frequency than comparable MOSFETs, thereby enabling wireless energy conversion at even higher frequencies such as the 13.56 MHz wireless power transfer standard. Furthermore, the lower gate charge also leads to lower power dissipation in the gate driver itself, thereby eliminating the need for heatsinks.

10.7 LiDAR and Pulsed Laser Applications

Another application for which the unique properties of GaN transistors are ideally suited is pulsed laser applications such as LiDAR (light distancing and ranging). LiDAR uses pulsed lasers to rapidly create a three-dimensional image of a surrounding area. This technique is widely used for geographic mapping functions such as those used by Google Maps. The higher switching speed of GaN transistors drive superior resolution and response time that enable LiDAR applications to go beyond simple mapping functions to applications such as real-time motion detection for video gaming, computers that no longer require touch screens, and fully autonomous vehicles.

In LiDAR, as well as other pulsed laser applications, the need exists for the generation of either very short duration current pulses, or finite pulses with accurate constant amplitude and pulse width. In both cases, the generation of a fast rising and falling edge is essential. The high speed switching of GaN devices is ideally suited for this application. LiDAR, in particular, needs shorter and shorter duration pulses at higher power, as spatial resolution is directly related to the pulse width:

$$\Delta R = \frac{c}{2}\Delta t \tag{10.3}$$

In Equation 10.3, ΔR is the spatial resolution, c is the speed of light, and Δt is the full width of the light pulse, measured at half the maximum amplitude – also known as full width at half maximum (FWHM). Thus, the spatial resolution is equal to half the distance from leading to trailing edge of the light pulse as it travels at the speed of light (about 0.3 m per ns). Furthermore, since LiDAR systems rely on the often faint reflection of the light returning to the detector to be above the background light levels, increasingly-higher-power laser pulses are required to improve performance.

To understand the practical constraints of pulsed laser applications, consider the simple pulsed laser drive circuit shown in Figure 10.63 [53]. The laser diode is in series with a switching device and the voltage source or capacitance energy storage. At the switching device turn-on, the current will be zero and the initial rate of current increase will be limited by the overall loop inductance as determined by Equation 10.4.

$$\frac{di}{dt} = \frac{V_{Supply} - V_{Diode}}{L_{Loop}} \tag{10.4}$$

In this equation $\frac{di}{dt}$ is the maximum current slew rate, V_{Supply} is the DC supply voltage, V_{Diode} is the laser diode operating voltage drop, and L_{Loop} is the overall loop inductance formed between the supply, laser diode, and switching device. To maximize the current slew rate, the overall loop inductance needs to be minimized. This is done through the use of low-inductive package laser diodes [54], and low inductance switching devices, for which the use of the enhancement mode LGA package GaN transistors are ideally suited. Furthermore, high-speed GaN transistors can significantly increase switching speeds, further minimizing the achievable pulse width.

As an example of the capabilities of GaN transistors for pulsed laser applications, the current pulse waveforms for a laser diode using GaN transistors, compared to a similar result using silicon MOSFETs, are shown in Figure 10.64. Comparing the rising edges of these waveforms, it can be seen that the GaN transistors are capable of switching 400 A in less than 50 ns (>8 A/ns), compared to the MOSFET switching 300 A in less than 70 ns (>4.3 A/ns). Not only are the GaN transistors able to almost double the current slew rate, but they

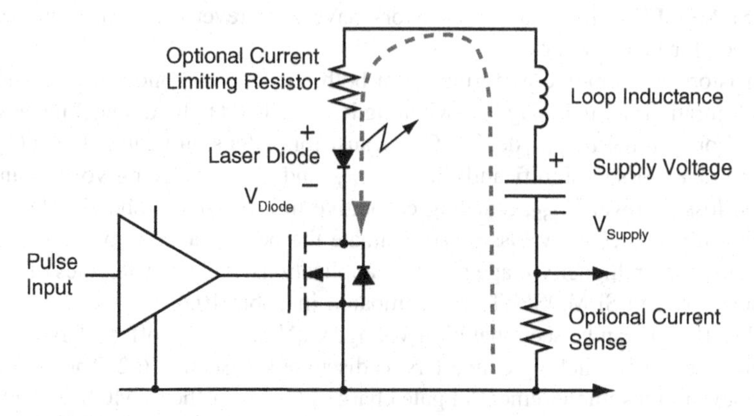

Figure 10.63 Simple pulsed laser drive circuit implementation. Pulsed current path shown as dashed line

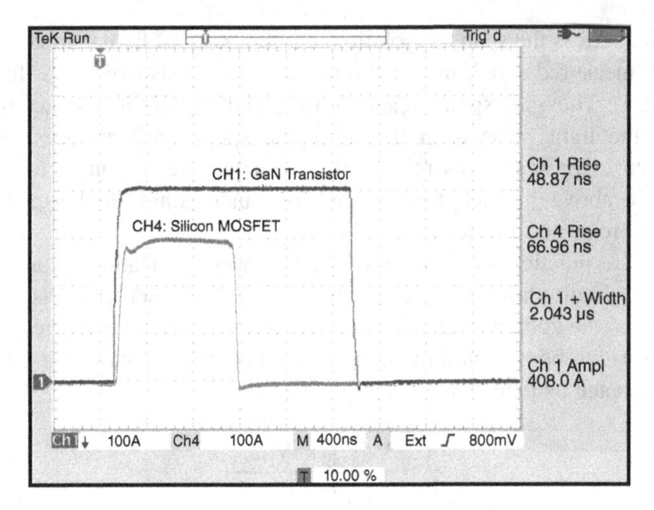

Figure 10.64 Laser diode current pulses using silicon MOSFET and GaN transistor respectively

are also able to increase the overall current capability by virtue of a lower device on-resistance, enabling significant performance improvements in LiDAR and other pulsed laser applications.

10.8 Power Factor Correction (PFC)

PFC circuits are required in modern power supplies to maximize real power and minimize THD, to improve the efficiency of the power grid. The conventional PFC boost rectifier, shown in Figure 10.65(a), is prevalent in current designs. To improve PFC performance, bridgeless topologies are being considered to eliminate one diode from the current conduction path, reducing power loss [55]. To enable improved PFC topologies, better performing devices are required. The bridgeless totem pole PFC boost rectifier, shown in Figure 10.65(b), is a promising topology, but its operation has been limited by the high reverse recovery of traditional Si MOSFETs [56]. GaN transistors have zero reverse recovery and thus make an ideal choice for this topology.

GaN transistors can improve performance in both conventional and bridgeless PFC boost rectifier configurations by reducing the switching losses related to hard-switching as discussed in Chapter 6. For a similar $R_{DS(on)}$ device, GaN transistors offer significantly lower Q_G, thereby reducing gate drive losses, significantly lower Q_{GD} and Q_{GS2}, reducing voltage and current commutation losses, lower E_{OSS}, reducing capacitive turn-on losses, and significantly lower Q_{RR}, reducing reverse recovery losses. To evaluate the performance improvements possible with GaN transistors in higher voltage resonant applications, the characteristics of 600 V GaN cascode transistors and Si MOSFETs are compared in Table 10.4.

From Table 10.4, it can be seen that high-voltage GaN transistors offer a large reduction in the key FOMs for hard-switching converters as discussed in Section 6.2. The GaN transistor has an over seven times improvement in gate charge FOM over the Si MOSFET, resulting in lower gate driving losses in high-frequency designs. The hard-switching FOM_{HS} is almost eight times lower for a GaN transistor than an Si MOSFET, and this will translate directly into

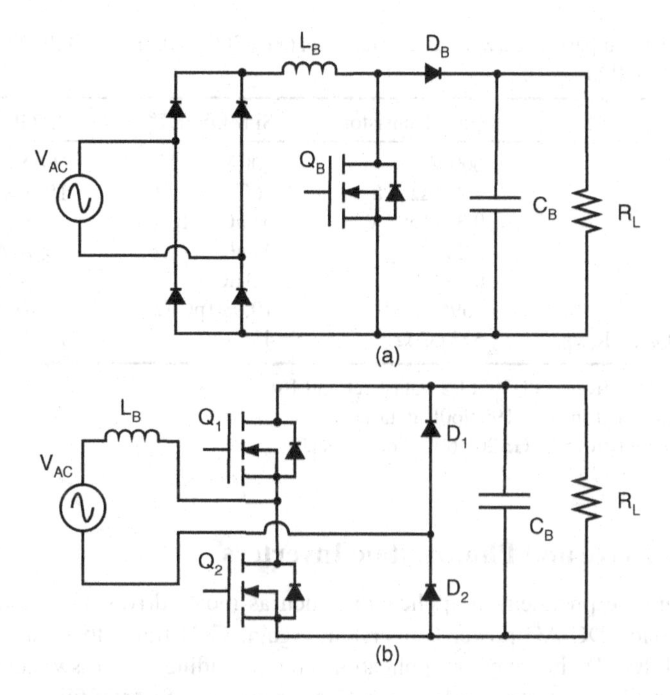

Figure 10.65 (a) Conventional PFC boost rectifier (b) bridgeless totem pole PFC boost rectifier

lower losses and improved high-frequency PFC performance. For a 230 V_{AC_IN} to 400 V_{OUT}, 1500 W conventional PFC converter operating at 500 kHz, GaN transistors have demonstrated a 1% gain in full load efficiency and a 0.25% gain at 20% load when compared to state-of-the-art Si MOSFETs. This leads to a reduction in full-load power loss from 42 W to 26 W [57]. For the bridgeless totem-pole PFC rectifier shown in Figure 10.65, efficiencies over 98.5% were achieved with GaN transistors for a 230 V_{AC_IN} to 400 V_{OUT} converter operating at 50 kHz [56].

Table 10.4 Device comparison between GaN transistor (TPH3006PS) and Si MOSFET (FCP190N60E)

Parameter	GaN transistor[a]	Si MOSFET[b]	FOM ratio
Voltage Rating (V_{DSS})	600 V	600 V	
$R_{DS(on)}$	150 mΩ at 8 V	160 mΩ at 10 V	
Q_G	9.3 nC at 8 V[a]	63 nC at 10 V	
Q_{GD} at 380 V	3.2 nC[a]	24 nC	
Q_{GS2} at 10 A	0.5 nC[a]	3.3 nC[b]	
E_{OSS} at 380 V	4.0 μJ[a]	5.5 μJ	
Q_{RR}	54 nC	4800 nC	
$Q_G \cdot R_{DS(on)}$	1395 pC·Ω	10,080 pC·Ω	7.23 × reduction
FOM_{HS} $(Q_{GD} + Q_{GS2}) \cdot R_{DS(on)}$	555 pC·Ω	4368 pC·Ω	7.87 × reduction

[a] All values calculated from TPH3006PS datasheet and [6].
[b] All values calculated from FCP190N60E datasheet.

Table 10.5 Device comparison between a GaN transistor (TPH3006PS), an Si MOSFET (FCP190N60E), and an Si IGBT (STGB20V60DF)

Parameter	GaN Transistor[a]	Si MOSFET[b]	Si IGBT[c]
Voltage rating (V_{DSS})	600 V	600 V	600 V
$R_{DS(on)}$	150 mΩ at 8 V	160 mΩ at 10 V	150 mΩ at 15 V, 10 A[c]
Q_G	9.3 nC at 8 V[a]	63 nC at 10 V	116 nC at 15 V
Q_{GD} at 380 V	3.2 nC[a]	24 nC	(Q_{GC}) 45 nC[c]
Q_{GS2} at 10 A	0.5 nC[a]	3.3 nC[b]	(Q_{GE2}) 3.8 nC[c]
$Q_G \cdot R_{DS(on)}$	1395 pC·Ω	10,080 pC·Ω	17,400 pC·Ω
FOM_{HS} ($Q_{GD} + Q_{GS2}$)·$R_{DS(on)}$	555 pC·Ω	4368 pC·Ω	7320 pC·Ω

[a] All values calculated from TPH3006PS datasheet and [6].
[b] All values calculated from FCP190N60E datasheet.
[c] All values calculated from STGB20V60DF datasheet.

10.9 Motor Drive and Photovoltaic Inverters

Voltage inverters are prevalent in applications such as motor drives and photovoltaic (PV) inverters to provide DC-AC power conversion. Again, GaN transistors can outperform Si MOSFETs and IGBTs in inverter applications by providing faster switching speeds. A comparison of device performance between GaN transistors, Si MOSFETs, and Si IGBT's can be seen in Table 10.5. The benefits of GaN transistors were demonstrated in a 4 kW, 50 kHz PV inverter, where the GaN transistor-based design was able to achieve higher efficiency than a 16 kHz IGBT-based solution, with the higher frequency, allowing for an almost 45% reduction in converter volume [58]. In motor drive applications, GaN transistors have demonstrated improved system performance in a 2.2 kW motor drive system where the GaN transistors operated at 100 kHz, an over six-fold increase over the 15 kHz IGBT-based design, while providing a 2% increase in efficiency [59].

10.10 Summary

Highlighted in this chapter were existing applications, such as point-of-load converters, Class-D audio amplification, motor drives, and power factor correction, which benefit from the superior performance of GaN transistors compared with their silicon MOSFET and IGBT counterparts. Additionally, GaN transistors enable applications that are not possible with the aging silicon transistor technology. These emerging applications include envelope tracking, wireless power transfer, and LiDAR, just to name a few. Improvements in switching losses are complemented by improvements in gate drive losses and layout parasitics to demonstrate that GaN can outperform silicon in almost any power conversion application.

In the final chapter, key factors that determine the displacement rate of GaN technology into the world of silicon transistors will be considered.

References

1 International Rectifier, J. B. Forsythe, "Paralleling of Power MOSFETs for High Power Output," Appl. Note. Available from http://www.irf.com/technical-info/appnotes/para.pdf.

2. Wu, Y.F. (2013) "Paralleling high-speed GaN power HEMTs for quadrupled power output," Twenty-Eighth Annual IEEE Applied Power Electronics Conference and Exposition (APEC), Long Beach, CA, 16–21 March 2013, pp. 649–655.
3. IEEE 802.3 at™-2009 Ethernet standard, Available from http://standards.ieee.org/about/get/802/802.3.html.
4. Lu, B., Liu, W., Liang, Y., Lee, F.C., and Wyk, J.D. (2006) "Optimal design methodology for LLC resonant converter," Applied Power Electronics Conference and Exposition (APEC) 2006 Twenty-First Annual IEEE, Dallas, TX, 2006, pp. 533–538.
5. Yang, B., Lee, F.C., Zhang, A.J., and Huang, G. (2002) "LLC resonant converter for front end DC/DC conversion," Applied Power Electronics Conference and Exposition (APEC) 2002 Seventeenth Annual IEEE, Dallas, TX, 2002, pp. 1108–1112.
6. Mishra, U. and Wu, Y. (2013) "Latest high voltage GaN devices for inverters," PCIM Europe 2013, pp. 724–729.
7. Briere, M. (2013) "The Status of 600 V GaN on Si-based Power Device Development at International Rectifier," PCIM Europe 2013, pp. 735–742.
8. Huang, X., Liu, Z., Li, Q., and Lee, F.C. (2013) "Evaluation and application of 600 V GaN HEMT in cascode structure," Applied Power Electronics Conference and Exposition (APEC) 2012 Twenty-Seventh Annual IEEE, Orlando, FL, 2012, pp. 1279–1286.
9. International Rectifier Corporation, "Class-D Amplifier Design Basics II," Rev. 1, 19 Feb. 2009. Available from http://www.irf.com/product-info/audio/classdtutorial2.pdf.
10. International Rectifier Corporation, "Protected digital audio amplifier," IRF2092 datasheet, 2007 [Revised 18 Oct. 2010]. Available from http://www.irf.com/product-info/datasheets/data/irs2092.pdf.
11. J. Strydom, "eGaN® FET- Silicon Power Shoot-Out Volume 11: Optimizing FET On-Resistance," *Power Electronics Technology*, Oct. 2012. Available from http://powerelectronics.com/discrete-semis/gan_transistors/egan-fet-silicon-power-shoot-out-volume-11-optimizing-fet-on-resistance-1001/.
12. Efficient Power Conversion Corporation, "Development board EPC9106 quick start guide – 150 W/8 Ω class-D amplifier," Available from http://epc-co.com/epc/documents/guides/EPC9106_qsg.pdf.
13. Wimpenny, G. (Jan. 2012) "Understand and characterize envelope-tracking power amplifiers," EE Times Magazine.
14. OpenET alliance, "Introduction to envelope tracking," Available from www.open-et.org/Intro-to-ET-pa-712.php.
15. Staudinger, J., Gilsdorf, B., Newman, D., Norris, G., Sadowniczak, G., Sherman, R., and Quach, T. (2000) High efficiency CDMA RF power amplifier using dynamic envelope tracking technique. *IEEE Microwave Symposium Digest*, **2**, 873–876.
16. Baker, S. (14 July 2011) "Applying envelope tracking to high-efficiency power amplifiers for handset and Infrastructure transmitters," *Cambridge Wireless Radio SIG*, Available from http://www.cambridgewireless.co.uk/Presentation/Steven%20Baker.pdf.
17. Chan, K. (Jan. 2010) "GC5325 Envelope Tracking," Texas Instruments, Application Report SLWA058.
18. Hendy, J. (2009) Transmitter power efficiency. *Broadcast Engineering Magazine*, **51** (11), 8–13.
19. Vasić, M., Garcia, O., Oliver, J.A., Alou, P., and Cobos, J.A. (2012) Survey of architectures and optimizations for wide bandwidth envelope amplifier, Power Electronics and Motion Control Conference, EPE/PEMC 4–6 Sep. 2012, pp. LS8d.1-1-LS8d.1-7.
20. Jeong, J. and Kimball, D.F. (2009) Wideband envelope-tracking power amplifier with reduced bandwidth power supply waveform. *IEEE Transactions on Microwave Theory and Techniques*, **52** 1381–1384.
21. Strydom, J.T. (April 2012) "eGaN® FET- Silicon Power Shoot-Out Volume 8: Envelope Tracking," *Power Electronics Magazine*, Available from http://powerelectronics.com/gan-transistors/egan-fet-silicon-power-shoot-out-volume-8-envelope-trackingShootout 8.
22. Cucak, D., Vasić, M., Garcia, O., Oliver, J.A., Alou, P., and Cobos, J.A. (2012) "Application of eGaN FETs for highly efficient radio frequency power amplifier," Integrated Power Electronics Systems, CIPS March 2012, pp. 1–6.
23. Efficient Power Conversion Corporation, Appl. Note AN015, "Introducing a Family of eGaN FETs for Multi-Megahertz Hard Switching Applications," Available from http://epc-co.com/epc/documents/product-training/AN015 eGaN FETs for Multi-Megahertz Applications.pdf.
24. Efficient Power Conversion Corporation, "EPC8005 – Enhancement-mode Power Transistor," EPC8005 datasheet, Sep. 2013 [Revised Sept. 2013]. Available from http://epc-co.com/epc/documents/datasheets/EPC8005_-datasheet.pdf.
25. Wu, T., "Cdv/dt induced turn-on in synchronous buck regulators," white paper, International Rectifier Corporation.
26. Reusch, D. and Strydom, J. (2014) Understanding the effect of PCB layout on circuit performance in a high frequency gallium nitride-based point of load converter. *Power Electronics, IEEE Transactions*, **29** (4), 2008–2015.

27. Strydom, J. and Reusch, D. (2014) "Design and evaluation of a 10 MHz gallium nitride-based 42 V DC-DC converter," Applied Power Electronics Conference (APEC) 2014 Twenty-Ninth Annual IEEE Session T30, Fort Worth, TX, 16–20 March 2014.
28. http://www.wirelesspowerconsortium.com/
29. Tseng, R., von Novak, B., Shevde, S., and Grajski, K.A. (2013) "Introduction to the alliance for wireless power loosely-coupled wireless power transfer system specification version 1.0," IEEE Wireless Power Transfer Conference 2013, Technologies, Systems and Applications, May 15–16, 2013.
30. "Industrial, Scientific, and Medical Equipment," FCC regulation: CFR 2010, title 47, vol. 1, Part 18. Available from http://www.ecfr.gov/cgi-bin/text-idx?SID=abcb16dbad4883038a13c0a1a0e28df0&node=47:1.0.1.1.18&rgn=div5.
31. El-Hamamsy, S.-A. (1994) Design of high-efficiency RF class-D power amplifier. *IEEE Transactions on Power Electronics*, **9** (3), 297–308.
32. Hung, T.-P. (Jan. 2008) "High efficiency switching-mode amplifiers for wireless communication systems," Ph.D. dissertation, University of California, San Diego, CA.
33. Choi, D.K. (March 2001) "High efficiency switched-mode power amplifiers for wireless communications," Ph.D. dissertation, University of California, Santa Barbara, CA.
34. Gerrits, T., Duarte, J.L., and Hendrix, M.A.M. (2010) "Third harmonic filtered 13.56 MHz push-pull class-E power amplifier," IEEE Energy Conversion Congress and Exposition (ECCE) conference, September 2010, pp. 742–749.
35. Siddabattula, K.,"Wireless Power System Design Component and Magnetics Selection," Texas Instruments, presentation. Available from http://e2e.ti.com/support/power_management/wireless_power/m/mediagallery/526153.aspx.
36. Kurs, A., Karalis, A., Moffatt, R., Joannopoulos, J.D., Fisher, P., and Soljačić, M. (2007) Wireless power transfer via strongly coupled magnetic resonances. *Science Magazine*, **317** (5834), 83–86.
37. Qualcomm (Sep. 2011) Wireless power transfer enabling the mobile charging ecosystem. *Darnell Power Forum.*
38. de Rooij, M.A. and Strydom, J.T. (2012) eGaN® FET–silicon shoot-out Vol. 9: wireless power converters. *Power Electronics Technology*, 22–27. Available from http://powerelectronics.com/discrete-power-semis/egan-fet-silicon-shoot-out-vol-9-wireless-power-converters.
39. De Rooij, M.A. and Strydom, J.T. (2012) eGaN® FETs in low power wireless energy converters. *Electro-Chemical Society Transactions on GaN Power Transistors and Converters*, **50** (3), 377–388.
40. de Rooij, M.A. and Strydom, J.T. (August (2013)) "High Efficiency Voltage Mode Class-D topology," U.S. patent pending.
41. Efficient Power Conversion Corporation, "EPC2007 – Enhancement-mode Power Transistor," EPC2007 datasheet, Sep. 2011 [Revised Jul. 2013]. Available from http://epc-co.com/epc/documents/datasheets/EPC2007_datasheet.pdf.
42. Witricty Corp., coil set part numbers 190-000037-01 and 190-000038-01, Available from www.witricity.com.
43. Raab, F.H. (1977) Idealized operation of the class-E tuned power amplifier. *IEEE Transactions on Circuits and Systems*, **24** (12), 725–735.
44. Kazimierczuk, M. (1984) Collector amplitude modulation of the class-E tuned power amplifier. *IEEE Transactions on Circuits and Systems*, **31** (6), 543–549.
45. Chen, K. and Peroulis, D. (2011) Design of highly efficient broadband class-E power amplifier using synthesized low-pass matching networks. *IEEE Transactions on Microwave Theory and Techniques*, **59** (12), 3162–3173.
46. Chen, W., Chinga, R.A., Yoshida, S., Lin, J., Chen, C., and Lo, W. (2012) "A 25.6 W 13.56 MHz wireless power transfer system with a 94% efficiency GaN class-E power amplifier," IEEE MTT-S International Microwave Symposium Digest (MTT), June 2012, pp. 1–3.
47. Texas Instruments, "LM5113 5A, 100V Half-Bridge Gate Driver for Enhancement Mode GaN GETs," LM5113 datasheet, June 2011 [Revised Apr. 2013].
48. Sokal, N.O. and Sokal, A.D. (1975) Class-E – A new class of high-efficiency tuned single-ended switching power amplifiers. *IEEE Journal of Solid-State Circuits*, **SC-10** (3), 168–176.
49. Sokal, N.O. (2001) Class-E RF power amplifiers. *QEX Magazine*, (204), 9–20.
50. Efficient Power Conversion Corporation, "EPC2012 – Enhancement-mode Power Transistor," EPC2012 datasheet, Aug. 2011 [Revised Oct. 2012]. Available from http://epc-co.com/epc/documents/datasheets/EPC2012_datasheet.pdf.
51. Fairchild Semiconductor, "FDMC8622–100 V N-Channel Shielded Gate Power Trench® MOSFET," FDMC8622 datasheet, Rev. C5, Oct. 2013. Available from http://www.fairchildsemi.com/pf/FD/FDMC8622.html.
52. Fairchild Semiconductor, "FDMC86248–150 V N-Channel Shielded Gate Power Trench® MOSFET," FDMC86248 datasheet, Rev. C3, Sept. 2012. Available from http://www.fairchildsemi.com/pf/FD/FDMC86248.html.

53. IXYS-Colorado, "PCO-7120 Laser Diode Driver Module," Installation and operation notes, Available from http://www.ixysrf.com/index.php?option=com_joomdoc&task=document.download&path=dei-scientific/operating-manuals/PCO-7120_Manual_RevA.pdf.
54. PicoLAS, Application Note # 02: "Impedance of Laser Diodes, Inductive Behaviour," Available from http://www.lasercomponents.com/fileadmin/user_upload/home/Datasheets/lc/veroeffentlichung/treiberelektronik-pulslaserdioden.pdf.
55. Huber, L., Jang, Y., and Jovanovic, M.M. (2007) Performance evaluation of bridgeless PFC boost rectifiers," Applied Power Electronics Conference and Exposition (APEC), 2007 Twenty-Second Annual IEEE, Anaheim, CA, pp. 165–171.
56. Zhou, L., Wu, Y.F., and Mishra, U. (2013) "True-bridgeless totem-pole PFC ased on GaN HEMTs," PCIM Europe 2013, pp. 1017–1022.
57. Wu, Y.F., Gritters, J., Shen, L., Smith, R.P., McKay, J., Barr, R., and Birkhan, R. (2013) "Performance and robustness of first generation 600 V GaN-on-Si power transistors," IEEE Workshop on Wide Bandgap Power Devices and Applications.
58. Blake, C. (April 2013) "GaN power devices slash size, raise efficiency of 4 kW solar inverter," *How2Power Today*, exclusive technology feature.
59. Wu, Y.F., Kebort, D., Guerrero, J., Yea, S., Honea, J., Shirabe, K., and Kang, J. (2012) "High-frequency, GaN diode-free motor drive inverter with pure sine wave output," PCIM Europe 2012, pp. 76–83.
60. Charged, Electric Vehicles Magazine, Available from http://chargedevs.com/features/whats-wireless-ev-charging/, issue 9-Aug 2013, [accessed Jan. 2014].

11

Replacing Silicon Power MOSFETs

11.1 What Controls the Rate of Adoption?

The silicon power MOSFET journey, spanning more than 30 years, taught us that there are four key variables controlling the adoption rate of a disruptive power management technology [1].

1. Does it enable significant new capabilities?
2. Is it easy to use?
3. Is it *very* cost effective for the user?
4. Is it reliable?

In the next few sections, and using the information from the previous chapters, the readiness of the GaN transistor to replace the silicon power MOSFET, based on these four criteria, will be reviewed.

11.2 New Capabilities Enabled by GaN Transistors

The most significant new capabilities enabled by GaN transistors stem from the disruptive improvement in switching speed (see Figure 11.1). As discussed in Chapter 1, GaN transistors also have a much higher critical electric field capability than silicon, which enables this new class of devices to withstand much greater voltage from drain to source, with much less penalty in on-resistance. This capability, coupled with higher electron mobility and innovative device packaging, has created a class of devices that are significantly smaller and faster than their silicon predecessors.

As GaN transistors gain wider acceptance as the successor to the power MOSFET, designers have been able to leverage the increased switching performance to improve power conversion system efficiency, size, and cost. Examined in Chapters 6–8 were the advantages of GaN transistors in hard- and resonant-switching topologies from hundreds of kHz up to hundreds of MHz, and even RF amplifiers into the multi-GHz range. Chapter 9 explored the exceptional capabilities of enhancement-mode transistors to withstand large amounts of radiation exposure.

GaN Transistors for Efficient Power Conversion, Second Edition.
Alex Lidow, Johan Strydom, Michael de Rooij, and David Reusch.

Companion Website: http://www.wiley.com/go/gan_transistors

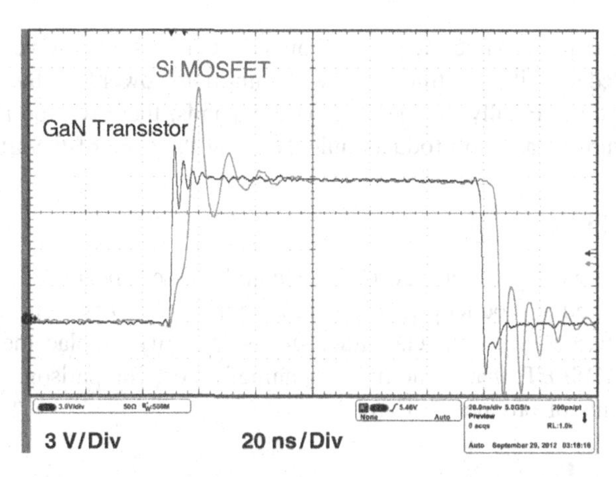

Figure 11.1 Switching speed comparison between a GaN transistor (EPC2015) and a MOSFET (BSZ040N04) in a buck converter switching at 1 MHz with a $V_{IN} = 12$ V, $V_{OUT} = 1.2$ V, $I_{OUT} = 20$ A

In Chapter 10, early-adopter applications such as wireless power transfer, envelope tracking, LiDAR, power factor correction, and Class-D audio were described in detail and compared against state-of-the-art power MOSFETs in the same circuit topologies. In all the examples, GaN transistor technology can shrink system size, enhance system efficiency, and increase power density.

11.3 GaN Transistors are Easy to Use

How easy a device is to use depends on the skill of the user, the degree of difficulty of the circuit under development, how different the device is compared with devices that are within the experience of the user, and the tools available to help the user apply the device. GaN transistors are very similar in their behavior to existing power MOSFETs and, therefore, users can greatly leverage their past design experience.

One key difference, relatively high frequency response, is both a step function improvement over any prior silicon device and an added consideration for the user when designing a circuit. As described in Chapter 4, small amounts of stray parasitic inductance can cause increased power losses and a large overshoot in the gate-to-source voltage that could potentially damage devices.

On the other hand, there are several characteristics that render these devices *easier* to use than their silicon predecessors. For example, the threshold voltage is virtually independent of temperature over a wide range for enhancement-mode devices (cascode devices have a threshold voltage set by a silicon MOSFET and, therefore, have the same dependence on temperature as a MOSFET), and the on-resistance also has a significantly lower temperature coefficient than silicon.

User-friendly tools can make a big difference to how easy it is to apply a new type of device. SPICE device models, as well as thermal models, are widely available for user download [2]. Pre-assembled circuit kits are available from GaN transistor manufacturers such as Efficient Power Conversion Corporation, Transphorm, and GaN Systems. In addition, ICs designed specifically to drive GaN transistors, make the designer's job easier by compressing designs, and thus reducing common source and power loop inductances. Driver ICs can also manage the higher speeds needed to switch the transistors with minimum power loss, but without excessive overshoot or dead-time. Finally, as the user base expands, there will be many experienced designers able to turn ideas into products quickly using the state-of-the-art GaN transistors.

11.4 Cost vs. Time

Cost comparisons between products of different technologies can be difficult. In addition, costs are not always reflected in product prices if there is an imbalance between supply and demand in the consuming market. Since the GaN transistor is primarily a replacement for the mature, but aged, power MOSFET, that is the most meaningful cost comparison.

The basic elements of product cost are:

- starting material
- epitaxial growth
- wafer fabrication
- test and assembly.

Other elements that contribute to costs such as yield, engineering costs, packing and shipping costs, and general overhead costs are assumed to be similar between the technologies for the purpose of this analysis.

11.4.1 Starting Material

Today, GaN transistors are produced on 150 mm substrates with a migration to 200 mm diameter likely in the near future. Power MOSFETs are produced on anything from 100 mm through 200 mm substrates by the many manufacturers in this business. Because the most cost-competitive GaN transistors use standard silicon substrates, there is no cost penalty compared with power MOSFETs fabricated on similar diameter starting material. In fact, there is little cost difference per unit area between 150 mm and 200 mm silicon wafers and, therefore, as far as starting material is concerned, there is no real cost difference per wafer. If one considers the fact that the GaN transistor has less device area than a silicon device with similar current-carrying capability, then the cost per function is lower for GaN transistors than for comparable Si MOSFETs.

11.4.2 Epitaxial Growth

Silicon epitaxial growth is a mature technology with many companies making highly efficient and automated machines. MOCVD GaN equipment is available from at least two sources, Veeco in the US, and Aixtron in Germany. Both make capable and reliable machines whose primary use has been the growth of GaN epitaxy used in the fabrication of LEDs. None of the machines are optimized for GaN on silicon epitaxy, nor do they have levels of automation that are common on silicon machines. As a result, GaN epitaxy on silicon is significantly more expensive than silicon epitaxy today. But this is not fundamental. Processing times and temperatures, wafer diameter, materials costs, and machine productivity are all on a fast track of improvement and there is no fundamental limit preventing costs from being similar to silicon epitaxial costs. Within the next few years, assuming widespread adoption of GaN transistors as a replacement for silicon power MOSFETs, it is expected that the cost of the GaN epitaxy will approach that of silicon epitaxy.

11.4.3 Wafer Fabrication

The simple structure of a GaN transistor, depicted in Chapter 1, is not difficult to build in a standard low-cost silicon wafer fabrication facility. Processing temperatures are similar to silicon BCDMOS, and cross-contamination can be managed easily. Today, GaN transistors are processed in standard silicon wafer fabrication facilities side-by-side with silicon ICs and power MOSFETs. In addition, the simple GaN transistor structure has many fewer processing steps than a comparable state-of-the-art power MOSFET. As GaN transistor volumes grow, the cost should become less than that of a leading-edge power MOSFET. The cascode GaN transistor's cost, however, is the sum of the cost of a GaN transistor and a silicon MOSFET, and therefore, may remain higher than the wafer fabrication costs of either a monolithic GaN transistor or a silicon MOSFET.

11.4.4 Test and Assembly

Although in the assembly process, there are significant differences favoring the cost structure of GaN transistors, the testing costs are equivalent.

Silicon power MOSFETs need a surrounding package, typically made of a copper leadframe, some aluminum, gold, or copper wires, all in a molded epoxy envelope. Connections need to be made to the top and bottom of the vertical silicon device; the plastic molding is needed to keep moisture from penetrating the active device, and there needs to be a means of getting the heat out of the part. Traditional power MOSFET packages such as the SO8, TO220, or DPAK, add cost, parasitic inductance, electrical and thermal resistance, and increase reliability and quality risks to the product. As discussed in Chapters 1–4, GaN transistors with a rated voltage of 200 V or less come in an LGA (WLCSP) format and can be used as a "flip chip" without compromising the electrical, thermal, or reliability characteristics. Assembly costs for these devices, therefore, would be lower than their MOSFET counterparts. Cascode devices discussed in Chapters 1, 2 and 10 are in more traditional power packages such as TO-220 or PQFN package styles, and, as a result of the dual-chip packaging required to accommodate both a GaN transistor and a MOSFET, are expected to cost somewhat more than a MOSFET to assemble.

Table 11.1 summarizes the cost comparison and anticipates that costs for GaN transistors will be lower in the case of a monolithic enhancement-mode transistor, and somewhat higher for a cascode transistor. This comparison, however, does not take into account any differences in yields. Manufacturing yields of GaN transistors are already close enough to MOSFET yields such that, when considered in conjunction with the relative die size comparison presented in Chapter 1, GaN transistors will most likely cost less to manufacture within a few years.

11.5 GaN Transistors are Reliable

The cumulative reliability information available on silicon power MOSFETs is staggering. Many years of work have been devoted to understanding failure mechanisms, controlling and refining processes, and designing products that have distinguished themselves as the highly reliable backbone of any power conversion system. GaN transistors are just a few years into this

Table 11.1 Summary of the cost difference between GaN transistor costs compared with silicon power transistors in 2013 and as anticipated in 2016. (a) monolithic enhancement-mode transistors, and (b) cascode GaN transistors

Monolithic enhancement-mode			Cascode GaN		
	2013	2016		2013	2016
Starting material	lower	lower	Starting material	same	same
Epi growth	higher	same	Epi growth	higher	same
Wafer fab	same	lower	Wafer fab	higher	higher
Assembly	lower	lower	Assembly	higher	higher
Overall	higher	lower	Overall	higher	higher

journey. Results to date are excellent. Manufacturers have published results from their qualification tests [3–7], and devices have been applied successfully to many RF and power applications with good results. Some GaN transistors have also proved reliable under harsh radiation exposure as discussed in Chapter 9. Reliability testing is rapidly demonstrating that GaN technology is robust under a wide range of accelerated life-test conditions.

11.6 Future Directions

The GaN technology journey is just beginning. There are profound improvements that can be made in basic device performance as measured by figures of merit, Miller ratio, $R_{DS(on)} \times$ area, and cost. Still, we are far away from theoretical performance limits. It is quite reasonable to expect factors of two or more reductions in the key figures of merit every two to four years.

Perhaps the greatest opportunity for GaN technology to impact the performance of power conversion systems comes from the intrinsic ability to integrate both power-level and signal-level devices on the same substrate. As an example, Figure 11.2 is a top view of a monolithic full-bridge power device in development. GaN technology, much like silicon-on-insulator (SOI) technology, has no significant parasitic interaction between components, allowing designers to easily develop monolithic power systems on a single chip.

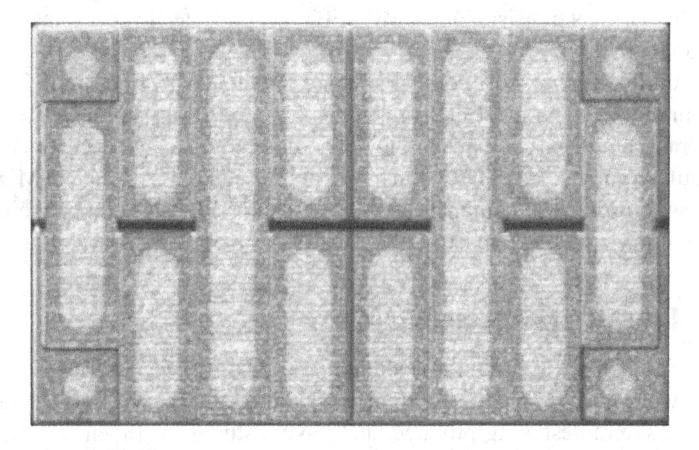

Figure 11.2 Monolithic full-bridge power device from EPC

11.7 Conclusion

In the late 1970s, power MOSFET pioneers believed that they had a technology that would displace bipolar transistors completely. Thirty-plus years later, plenty of applications remain that prefer bipolar transistors over power MOSFETs, but the size of the power MOSFET market is many times larger than the bipolar market. This is due to all the new applications and new markets enabled by that breakthrough technology. Today, GaN technology is at that same threshold. Like the power MOSFET of 1976, GaN manufacturers are at the beginning of an exciting journey with new products and breakthrough capabilities almost monthly [8].

The power MOSFET is not dead, but it is nearing the end of the road for major improvements in performance and cost. As more and more GaN transistor-based designs come to market and GaN-on-silicon epitaxial growth technology matures, this technology will most probably become dominant over the next decade due to its large advantages in both performance and cost – advantage gaps that promise to widen as we quickly climb the learning curve.

References

1. Lidow, A. (2010) "Is it the end of the road for silicon in power management?" *CIPS 2010 Conference*, Nuremburg, Germany, March 2010.
2. Efficient Power Conversion Corporation, eGaN FET SPICE Models, Available from http://epc-co.com/epc/DesignSupportbr/Device models.aspx.
3. Yanping, Ma (2011) "EPC GaN Transistor Application Readiness: Phase One Testing," reliability report, Available from http://epc-co.com/epc/documents/product-training/EPC_relreport_030510_finalfinal.pdf.
4. Yanping, Ma (2011) "EPC GaN Transistor Application Readiness: Phase Two Testing," reliability report, Available from http://epc-co.com/epc/documents/product-training/EPC_Phase_Two_Rel_Report.pdf.
5. Yanping, Ma (2011) "EPC GaN Transistor Application Readiness: Phase Three Testing," reliability report, Available from http://epc-co.com/epc/documents/product-training/EPC_Phase_Three_Rel_Report.pdf.
6. Yanping, Ma (2011) "EPC eGaN FETs Transistor Application Readiness: Phase Four Testing," reliability report, Available from http://epc-co.com/epc/documents/product-training/EPC_Phase_Four_Rel_Report.pdf.
7. Yanping, Ma (2011) "EPC eGaN FETs Transistor Application Readiness: Phase Five Testing," reliability report, Available from http://epc-co.com/epc/documents/product-training/EPC_Phase_Five_Rel_Report.pdf.
8. Pitcher, G. (27 January, 2009) "Power to change – how GaN might revolutionize embedded power device design," *New Electronics*, Available from http://www.newelectronics.co.uk/electronics-technology/power-for-change/16806/.

Appendix
Glossary of Terms

Term	Symbol	Units	Definition	Chapter
Band gap energy	E_G	eV	A band gap, also called "an energy gap" or "band gap energy," is an energy range in a solid where no electron states can exist. The band gap is a major factor determining the electrical conductivity of a solid. Substances with large band gaps are generally insulators; those with smaller band gaps are semiconductors, while conductors either have very small band gaps or none, because the valence and conduction bands overlap	1
Capacitance (drain-source)	C_{DS}	farads	The capacitance between the drain terminal and the source terminal	2, 3, 5
Capacitance (gate-drain)	C_{GD}	farads	The capacitance between the gate terminal and the drain terminal	2, 3,5
Capacitance (gate-source)	C_{GS}	farads	The capacitance between the gate terminal and the source terminal	2, 3, 6, 7,10
Capacitance (input)	C_{ISS}	farads	The input capacitance of the device is the sum of the C_{GD} and C_{GS}	2, 6, 7, 10
Capacitance (output)	C_{OSS}	farads	The output capacitance of the device is the sum of the C_{GD} and C_{DS}	2, 3, 5, 6, 7, 8, 10
Carrier mobility	μ	$cm^2/V{\cdot}s$	The electron mobility characterizes how quickly an electron can move through a metal or semiconductor, when pulled by an electric field	1, 2, 6,10
Common source inductance	CSI	henrys	Inductance shared by the drain-to-source power current path and gate driver	4, 6, 10
Critical electric field	E_{crit}	MV/cm	When the electric field gradient exceeds the critical electric field, valence bonds between atoms are ruptured and current flows.	1, 2, 11
Device junction temperature	T_J	^oC	The temperature of the device junction during operation	2
Diode reverse recovery charge	Q_{RR}	coulombs	The amount of charge that needs to be pulled out of the body diode of a FET in order to turn the device OFF. Enhancement-mode GaN transistors with pGaN gates have zero Q_{RR}	2, 6, 10

Drain current at the quiescent operating point	I_{DQ}	amperes	Drain current in the transistor at the quiescent operating point in a linear amplifier	8
Drain efficiency	η_D	percentage	Ratio of P_{RFout}/P_{DC}	8
Effective dead-time	t_{eff}	seconds	The time from when the gate voltage of the device reaches the turn-off plateau voltage to when the other device's load current commutates from the diode	3, 6,7 10,11
eGaN® FET	eGaN FET		Trademarked symbol for enhancement-mode gallium nitride on silicon FET	1, 6, 8, 9, 10, 11
Figure of Merit	FOM	mΩ·nC	The FOM is a way to compare different device technologies. The most common FOM is the calculated by multiplying the $R_{DS(on)}$ or a given device times the Q_G	1, 3, 6, 7, 8, 9, 10, 11
Figure of Merit–hard switching	FOM_{HS}	mΩ·nC	The hard switching FOM is a way to compare different technologies in hard switching applications such as buck converters. $FOM_{HS}=(Q_{GD}+Q_{GS2})\cdot R_{DS(on)}$	6,10
Figure of Merit–soft switching	FOM_{SS}	mΩ·nC	The soft switching FOM is a way to compare different technologies in soft switching and resonant applications. $FOM_{SS}=(Q_{OSS}+Q_G)\cdot R_{DS(on)}$	7,10
Gate charge required to increase gate voltage to threshold voltage	Q_{GS1}	coulombs	Charge required to increase gate voltage from zero to the stated threshold voltage of the device	6
Gate charge for the current transition interval	Q_{GS2}	coulombs	Charge required to increase gate voltage from the stated threshold voltage of the device to the plateau voltage (current conduction interval)	6
Gate charge from zero to the onset of the plateau voltage	Q_{GS}	coulombs	Charge required to increase gate voltage to the plateau voltage	6
Gate charge during the voltage transition interval	Q_{GD}	coulombs	The charge transferred between the gate terminal and the drain terminal when the drain voltage changes	2, 3, 5

Gate charge (total)	Q_G	coulombs	Total gate charge required to drive a device from zero to rated gate voltage (fully enhanced)	2, 3, 6, 7, 10
Gate driver output voltage	V_{DR}	volts	On-state output voltage of gate driver	6, 7
Gate plateau voltage	V_{pl}	volts	Gate voltage at which the drain-source voltage transition occurs during a hard-switching event	3, 6
Gate threshold voltage	V_{th}	volts	Gate voltage that needs to be applied to initiate drain-source conduction	2, 3, 6
Gate-drain charge	Q_{GD}	coulombs	The charge transferred between the gate terminal and the drain terminal when the drain voltage changes	2, 3, 5
High electron mobility transistor	HEMT		High electron mobility transistor (HEMT), also known as heterostructure FET (HFET) or modulation-doped FET (MODFET), is a field effect transistor incorporating a junction between two materials with different band gaps (i.e., a heterojunction) as the channel instead of a doped region, as is generally the case for MOSFET	1, 2, 3, 4, 6, 8, 9, 10
Internal gate resistance	R_G	ohms	The resistance inside the transistor gate that limits how fast charge can be pulled out of, or pushed into the gate electrode	3, 7
Land grid array package	LGA		The land grid array (LGA) is a type of surface-mount packaging for integrated circuits (ICs) and eGaN FETs. An LGA can be electrically connected to a printed circuit board (PCB) by soldering directly to the board	1, 2, 4, 5, 6, 7, 10,11
Linear energy transfer	LET	$MeV{\cdot}cm^2/mg$	Linear energy transfer (LET) is a measure of the energy transferred to material as an ionizing particle travels through it. Typically, this measure is used to quantify the effects of ionizing radiation on electronic devices	9

Metal oxide chemical vapor deposition	MOCVD		MOCVD is a chemical vapor deposition method of epitaxial growth of materials, especially compound semiconductors from the surface reaction of organic compounds or metalorganics and metal hydrides containing the required chemical elements	1, 11
Miller ratio	Q_{GD}/Q_{GS}		The Miller Ratio is a gauge of how much the gate terminal may be susceptible to false turn-on when a high dv/dv is applied at the drain terminal	2, 3, 10, 11
On-resistance	$R_{DS(on)}$	ohms	When the eGaN FET is in the on-state, it exhibits a resistive behavior between the drain and source terminals. This resistance is called $R_{DS(on)}$ for "drain-to-source resistance in on-state" and is the sum of many elementary contributions	1, 2, 3, 4, 5, 7, 8, 9, 10, 11
Output Charge	Q_{OSS}	coulombs	The charge required to be supplied to the drain terminal to achieve a certain voltage on the drain relative to the source	6, 7, 9, 10
Power (DC power to the RF transistor)	P_{DC}	watts	DC power delivered to the RF transistor	8
Power (output RF)	P_{RFout}	watts or dBm	RF power of the amplifier	8
Power losses (turn-off)	P_{off}	watts	Power losses due to the turn-off switching transition	6
Power losses (turn-on)	P_{on}	watts	Power losses due to the turn-on switching transition	6
Power over Ethernet – power sourcing equipment	PoE-PSE		Power sourcing equipment (PSE) is a device that provides (sources) power on the Ethernet cable	10
Reflection coefficient (input port)	s_{11}	percentage	The percentage of the incident wave that is reflected back from the input port	8
Reflection coefficient (input power forward gain)	s_{21}	percentage	The percentage of the input port incident wave that is reflected to the output port	8

Reflection coefficient (output port reverse gain)	s_{12}	percentage	The percentage of the output port incident wave that is reflected to the input port	8
Reflection coefficient (output port)	s_{22}	percentage	The percentage of the incident wave that is reflected back from the output port	8
Rollett stability factor	K		Rollett stability factor is a test for unconditional stability	8
Single event burnout	SEB		Single-event burnout (SEB) is typically a latch-up event that is possible with power eGaN FETs. It may be triggered by the passage of a heavy-ion. If unmitigated, SEB can be destructive. SEB can be prevented by limiting drain current or switching on the eGaN FET when SEB is detected	9
Single event effects	SEE		Single-event effects (SEE) occur when a high-energy particle travels through a semiconductor; it leaves an ionized track behind. This ionization may cause a highly localized effect similar to the transient dose one–a benign glitch in output, a less benign bit flip in memory or a register, or, especially in high-power transistors, a destructive latch up and burnout. Single event effects have importance for electronics in satellites, aircraft, and other both civilian and military aerospace applications	9
Single event gate rupture	SEGR		Single-event gate rupture (SEGR) is a destructive event that can occur when an energetic ion passes through the gate of a power transistor. A high field is created across the oxide causing permanent failure. SEGR cannot be mitigated using circuit level approaches	9
Smith chart			The Smith chart is a graphical aid designed for electrical engineers specializing in radio frequency (RF) engineering to assist in solving problems with transmission lines and matching circuits. The Smith chart can be used to represent many parameters including impedances, admittances, and reflection coefficients	8

Switching frequency	f_{sw}	hertz	Operating frequency of the transistor	6, 7, 10
Thermal impedance–junction-to-board	$Z_{\Theta JB}$	°C/W	Thermal impedance from the device junction to the bottom of the solder bumps, without consideration of the type or size of the mounting circuit board. The impedance is typically used when the device is not in steady state	2
Thermal resistance–junction-to-ambient	$R_{\Theta JA}$ or R_{THJA}	°C/W	Thermal resistance - junction to ambient, is measured with the DUT. is mounted onto a single sided 2-ounce FR-4 circuit board whose area is 1 square inch (645.16 square millimeters)	2, 5
Thermal resistance–junction-to-board	$R_{\Theta JB}$ or R_{THJB}	°C/W	Thermal resistance from the device junction to the bottom of the solder bumps, without consideration of the type or size of the mounting circuit board	2
Thermal resistance–junction-to-case	$R_{\Theta JC}$ or R_{THJC}	°C/W	Thermal resistance from the device junction to the silicon backside of the transistor	2
Thermal resistance (effective)	$R_{\Theta(Effective)}$	°C/W	$R_{\Theta(Effective)}$ is the effective thermal resistance when all resistance paths are added together	2, 5
Threshold voltage	$V_{GS(TH)}$ or V_{TH}	volts	In a GaN transistor the threshold voltage is the voltage applied between gate and source that enhances the 2DEG under the gate enough to begin conducting current from drain-to-source	1, 2, 4, 6, 9, 11
Transconductance	g_m	siemens	Transconductance is the ratio of the current change at the output port to the voltage change at the input port	6
Turn-off current fall time	t_{CF}	seconds	Hard-switching fall time for current in the transistor	6
Turn-on voltage fall time	t_{VF}	seconds	Hard-switching fall time for voltage across the transistor	3, 6
Turn-on current rise time	t_{CR}	seconds	Hard-switching rise time for current in the transistor	6
Turn-off voltage rise time	t_{VR}	seconds	Hard-switching rise time for voltage across the transistor	6

Two-dimensional electron gas	2DEG		A two-dimensional electron gas (2DEG) is a gas of electrons free to move in two dimensions, but tightly confined in the third. This tight confinement leads to quantized energy levels for motion in that direction, which can then be ignored for most problems. Thus the electrons appear to be a 2D sheet embedded in a 3D world	1, 2, 5, 9
Zero voltage switching time	t_{ZVS}	seconds	The time required to discharge a transistors output capacitance to achieve a zero voltage switching transition	7

Index

GaN Transistors for Efficient Power Conversion, Second Edition.
Alex Lidow, Johan Strydom, Michael de Rooij, and David Reusch.

Companion Website: http://www.wiley.com/go/gan_transistors

Printed in the United States
By Bookmasters